Building a GIS

System Architecture Design Strategies for Managers

Dave Peters

ESRI PRESS

REDLANDS, CALIFORNIA

ESRI Press, 380 New York Street, Redlands, California 92373-8100

Copyright © 2008 ESRI

All rights reserved. First edition 2008
10 09 08   2 3 4 5 6 7 8 9 10

Printed in the United States of America

Library of Congress Cataloging-in-Publication Data
Peters, Dave, 1948-
Building a GIS : system architecture design strategies for managers / Dave Peters.
      p.   cm.
   Includes index.
      ISBN 978-1-58948-159-6 (paper back : alk. paper) 1. Geographic information systems–Design. I. Title.
   G70.212P48   2008
   910.285–dc22   2008021553

Ask for ESRI Press titles at your local bookstore or order by calling 1-800-447-9778. You can also shop online at
www.esri.com/esripress. Outside the United States, contact your local ESRI distributor.

ESRI Press titles are distributed to the trade by the following:

In North America:
Ingram Publisher Services
Toll-free telephone: (800) 648-3104
Toll-free fax: (800) 838-1149
E-mail: customerservice@ingrampublisherservices.com

In the United Kingdom, Europe, and the Middle East:
Transatlantic Publishers Group Ltd.
Telephone: 44 20 7373 2515
Fax: 44 20 7244 1018
E-mail: richard@tpgltd.co.uk

On the cover
*Cover images by Kim Steele/PhotoDisc/Getty Images,
Keith Brofsky/PhotoDisc/Getty Images, Chad Baker/PhotoDisc/Getty Images*

*Special thanks to Roger Tomlinson for allowing use of his materials from chapter 10
of* Thinking About GIS: Geographic Information System Planning for Managers, *third edition.*

Contents

Part I: Understanding the technology

List of figures

Foreword

Why would a software company publish a book that talks so much about hardware? Because we're in the business of supporting what our clients do. Whether you are implementing GIS for the first time or expanding it further into your organization, this change may require shifts in network communications and hardware platforms in order to support GIS and everything it's connected to. Part of achieving the potential of GIS software involves properly designing the hardware infrastructure needed to support it. We realized this early on and, almost 20 years ago, put together a team at ESRI to focus on system design for GIS.

We got Dave Peters on board—a physicist, pilot, and engineer who had been putting systems together for the U.S. Air Force—and installed him in a small office next to mine so I could keep an eye on him. He became the lead systems architect for ESRI and the author of an invaluable technical reference document, *System Design Strategies*, on which this book is based.

Dave Peters now has more than 27 years of systems integration experience, yet he insists that we have our customers to thank for the wisdom in his book. He's right. Through observing and listening to their experience with ESRI software technology, we have documented each new lesson learned from consulting and customer feedback. The system design models and tools in this book owe their reliability and usefulness to that background.

We have offered consulting services long enough to know that there is no one way to configure a system that fits every organization. In fact, there are probably as many valid ways to go about designing a system as there are organizations that need one. The approach detailed in this book isn't the only way to do the job, but it's an approach based on system design experience spanning three decades—the logic and principles of system design discussed here are time-tested.

Like its author, who is both a teacher and an inventor of creative solutions, this book does double duty: it offers novices a handle on the classic, technical fundamentals while lending experts the latest, state-of-the-art planning tools and models. This book is unique in its field and stands alone, but it is also the second of two ESRI Press books that focus on helping organizations and their teams of planners implement GIS. The first book, *Thinking About GIS: Geographic Information System Planning for Managers*, was written by Roger Tomlinson, and Dave worked with Roger on that book's system design chapter. Roger found Dave's advice so sound and his ideas so groundbreaking that he suggested to ESRI Press that Dave write a book of his own.

Dave Peters is a very pragmatic visionary—the best kind in the computer industry. He is constantly applying his foresight and imagination toward developing simple, universal tools that solve complex, real-world problems. His solutions, always on the cutting edge of change, take into account new and improved Web mapping services software, multicore server technology, and integrated system performance. Dave's focus is on how to take advantage of these advancements in technology, which promise reduced costs as well as improved user productivity. The platform sizing models in this book reflect the recent advances in technology; other tools are here to help you configure your network to handle the increasing volume of traffic incumbent with making good use of expanding Web services.

Today, more than one million people worldwide use GIS for thousands of different reasons. And the technology is poised to make an even greater impact in society as more organizations begin to find GIS as vital as their workplace computers. Therefore, planning for GIS has never been more important than now, and this book offers an easy-to-use format everyone on the planning team can follow. Also, in teaching ESRI's class "System Architecture Design for GIS" in many countries over the past two years, Dave has found that the international demand for the state-of-the-art Capacity Planning Tool used in his class has grown so widespread that offering it with this book seemed the best way to meet the need.

We want you to have a pretty good idea of what you might be getting into when you implement a GIS in your organization, not only the potential of GIS software but also the hardware infrastructure needed to support it. Depending on how you use it, GIS can put a fairly heavy load on a system, and we want you to be prepared for that. Of course, there are positive reasons for this: GIS is capable of handling a great deal of work, so it needs a system adequate to the task of supporting all that accomplishment.

In the final analysis, you will be the one designing a system for your organization that will result in successful GIS operations. We hope this helps.

Jack Dangermond

Preface

Before I was a systems architect, I was a physicist and engineer, and before that, a pilot. I remember more than one mission when I found myself flying in the clouds. What did I do? I learned to trust my instruments. It's very dangerous to fly by the seat of your pants in that situation; when the stakes are high, you have to trust your control panel. The stakes are very high when designing a system for technology, too, yet you'd be surprised by how many people think they can "feel their way" to a proper system design. In contrast to that, I use the performance models in this book to guide my design decisions. As a systems architect, I've created an instrument panel from these models, the Capacity Planning Tool, or CPT (on the CD accompanying this book), to automate many of the system design analysis tasks. These models—developed over many years of experience and observation, learning from geographic information system (GIS) users what worked and what didn't—form the methodology offered to you here, as you seek to mitigate risk and ensure the success of your GIS implementation. The system architecture design methodology in this book has already led to thousands of successful GIS deployments by ESRI customers over the last two decades.

This book is the second in a growing series establishing the fundamentals of GIS planning. The series' mission is to promote more effective GIS planning, especially for large-scale projects, in which planning will significantly improve, if not guarantee, the chances of implementing and maintaining successful GIS enterprise operations. In addition to its focus on the system design process itself, this book puts its emphasis on integrating planning with system implementation and performance tuning. Geared toward the same broad readership as that of Roger Tomlinson's classic, *Thinking About GIS*, this companion book further bridges the communication gap between those who bring different specialties to the planning effort: CEOs and CIOs who are not expert in technology still need to assess what's cost-effective; IT directors and systems administrators need to understand how GIS processes may be more demanding on a system than other software; and GIS managers, specialists, cartographers, and developers may be surprised to learn how many different ways their work factors in to building a high-performance and scalable platform for GIS.

The book is structured to reach all the various stakeholders with the information pertinent to their role. Each chapter begins by emphasizing the basics in an introduction that, like an executive summary, is intended to make the kind of information decision makers require more accessible to them. Upper managers trust their hands-on people to handle the details, but they still need a grasp on the fundamentals in order to monitor the project and direct those charged with moving it forward to successful completion. Information grows progressively more detailed as you move on through each chapter, with the numbers you need charted in tables and plenty of graphics to illustrate the specific topic or component and its relationship to system design. You can use the Capacity Planning Tool on the CD to support any system design project. Also on the CD are PDFs of the chapters describing the fundamentals on which the CPT is based (chapters 7-9), how to use it (chapters 10-11), and classroom exercises, useful in reviewing the issues of primary importance chapter by chapter.

As a physicist with a specialty in engineering, I've been putting systems together for 27 years, 18 of those years as systems consultant and lead technical architect for ESRI. This book is a product of both types of experience and describes an approach to system architecture design specifically intended to support successful GIS operations. With it in hand, you have a time-tested process and a practical tool to help you plan the infrastructure—system configuration, platform sizing/selection, and network bandwidth—to sustain the GIS specific to your needs. My goal is to provide clarity on the fundamentals of system design as they relate to GIS, and specifically to GIS software by ESRI, the world leader in developing geographic information systems and my employer. While the latter two are reasons enough for the book's examples to be ESRI-software-centric, there's a stronger rationale for this focus. ESRI started developing and testing models almost 20 years ago to figure out how to put systems together to support what our clients do. Working with colleagues, customers, and test teams at ESRI, through trial and error and benchmark and performance testing, I began documenting what we were learning in a technical reference document called *System Design Strategies*, updating it at least twice a year since 1992 in order to share with our customers over the Web what we were learning.

Flying in the clouds was good preparation for a systems engineer starting to work in a software company. When I first got here, users were asking, What type of hardware should we get? How should we configure it? We were fortunate that customers asked these questions early on because it got us thinking about it and testing our ideas in the lab as early as 1991. By 1992, we started using the Standard Performance Evaluation Corporation (SPEC) benchmarks, published by the hardware vendors, to adjust for hardware performance changes. SPEC benchmarks are results of tests set up by the Standard Performance Evaluation Corporation, used to compare the performance of hardware. Over the years, this has given us numerical values to reference from the past to compare to present numbers. Since 2000, we've joined with our development staff

to conduct controlled performance validation testing to better quantify performance and scalability of each software release. Now we can demonstrate what we've been talking about for years in terms of what can be done in capacity planning, using the models and the Capacity Planning Tool in this book. The CPT includes templates for defining enterprise workflow requirements and calculating peak system capacity based on vendor platform selection and system configuration. It can be a tremendous aid in planning a successful GIS implementation—at first launch and at the least cost—when you use it beforehand to model the various system options under consideration.

But a word of caution: powerful as it is, the CPT is no substitute for understanding the fundamentals of system design, which are also explained in this book. I know from our customers and from students in my classes that it's very tempting to think you can just plug in the numbers and let the CPT do the rest. It is quite a useful tool insofar as you can try out various configurations in theory before buying and implementing the real thing. As a realistic model based on real-world experience, it can get you within range of the specs you need. But a model is only as good as the numbers you put into it. And what if the real world changes, as it inevitably does—we have seen big, rapid changes in the last few decades—how are you going to tweak the model so that it faithfully conforms to the new realities? We need to talk about the real world the models represent, as we do here, because models are just tools. A tool is good, but only in the hands of a person who knows how to use it. And a tool does its best work when used exactly as intended. You wouldn't use a hammer to drive in pilings; and you wouldn't start to build without a blueprint either. You need to know what you need out of the GIS and to describe the information products that will provide it, before you are ready to develop your design solution. This is why a user needs assessment—documenting exactly what your organization requires of a GIS—is prerequisite to designing the system for it.

In any cost-benefit analysis, the advantages of modeling a technology solution before committing resources to it are enormous and manifold. And the present time in particular is a strike-while-the-iron-is-hot moment in the evolution of computer technologies. Technologic advancements in software, hardware, operating systems, and the Web have recently converged to offer opportunities for more functionality and efficiency at less cost. With the new multicore hardware technology, for example, a dollar buys twice the processing power it bought before. Software offers more functionality and interoperability for the same licensing fees.

We have developed models over time to support the systems integration of these advanced and evolving technologies. And now, what we anticipated in the last decade as the next generation of GIS has arrived. The future is here, and not only that, the wherewithal to grow into—and seamlessly adjust to—future changes is here as well, in the scalability of both software and hardware architecture. Future changes will enable more and more GIS sharing of a variety of geographic tools and information layers for use in the new Web 2.0 space. This, in turn, will enable building more and more federated solutions that share pieces of information for use in user workflow collaborations. ArcGIS 9.3 opens the door for a bright and challenging future with a great many new opportunities. But in order to take best advantage of the opportunities incumbent in all these advancements, you must understand how all these components work together. And you must work together well with colleagues, who bring different but necessary expertise, in order to plan and maintain systems that work well together, too. This book is intended to help you do both.

Acknowledgments

This is only the beginning of the story—technology is changing every day, and we have only scratched the surface. The practical view of technology shared in this book evolved over a period of more than 18 years with the help of many people. It is an effort to share what I have learned from others—those things, large and small, that help us build a successful GIS. The special people include our system design consultants, my associates Tom Pattison and Ty Fabling, and our Enterprise Test Lab team led by Jeff DeWeese. These dedicated professionals worked with me for many years to test and document what we learned. A special thanks to Jack Dangermond for his trust, encouragement, and belief that the right system design can make a difference for our customers—and for taking a chance by letting us share what we learn about the technology. This book would not be possible without the vision and encouragement of Dr. Roger Tomlinson, who recognized and accepted me as a partner in sharing a great technology. His fatherly advice and attention to detail have been a positive challenge that kept me honest over the years. Thanks to the ESRI Development team, who worked with us to understand the technology. Thanks to ESRI Professional Services, who accepted us as part of their implementation team. Thanks to the ESRI Learning Center for encouraging us to share our experience as part of their training program. And thanks to the hundreds of ESRI customers who invite us to work with them to design and implement successful GIS operations. Thanks to Candace Hogan, who put the words to my ideas and who gave this book a life—her patience and encouragement made this book possible. Thanks to the years of love and respect from my wonderful wife Julie—one who believes anything is possible if we do it right. Thanks to my daughters, Jennifer and Tera, who never lost faith in their dad; and a special thanks to Sheryl and to Steve for adopting me as theirs. Thanks to a faith beyond understanding that can guide us to the truth. There are many others that have contributed in their own special way, and I am thankful for them all.

Dave Peters

Part I

Understanding the technology

1   System design process

This book describes an approach to system architecture design specifically intended to support successful geographic information system (GIS) operations. With it in hand, you have a process—not a prescription—and a practical tool (see CD) to help you plan the infrastructure for a GIS specific to your needs. Knowing what you need is all-important to this process, and it is the assumption of this book that you have already identified what your organization requires from a GIS. This is because system architecture design defines the relationship between peak user workflow performance needs and the hardware and network infrastructure required to meet them.

The purpose of this book is to provide managers with not only the tools but also the information to help them build a successful infrastructure for GIS. A fundamental understanding of the supporting information technology, GIS performance and scalability, and the models and methods in this book fosters the type of management framework that leads to success. These fundamentals of system architecture design could help in building a system for any technology. But we consultants at ESRI have applied them to guide our customers to successful GIS deployments.

We are sharing our understanding of the technology (chapters 1–6), how the technology comes together to fulfill user performance requirements (chapters 7–9), and the process for completing a system design (chapters 10–12) in which user requirements lead to hardware specifications. If you are a system architect, you may be tempted to jump to chapter 11 where the system design process itself is spelled out step by step. Don't do it; rather, look before you leap. Even system architects are not ready to begin until they understand the technology and the Capacity Planning Tool (CPT), within the first 10 chapters, because the technology has changed so dramatically, so recently. (We do not send our system design consultants to a customer site until they understand what's inside the first 10 chapters of this book.)

There used to be a time when it was possible to simply define user requirements, select a software solution of choice, and then identify hardware and network specifications to fulfill system capacity needs. Nowadays it's more complex. Selecting the right technology requires a more specific understanding of GIS user needs and system infrastructure limitations. Enterprise GIS operations today are supported by a variety of software and hardware technologies that must work together to meet user performance needs when operating at peak system capacity. Technology selection must take into consideration performance and scalability requirements and infrastructure limitations. An integrated business needs assessment—evaluating GIS user needs along with projected system performance and scalability throughout the requirements definition process—is the recommended way to ensure deployment success.

For this reason, system architecture design—better described as an integrated business needs assessment—is not at heart a step-by-step process anymore. It is more like working on a puzzle. There are procedures within the process, such as those that must be followed during a project implementation, but first, as in a puzzle, a system must be understood as a whole before it can be properly put together. Like the pieces of a puzzle, many interacting components compose the whole, each related to another in a very special way. Each special relationship emerges during the system architecture design process to reveal to you its significance to your design. The components of this puzzle are what we examine in *Building a GIS*, and our discussion is divided into three parts.

Part I: Understanding the technology
Chapter 1 presents the very high-level view, such as identifying the pieces of the system design puzzle (figure 1-1). The next five chapters, which complete the first section, describe these puzzle pieces, answering the following questions:

Chapter 2. Software technology—what are the software technology options?

Chapter 3. Network communications—what effect will a given technology solution exert on the existing infrastructure?

Chapter 4. GIS product architecture—what configuration options match up with your availability and scalability needs?

Chapter 5. Enterprise security—how do you adjust your architecture solution to accommodate your security needs?

Chapter 6. GIS data administration—what are your options for maintaining and providing the required GIS data resources?

Part II: Understanding the fundamentals

The relationships between the pieces of the puzzle are expressed in terms of system performance and cost. The system performance relationships are presented in the second part of the book, chapters 7–9. You can obtain information about cost from the software and hardware vendors, once you identify the right solutions for your organization.

Chapter 7. Performance fundamentals—an overview of the performance relationships between the pieces of the system design puzzle.

Chapter 8. Software performance—an overview of software performance considerations: how will your technology decisions affect system performance and scalability?

Chapter 9. Platform performance—what do the vendors have to offer, and how do you select the hardware that will provide the required processing power?

Part III: Putting it all together

Putting all the pieces together completes the puzzle. The CPT and the actual system design process itself are presented in the final three chapters.

Chapter 10. Capacity planning—a fully illustrated and detailed description of how to use the Capacity Planning Tool (a composite of several integrated information management tools that identify user workflow requirements and select the right hardware and network solution), including how to customize it for your own situation.

Chapter 11. Completing the system design—a walk through the system design process, using a case study to show how to bring the pieces of the puzzle together in the system architecture.

Chapter 12. System implementation—guidelines for implementation of the selected design solution, which take into account the maintenance and tuning that follow.

The CPT on the CD

The performance components that contribute to success are clearly defined and modeled in the CPT, which is provided on the accompanying CD and thoroughly described in the book. Templates are provided in the CPT for collecting user requirements, as are standard workflow models that translate peak user loads to selected platform processing environments. The CPT translates the peak user workflows for you as specific platform and network specifications, applying the performance models presented in chapter 7 so that you can use it to evaluate which mixes of technology will best meet your needs. In doing so, you may find the selected software technology will not support users over the available network infrastructure. You may find your favorite vendor does not provide the best hardware for your preferred solution. You may find you need to upgrade your network infrastructure, or change your business processes to conform to a more distributed architecture. As you change each piece of the puzzle, its relationship with the other pieces of the puzzle will change and the overall solution will be reevaluated. The CPT is intended to make this process of reevaluation easier for you. For each solution, the CPT takes each component into consideration and evaluates performance of the integrated system: Does the selected solution meet your organization's needs? If not, what other alternatives will better support your needs?

Also on the CD, you'll find teaching and learning aids, including a list of multiple-choice questions associated with each chapter. These might be useful for teachers and students, or simply for readers interested in reviewing the main points of discussion.

Asystem is only as strong as its weakest link. Enterprise GIS operations are supported by a large number of infrastructure components, wherein the weakest component will limit system capacity and user productivity. Selecting the right software technology, building proper applications, establishing an effective database design, and procuring the right hardware all play a critical role in fulfilling system performance and scalability expectations. Understanding your infrastructure needs before you buy can make the difference between success and failure and significantly reduce the overall system deployment cost. The key to system design success is understanding what you need, establishing measurable performance milestones, and managing incremental progress toward performance and scalability goals throughout the development and deployment process.

Computers are part of our life because they save people—and organizations—time and money. Tasks that used to take hours or weeks can now be done in minutes, even seconds. People use GIS for the same reason they use computers: because it makes them more productive, more efficient at work. Anything that can improve efficiency and productivity is welcome. System performance—how fast and reliably the components of a computer system can process an application and display the results for you—directly contributes to user productivity. You need to configure all the elements in a system—software, hardware, network communications—so that they work well and efficiently together in doing what you need to get done.

Technology is changing, and unless people take the time to understand it, they can get in over their heads and suffer performance problems. Have you ever experienced a problem with a system at point A, only to discover that the actual cause of the problem was located way over at point E, a seemingly unrelated area? Usually these realizations come after spending quite a while exploring a lot of dead-ends trying to get to the root of the matter, often settling for a temporary workaround. No one needs to tell you, these times spent chasing symptoms interrupt the workflow, and if they happen over and over again, can significantly reduce productivity. They also could be a sign that there may be something wrong with the system design.

Far and away better, of course, is to get the system right the first time around. Then you can spend your tinkering time doing serious planning and maintenance, to keep your system at the ready to adjust, as advancements in technology and organizational goals present new opportunities for growth through change. Getting the system right means achieving *performance*: a proper system is one that works the way you want and as fast as you need. A system able to take advantage of opportunities and grow is a *scalable* one.

The key to achieving both performance and scalability —the prerequisite for getting the system right and for keeping it right as the demand on it grows—is the same task: understanding the nature of each component of a system, the interrelationships between components, and how changes in one affect the other. Each component technology does have an affect on overall system performance. The reason you might not have known right away the symptomatic connection between point A and point E in the example above is no doubt due to the intricacies of these interrelationships. They seem complex, but after we've sorted them out, you will see how simple system design really is.

System architecture design for GIS

Computer systems were first used to automate cartographic map production in the late 1960s. Before that, geographic analysis, a method for displaying the relationship of many layers of geospatial information, relied on more time-consuming methods. Traditional methods of creating Mylar representations of each dataset and overlaying the layers of spatial data could take weeks to complete a simple information product. Early computers were able to automate the analysis process and reduce map production timelines from weeks to hours. Current computer technology can complete this same type of analysis and generate dynamic map displays in less than a second.

The term *geographic information system* (GIS) was introduced by Roger Tomlinson, and was later promoted by professors at Harvard University in the 1970s, inspiring several geographic consulting companies to develop and expand GIS technology. One of those companies was ESRI, beginning in 1969. Consider that its software runs on more than one million desktops in more than 300,000 organizations now, and you can see that the evolution of GIS has followed the advance of computer technology.

Local government and business started deploying large GIS operational systems in the early 1990s, and soon it became clear that the success of such distributed GIS operations was strongly connected with an understanding of GIS performance and scalability in a distributed computer environment. Distributed GIS operations were characterized by high performance computer systems coupled with a high demand on network communications. GIS deployment required several years' investment in data migration and custom application development. That, along with the required system infrastructure investments, was expensive. An understanding of system design requirements was critical in supporting successful GIS deployments. This is why a software company became so interested in hardware system performance early on.

My initial responsibility when I joined ESRI was to develop a team that would support successful implementation of turnkey ("out of the box") GIS sales. Part of that was the acquisition and installation of hardware and software technology adequate to the task of supporting an organization's GIS workflows—or, as we call them, GIS peak operational system level requirements. We established a Systems Integration Department with four specific teams (system architecture design, project management, system installation, and project control). We helped complete several hundred successful GIS turnkey project implementations between 1992 and 1998, and learned a few things along the way. The best practices developed to facilitate implementation of turnkey GIS sales were also used to address failed system implementations. We learned from our mistakes and those of our customers.

Technology has changed tremendously, but the three fundamental building blocks required for each project implementation have not changed in all this time:

1. A clear definition of the peak user workflow requirements (user requirements analysis)
2. A clear understanding of infrastructure requirements to support the peak user workflows (system architecture design)
3. An implementation strategy that provides proper systems integration management from initial contract authorization to final system acceptance (project management)

A variety of best practices were soon established to facilitate a proper user requirements analysis. It had been evident during earlier implementations in the 1970s and 1980s, and then became abundantly clear, that identifying your user requirements—knowing exactly what you need to get out of a GIS (being able to describe the information products)—was essential for a GIS implementation to be successful. Implementations without a clear goal and purpose would fail.

Distributed processing systems introduced in the 1990s required a proper platform and network infrastructure to ensure the GIS benefits identified in the user requirements analysis could be achieved in a distributed environment. Many systems were deployed without a clear understanding of the system infrastructure requirements, and many of these early systems failed to meet user performance expectations.

In the early 1990s, I began to develop a system architecture design process for ESRI that would identify the proper hardware and network infrastructure required to support successful GIS implementations. Like a system itself, this process has been developed, maintained, and fine-tuned over the years, through working with colleagues and clients who have helped

System Architecture Design
What is System Architecture Design?

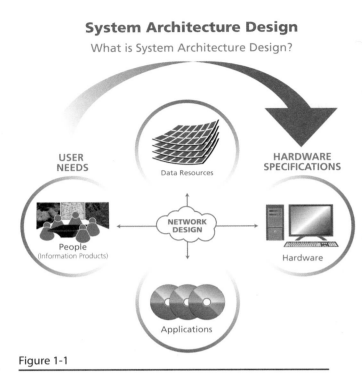

Figure 1-1

System architecture design process.

make it as productive as it is today. GIS is usually integrated into an existing system, so this design process takes into consideration an organization's infrastructure limitations. The process can be used to make specific recommendations for hardware and network solutions that are actually based on existing and projected user needs. (You'd be surprised how many projects just guess.) Based in reality, such system designs reduce implementation risk and are more likely to be approved. In any case, a system design is a prerequisite for gaining the go-ahead for your GIS project from upper management.

This system design methodology recognizes that people, application technology, and selected data sources are equally important in determining the optimum hardware solution, as shown in figure 1-1.

- **People**: Understanding user needs—information products and the procedures for making them—establishes a rationale by which to estimate peak system workflow loads.
- **Applications**: Software technology determines the processing requirements (system loads) that must be handled by the hardware solution.
- **Data**: The type of data source and what's required to access it (data access requirements) show you how the processing load is distributed across the system architecture.

What people need; what it takes to create what people need; and how all the components of a system work together—every day and reliably—these are the elements that comprise a balanced system design.

The ESRI system architecture design process provides specific deployment strategies and associated hardware specifications based on identified operational workflow requirements.

Why it's important

Vendor computer hardware and network infrastructure expenses represent a significant percentage of the overall system cost of deploying and maintaining distributed GIS operations. For many implementations, costs for hardware procurement and information technology (IT) administration and support exceed the cost of the GIS software. These costs must be identified during the project approval process and managed throughout system deployment to ensure that resources are available for implementation of the technology.

In order to define the overall project requirements, you must understand the relationship between GIS user needs, GIS software processing requirements, and computer hardware and network infrastructure performance and capacity. Understanding the technology (both GIS software and vendor hardware) is fundamental. To meet peak user performance requirements, a distributed computer environment must be designed properly.

Standard IT best practices confirm the importance of deploying a balanced system environment. The weakest link in the design will limit system level performance. User productivity can be improved by monitoring system performance, identifying system performance bottlenecks, and spending available system resources to upgrade and expand capacity of these "weak links" in the system design. Understanding the distribution of software processing loads and the network traffic generated by GIS workflows during peak operations is essential to selecting the right hardware and network bandwidth to handle and transport these workflow loads. That's why such understanding is the foundation for the system architecture design process.

Since 1992, ESRI has given my team the opportunity to maintain and continue to improve a system architecture design process that can provide specifications for a balanced IT solution. Investment in hardware and network components based on a balanced system load model provides the highest possible system performance at the lowest overall cost, as suggested by figure 1-2. In it, the chain represents the several factors that are linked in a system and therefore affect system performance. User workflows must be designed to optimize interactive client productivity; work request queues should be established to manage heavy batch

processing loads and enhance user productivity. Some information products are slower in the making than others, so user display requirements should be carefully evaluated: do you really need such a high-quality map when a simple one would do? Simple information displays promote quicker display performance and therefore a more productive user interface. The same holds true for the geodatabase design and database selection: how can they be optimized to make the best of user performance and productivity? If complex data models are needed for data maintenance, then possibly a simple distribution database replica could balance that out by providing high-performance view and query operations. The system platform components you select (servers, client workstations, storage systems) must perform adequately and provide the capacity to fulfill peak user workflow requirements. In addressing performance needs and bandwidth constraints over distributed communication networks, the system architecture design strategy should strike a balance between power (or quality) and economy (efficiency). The technology and configuration must be selected to conserve these shared infrastructure resources.

The weakest link (performance bottleneck) in this chain will determine the final system performance and capacity. That's why, in building a foundation for a productive operational environment, the system architecture design process must take into account every component of the system.

The primary infrastructure components contributing to system performance are identified in figure 1-3. These components include the user, application, and data source and how they are connected. What comes of their connections and relationships in the service of a system is the stuff of which performance models are made.

Almost every interrelationship can be factored in for use in your thinking about and modeling of a system. The peak number of concurrent users and their location represent the user access requirements. What's required

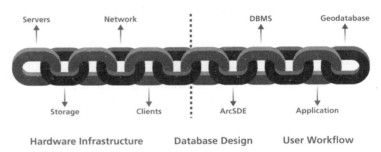

Figure 1-2

A balanced system load model leads to the highest system performance at the lowest overall cost.

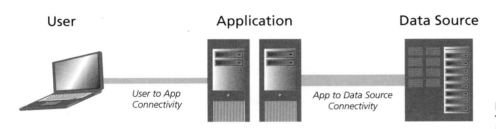

User   Application   Data Source

User to App Connectivity   *App to Data Source Connectivity*

Figure 1-3

Performance components.

in terms of display response time and user think time translates to the "user productivity needs." The application workflow establishes what you need in the way of platform capacity to satisfy the display service time requirements. The data source determines the database server capacity and storage requirements. The application and data architecture, combined with peak user workflow requirements, establish the network communication requirements.

Success with GIS

Selecting the right technology is the first step toward deployment success. Enterprise systems are composed of a variety of vendor components that must work together effectively to be productive. Vendor technologies are constantly evolving, introducing new functionality and hardware design enhancements. Software and hardware vendors develop their technology to support what they believe is generally accepted community interface standards. Ideally, all these vendor solutions will work together as a seamless system environment. In reality, both technology and standards are constantly changing. Technology that is used most together works best together. New technology introduces risk, and in most cases must be integrated with other system components over some period of time for interface anomalies to be identified and resolved. It can seem as if, just when you figure out one, another element comes along and you must figure it out all over again. Each component introduced, whether it's a first launch or just an upgrade, must be reevaluated in the context of the whole. So how often are you going to get lucky if you do this all by happenstance?

This is where planning comes in, as well as the obligation to keep abreast of—so that you can keep ahead of—change. Figuring out the design may be like a puzzle, but implementation should happen in stages, with careful planning.

There are several critical deployment stages that lead to a successful implementation. Understanding the importance of each stage and the key objectives within them that aim at success results in more effective enterprise implementations. Figure 1-4 shows the

different stages of GIS deployment and the advantages of managing implementation risk in stages. The cost of having to make a change in course increases exponentially as the project implementation proceeds. A change or correction made toward the end of the process can cost a thousand times more than if you'd identified the problem and made the change at the beginning.

Requirements stage
Understanding technology alternatives, quantifying user requirements, and establishing an appropriate system architecture design deployment strategy are critical during the requirements stage. Capacity planning during this phase can establish preliminary hardware performance specifications and validate that budget and performance expectations can be achieved. This is a planning stage where "getting it right" can save considerable effort and money by making choices that will lead to a successful deployment.

Design stage
System development and prototype testing validate functional interfaces and performance metrics. Functions must work and performance targets must be met to enable follow-on deployment. This is where time and money are invested to build and test the selected environment. Initial prototype testing demonstrates functionality and reduces system integration risk. Preliminary software performance testing can validate initial capacity planning assumptions and confirm that performance targets can be satisfied.

Construction stage
A successful initial system deployment can set the stage for success. This is where the solution comes together and final operational support needs are validated. This is an important time to demonstrate performance and capacity of the deployed system and validate that the selected hardware and infrastructure will support full production deployment.

Implementation stage
Final system procurement and deployment demonstrate operational success. Capacity planning metrics can be used to monitor and maintain system performance

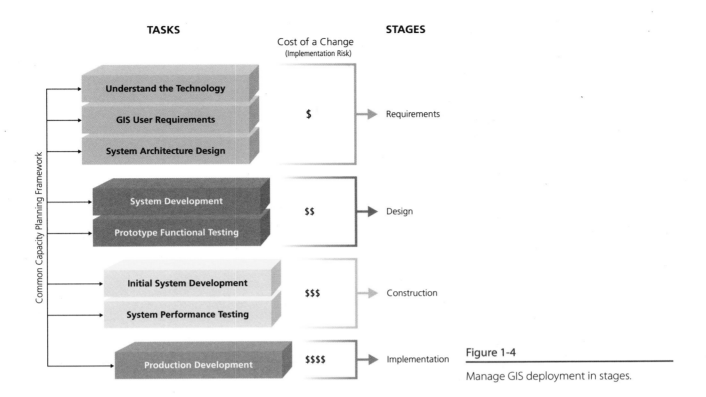

TASKS

STAGES

Cost of a Change
(Implementation Risk)

Common Capacity Planning Framework

Understand the Technology

GIS User Requirements

System Architecture Design

$

Requirements

System Development

Prototype Functional Testing

$$

Design

Initial System Development

System Performance Testing

$$$

Construction

Production Development

$$$$

Implementation

Figure 1-4

Manage GIS deployment in stages.

goals and guide system performance tuning. Good planning, development, and testing will result in a smooth deployment, productive operations, and satisfied users.

In the 1990s, much of the system cost was in data migration followed by a relatively high cost for the technology (software, hardware, etc.). Systems would be deployed over several years, which would spread costs over the system life cycle. The environment today has changed. Data is much more readily available (much faster acquisition time), and still leads as the highest overall GIS investment. Hardware and software costs have dropped by orders of magnitude and are approaching commodity prices.

Technology is moving much faster these days. Implementation timelines are short, and having to make a change in deployment strategy or having to deal with delivered systems that do not perform is an opportunity cost (lost revenue and additional labor) that can be significant. Managing implementation risk continues to be a top priority.

Getting it right from the start is best done by taking the time to understand the technology, quantify user requirements, select the right software technology, and deploy the right hardware. Not getting it right from the start will cost money to make it right later, either in system cost or in labor and time. Sometimes, if repeated failures erode upper management's confidence, the cost is that you don't get to have a GIS at all and the potential benefits are never realized.

What is the system design process?

The traditional GIS planning process includes a GIS needs assessment and a system architecture design. The GIS needs assessment identifies what you want out of the system, and describes specific information products, data requirements, and application needs that will improve business decisions and workflow productivity. The system architecture design uses system design guidelines and platform sizing models to establish vendor hardware specifications and identify peak network traffic requirements. Platform performance specifications are calculated based on user workflow requirements pinpointed in the GIS needs assessment.

Timing is important in the decision-making process, so the most effective system design approach is one that considers user needs and system architecture constraints throughout the design process—an integrated business needs assessment. It's a kind of holistic approach, if you want to look at it that way, but methodical nonetheless, and one intended to take into account everything that affects user productivity. Figure 1-5 provides an overview of the system design process as an integrated business needs assessment.

GIS needs assessment
The GIS needs assessment identifies how GIS technology can be leveraged to improve organizational or business operations. A fundamental objective is to clearly understand and identify what information products

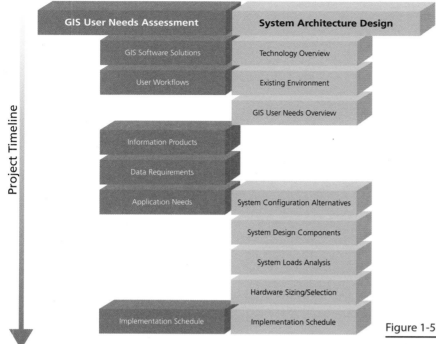

Project Timeline

Figure 1-5

System design process as an integrated needs assessment.

you want out of the GIS. Once you understand what information you want from the GIS, you are prepared to identify GIS application and data requirements and develop an implementation strategy for meeting these identified needs. The user requirements analysis is a process that must be accomplished by the user organization (improving user workflows, identifying more efficient business practices, and setting organizational goals and objectives). A GIS professional familiar with current GIS solutions and customer business practices can help facilitate this planning effort.

System architecture design

The system architecture design is based on user requirements identified by the GIS needs assessment. You must have a clear understanding of your GIS application and data requirements before you are ready to develop system design specifications. System implementation strategies should schedule hardware purchase requisitions "just in time" to support user deployment needs.

The system design normally begins with some level of what's called a "technology exchange"—in other words, people working together to understand and share what they need to know (the information in the first 10 chapters of this book). User participation is a key ingredient in the design process. Once the technology exchange is complete, the design process includes a review of the existing computer environment, GIS user requirements, and current GIS design alternatives.

The system design capacity planning tools translate peak user workflow requirements to specific platform specifications. An integrated implementation strategy is developed to achieve GIS deployment milestones.

Traditionally, the user needs assessment and the system architecture design were two separate efforts. But there are some key advantages in completing these efforts together. GIS software solutions should include a discussion of architecture options and system deployment strategies. The existing hardware environment and information on peak user workflows and user locations can be identified during the user needs workflow interviews. Technology selection should consider configuration options, required platforms, peak system loads for each technology option, performance and scalability, and overall system design costs. And finally the system implementation schedule must consider hardware delivery milestones. A primary goal of developing the new Capacity Planning Tool presented later in this document is to automate the system architecture design analysis in such a way that GIS professional consultants will be able to use the Capacity Planning Tool to complete an integrated business needs assessment, considering system architecture design implications throughout the user needs assessment process.

An integrated business needs assessment (user needs and system architecture design) is by far the best way to complete your planning process. You can easily integrate the system architecture design process with your user needs assessment by understanding part I and part

II of this book, and using the Capacity Planning Tool. You can use this tool—and the several modules within the tool—to document your workflow requirements and support your technology decisions during the user needs assessment process: this will be your integrated business needs assessment. In this step-by-step process for completing the integrated business needs assessment (in chapter 11), we will focus on the system architecture design steps, and touch lightly on the user requirements analysis. (Roger Tomlinson's book, *Thinking about GIS: Geographic Information System Planning for Managers*, provides the more complete step-by-step process for the user requirements analysis.)

Figure 1-6 provides an overview of the information management capabilities of the CPT that are available during the planning process. Step-by-step examples of how the CPT will empower your integrated business needs assessment is provided in the City of Rome case study in chapter 11. You need to complete part I and part II of this book to take full advantage of part III during your planning process.

You can use the CPT to provide an overview of your business needs, infrastructure specifications, and to establish your hardware vendor platform requirements. The CPT shows you the expected system performance for a variety of technology options. You can identify your network and platform environment and see if the technical solution you are considering can work

for you. Technology decisions can be made based on a full understanding of user workflow requirements and properly established system performance expectations. Once the system is operational in your environment, you can continue to base technology decisions on a complete understanding of your system performance and scalability. Establishing performance target milestones, based on credible information about the technology, will reduce implementation risk and build a framework to manage your implementation success.

Primary and secondary factors to consider

Distributed GIS solutions bring together a variety of vendor products. Each vendor technology is a part of the total enterprise that must be integrated into the existing system environment. Integration of any multivendor environment is made possible through voluntary compliance with generally accepted industry interface standards. When each new component is integrated into the system, the entire functionality, performance, and security of the system can be affected, and the following primary factors must be considered:

Functionality: Does the integrated system meet your functional workflow requirements?

Performance: Will the integrated system satisfy your performance needs during peak workflow operations?

Security: Does the final system environment support your enterprise security requirements?

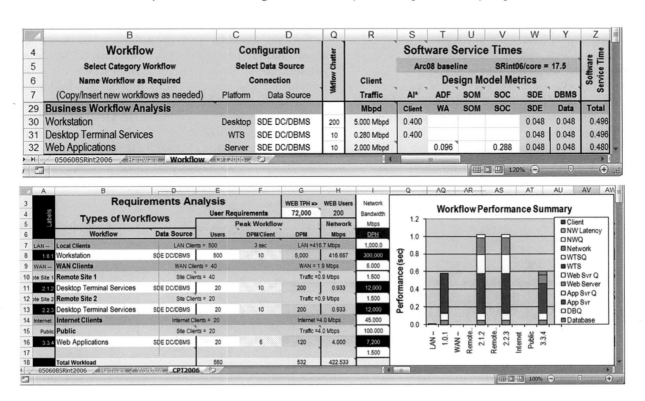

Figure 1-6

The CPT interface when using the Capacity Planning Tool to manage an integrated business needs assessment.

The primary design factors of functionality, performance, and security often dictate the initial system architecture design strategy. In many cases, policy issues, driven by established compliance factors or other technical or nontechnical issues, limit platform technology options and restrict deployment on older software versions. At first glance, the system architecture design can appear to be preordained, requiring only a simple implementation decision.

However, there are several secondary design factors that should be reviewed, since they often dictate implementation success or failure. These factors include the following:

- Cost considerations
- Scalability (ease of handling more users or higher volumes)
- Reliability (eliminating single points of failure)
- Mobility (support field editing or viewing)
- Availability (dependence on internal and external services)
- Quality of service or data
- Software stability
- Maintainability (centralized versus distributed architecture)
- Flexibility (adaptability to change)

A review and understanding of these secondary design factors in the context of your specific implementation strategy is necessary to ensure success. A complete design process will include a review of both the primary and secondary design factors to ensure proper technology selection and appropriate deployment strategy.

What system architects do

System architects establish the target architecture during the system design process; purchase and install the hardware needed to support the design; and resolve any performance issues during final implementation. Once the design is approved and the project is funded, the system architect is responsible for a final review

and update of the hardware specifications right before procurement. He or she is also responsible for scheduling the vendor installations and for participating in the monitoring and testing of system functionality and performance at each of the implementation milestones (see chapter 12).

Enterprise GIS environments include a broad spectrum of technology integrations; in other words, our systems include a lot of stuff that needs to play well together and the system architect has to make sure it does. Most environments today are composed of a variety of hardware vendor technologies, including database servers, storage area networks, Windows Terminal Servers, Web servers, map servers, and desktop clients—all connected by a broad range of local area networks, wide area networks, and Internet communications. All these technologies must function together properly to provide a balanced computing environment. So if the business did not begin with GIS, the system architect will be integrating new systems with existing business operations when you "go live" or launch a GIS. A host of software vendor technologies, including database management systems, ArcGIS Desktop, and ArcGIS Server software, Web services, and hardware operating systems—all must operate seamlessly with existing legacy applications. (Data and user applications are added to the integrated infrastructure environment to enable the final implementation.) The result is a very large, mixed bag of technology that must work together properly and efficiently to fulfill user workflow requirements.

Final purchase decisions are influenced by both operational requirements and budget limitations, introducing unique challenges for the system designer. Good leadership, qualified staff, and proven standard practices lead to successful deployments, so a wise system architect gets a team together for the GIS project from the start, to build from there. (Tomlinson's book, *Thinking About GIS*, describes the ideal GIS project team composition and methodology from the beginning stages of

Building a Solid Solution Strategy

Figure 1-7

Building blocks of the technical foundation supporting a distributed GIS environment.

planning for a GIS. The book you have in hand takes off from there.)

The building blocks

A general understanding of the fundamental technologies supporting a GIS enterprise solution provides a foundation for the system architecture design process described in this book. Figure 1-7 identifies the key technology building blocks underlying a distributed GIS environment. Each of these building blocks is a chapter in this book, which is divided into three parts, with the intention of first empowering you with the knowledge and tools to manage your own solution before you take action. The first part is comprised of chapters that describe the various components of a system. Part II describes the underlying principles of physics and engineering that determine how these pieces interact with each other. In Part III, having gained a deeper understanding of the technology relationships, we are ready to put the pieces together into a balanced system that will perform and scale, which is the goal of system architecture design for GIS.

Planning for success

As you know, technology is changing very rapidly. Most enterprise GIS deployments evolve over years of commitment, planning, and hard work. It is essential in today's world to plan for technology change and update these plans on an annual basis. GIS project planning should be scheduled to support the annual business cycle. Enterprise GIS is an evolving program that changes each year to support business objectives and keep pace with technology.

But the best plan is to understand what you're doing, every step of the way. Then, the technology won't get away from you and you won't run amok with it either. What do system architects do? They plan for success. They understand that computer systems are only as strong as their weakest link, and sometimes that weakest link is us. To understand a system as a whole, you have to know all its parts. To miss understanding any one of them is equivalent to missing more than one opportunity to save money and time. For the sake of simply making it work, you owe it to your GIS project to move forward with an intelligent plan. Planning ahead before investing in software and hardware saves money and reduces risk. GIS projects most often fail because system performance requirements are not satisfied. Setting appropriate performance targets, following development best practices, monitoring performance throughout deployment to ensure performance targets are met, and using the models and tools provided in this book increase your likelihood of success.

2 | Software technology

If system architecture design were a step-by-step process, the first step would be to review all your options before committing to any one of them. In other words, managers should focus on understanding all the pieces of the puzzle before making critical technology choices. A system is like a puzzle: it contains several interacting parts, each related to the other in a very special way. Each of those parts is also made up of components that carry on a special relationship with each other. Software technology is a good place to begin to understand these interacting parts and special relationships.

Software technology selection is a critical step in building a GIS. Within the past 15 years especially, GIS technology has grown tremendously, from serving a small group of users in a local office environment to integrating operations distributed on a global scale. Now the ArcGIS 9.3 release provides a new range of simple user methods for deploying a variety of mapping services, sharing GIS tools and information with Google and Microsoft Web 2.0 commercial and public users, and leveraging ArcGIS Explorer as a window to access information provided from every part of the world. All this promotes a growing level of information exchange and collaboration based in spatial thinking. What this means for your future and how can you use geography to make a difference in your organization will be a large factor in your decision making.

The GIS technologies you'll be choosing among include a variety of GIS data sources, ArcGIS Desktop applications for GIS power users, ArcGIS Server for supporting GIS business workflows and Web services, mobile client applications integrated with corporate business operations through wireless communications, and a variety of network services. Examining your GIS technology options is the first phase in building a GIS.

This chapter provides an overview of ESRI's GIS technology, past and present. Software solutions to serve your GIS needs could come from ArcGIS Desktop and ArcGIS Server technology, both of which serve a variety of GIS operations, involving local workstations, remote desktops and Web browsers, and mobile clients. Data can be accessed either from a centralized data center or in a distributed geodatabase architecture or in some combination of both. Understanding available technology options, in order to identify candidates that may satisfy your operational needs, establishes a baseline for making the best technology decision. Right now you're not buying, you're just window shopping; in other words, you're looking at the pieces of the puzzle, and your focus is on understanding the relationships between them, all of which will inform your decision making.

The histories of the various technologies reveal how they have influenced each other, another aspect of relationship to consider in designing a system that will accommodate change. Changes in platform processing performance and network communication bandwidth have already influenced the evolution of GIS technology, and are likely to continue to do so. Wireless communications has already opened a new frontier, enabling mobile users to be more tightly coupled with primary business workflow operations. GIS deployment strategies are changing from traditional department- and enterprise-level operations to emerging federated and service-oriented operations.

Selecting the right software is a primary system design task, as is choosing the most effective deployment architecture for that software. You should know there are many alternatives for both GIS software technology and architecture solutions. This chapter describes them (using ESRI examples) and what they may be good for, but you must make the decision as to what's best for your organization. As in all things GIS, you begin system architecture design by thinking about what you want to get out of the GIS. How will GIS optimize your organization's workflows? By understanding what you need a GIS to do, you are ready to identify the functions that will get it done. Software is the system component that provides this functionality. So, by defining the required functions, you have narrowed your software choices to those products that perform them.

You are likely to find the functionalities you need in the software available now—or in the near future—because traditionally their invention follows closely on the heels of the expressed user need. Historically, developers have responded to users' needs by creating software technology that meets the functional requirements identified by the GIS community. The history of GIS software products developed by ESRI follows the evolution of computer technology in general and the growing populations of GIS users in particular.

Understanding where the technology has been helps in anticipating where it is going. Over four decades, ESRI has developed GIS software to provide the functionality required by the GIS user community. Much has been learned over the years as technology has changed and deployment strategies have evolved to support user needs. Software development trends and enterprise architecture strategies provide clues about how GIS technology is changing.

This chapter begins with an overview of the ESRI software evolution as illustrated in figure 2-1. Understanding the target market for each member of the ESRI software family will help you identify your technology needs and develop a road map for migration to successful enterprise GIS operations.

ESRI software evolution

ARC/INFO software was installed on the Prime mainframe computer in 1982, offering developers and GIS professionals a rich toolkit for the geospatial query and analysis that would soon demonstrate the value of GIS technology. Most GIS work was performed by GIS professionals with extensive software training and programming experience. These GIS professionals developed information products (useful results of GIS applications), usually providing them in paper form, to support operational user needs. Then ArcView was introduced in 1992, an offering of easy-to-use commercial off-the-shelf (COTS) software that operational—not necessarily professional—GIS users could use directly. Thanks to ArcView, for the first time many kinds of GIS users could do their own project analysis and build their own information products without depending on a GIS professional to do their thinking for them. First-year ArcView sales totaled more than all the ARC/INFO sales over the previous 10 years put together. GIS users liked doing their own work, so an easy-to-use desktop software with which to perform GIS analysis and make basic geographic information products was a big hit. Early on in ESRI software's evolution, end users were looking toward simple software technology to provide access to and control of GIS data resources.

Since the 1980s, ESRI had been partnering with other businesses in using ARC/INFO technology to build applications for industry. This became the ESRI Business Partner (BP) program, a simple way to introduce a geographic representation into existing vertical market solutions. Using ArcView to better manage and display spatial relationships, this program grew in the 1990s. (There are many examples of BP ArcView applications still being used today in crime

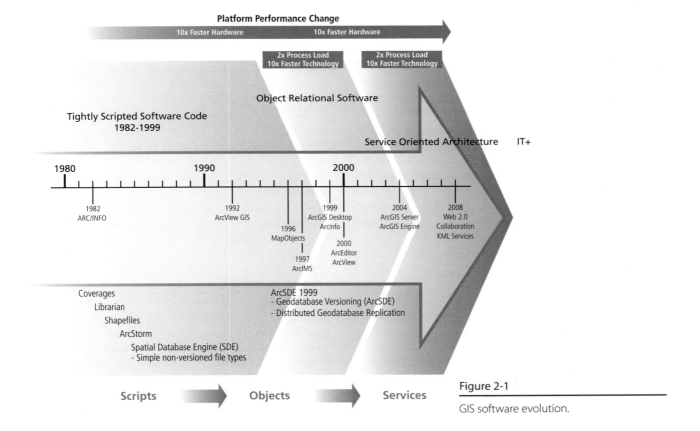

Figure 2-1

GIS software evolution.

analysis, emergency dispatch, and forestry, to name a few.) Developers found ArcView easy to integrate with existing applications, but they wanted the option of creating business solutions—and their accompanying geographic displays—without the ESRI user interface, so a new product, MapObjects, was introduced in 1996. Developed specifically for the ESRI Business Partner needs, MapObjects provided developers with a simple way to integrate map displays into their own application environments (the common standard desktop Visual Basic), without the ESRI user interface. Early on, developers were looking for standard software components that could be easily integrated with their own software solution.

Electric utility companies were using GIS in the early 1990s to manage their electrical power line infrastructure. Utility maintenance workflows were complex and required specialized power-flow and asset-management applications to buttress their distributed electrical facilities (telephone poles, capacitors, and power lines). GIS solutions were provided to support these operations. Compute-intensive GIS desktop software was used for the work management applications, and spatial data was integrated with enterprise facility and customer database solutions to support the operational workflow.

At the time, engineers located at remote field offices performed most of the facility maintenance tasks. To provide these field engineers with remote user access to centrally managed GIS desktop applications, centralized Unix Application Servers/Windows Terminal Servers were introduced. A relatively complex ArcStorm spatial database solution was introduced, to provide a scalable way to tightly integrate file-based spatial data with the tabular database systems of the facility. Soon enough, with enterprise GIS maintenance needs on the rise, simpler solutions for data management and application administration were needed. In response, ESRI worked with Oracle to develop the Spatial Database Engine (SDE). Deployed in 1997 as ArcSDE, it offered the first fully integrated, database-management solution at the enterprise level for the geographic database.

What about Web services? Initially, custom implementations of ArcView and MapObjects Internet mapping services handled the first GIS Web services. ArcIMS was introduced in 1997, providing a framework for publishing GIS information products to Web browser clients. Web technology introduced GIS information products to new users throughout the organization and to the public, rapidly expanding the population that uses GIS information products.

A boost from hardware

New opportunities for the advance of software technology came from computer hardware performance improve-ments in the late 1990s. In effect, faster hardware came along to compensate for the bigger processing loads incumbent with increasing software functionality. This is a story significant for system architects who need to be aware of the effect hardware can have on software, and vice versa. For ESRI, it happened this way: In the 1990s, ESRI software developers began to change the way they wrote the processing instructions or code underlying the software, moving from the traditional scripted sub-routine coding methods to object component coding techniques. Using common object interface standards, they could develop higher level code functions in a much more adaptable and expandable way. Component object functions could be used in many different software solutions, and could be combined with third-party components to enable much more rapid development. Exploring several of the new software object models (CORBA, JAVA, and COM), ESRI found the Microsoft Common Object Model libraries to be the most effective environment for the compute-intensive GIS application code. ESRI released the ARC/INFO GIS functions in the new ArcObject code base in 1999, for the ArcGIS Desktop software release.

This code base came to be known simply as ArcObjects, a library of software components that make up the foundation of ArcGIS. (ArcGIS Desktop, ArcGIS Engine, and ArcGIS Server are all built using the ArcObjects libraries. ArcGIS Desktop software provides the full set of ArcObjects technology, while focused subsets of the same ArcObjects support the ArcEditor, ArcView, ArcReader, and ArcExplorer desktop software solutions. See ESRI product family, page 23.)

ArcGIS Desktop software, providing GIS users with a simple and very powerful user interface supporting a full spectrum of standard GIS operations, came to be ESRI's flagship product. It started to lead the GIS industry with the most extensive set of geospatial functionality, with each release expanding that functionality. From the beginning, professional GIS users sought out the best tools available to advance GIS technology. Moving from the legacy ARC/INFO software to the new ArcObjects component development accelerated the advance of GIS software technology by orders of magnitude. Before the ArcObject code base, around the mid-1990s, GIS user requirements had been expanding faster than the ARC/INFO software release schedule could keep up with—new releases would fall short of what customers wanted. Yet within a couple of years after the ArcGIS Desktop release in 1999 (with its ArcObjects technology underpinnings), customers were asking developers to slow down—technology was advancing well beyond their needs and what they could absorb. There was a performance cost introduced with the new technology—ArcObjects applications required twice the processing resources as the older ARC/INFO code to produce the same information display. Computer hardware advances made up the difference the same year (1999), which made trading performance for much better technology a worthwhile exchange.

A period of adjustment
ArcGIS Server and ArcGIS Engine came along to provide GIS developers with access to the same rich ArcObjects software components available in the rest of ArcGIS Desktop software. ArcObjects—the library of software components that make up the foundation of ArcGIS—is now available as a desktop and server development framework, offering the full range of GIS functions for custom application development and deployment. Somewhere within that range of functionality, you can find the solutions that best fit your user needs; in system architecture design, it's important to know what those needs are before selecting the software, and that requires understanding your organization's users.

Not all GIS software users are looking for the same thing. Professional GIS users (geographers, scientists, architects, etc.) are looking for a simple, powerful, productive user interface they can use as a tool to support their work. ArcGIS Desktop provides these users with an open user interface to explore, build, study, analyze, and discover new things about the world, which in turn become tools that move GIS technology forward. GIS programmers and business partners are looking for software development environments that allow them to include ArcObjects functionality as part of their focused customer solutions, because it models the real world better. Many GIS users are not analysts or programmers; they want simple access to GIS information products and services—GIS makes their work more structured and productive. ArcGIS Server provides an optimum platform for professional GIS users to share information and services to these people along with colleagues worldwide who use GIS information in their work. IT professionals seek enterprise solutions that are stable, easy to support, dependable, scalable, and secure. Centralized application and data management—ArcSDE, geodatabase, distributed geodatabase synchronization, open standards and interoperability, data automation, ease of implementation, simple license management, data integrity, backup and disaster recovery, security—are primary concerns that affect supportability and maintainability, the cost factors managed by IT professionals. Because all the types of users have always been important to GIS, ArcGIS technology has grown and evolved to accommodate their varying needs.

Meanwhile, as we've seen, hardware technology has been evolving too, affecting software technology's own evolution. A maturing GIS technology took advantage of hardware performance gains to allow for more software functionality and adaptability. Faster hardware

encouraged software migration from the tightly scripted code of the 1980s and 1990s to the object relational ArcObjects code underlying ArcGIS software technology now. What this means to system architects is significant to know: ArcGIS Desktop generates twice the processing load of the older ARC/INFO scripted code to produce the same map display. In exchange for this performance cost, GIS technology has advanced at a much more rapid pace—with the functional benefits now far exceeding the performance trade-off.

We are seeing the same performance trade-off moving from ArcIMS to ArcGIS Server technology. In 1999, hardware technology performance gains absorbed the ArcGIS Desktop software performance cost in just one year. This pattern is repeating itself as customers migrate to ArcGIS Server technology. Any change requires a period of adjustment, but by recognizing a pattern of history repeating itself, the system architect can design to accommodate such change.

The factors of growth and change bring into play special considerations when designing a system specifically for GIS. As a data-intensive technology, GIS is a big consumer of server processing resources. It can generate a high level of network traffic as applications use large quantities of data from shared database-, file-, or Web-based data sources. If you don't get the hardware and network capacity right, GIS users will not be able to perform their work. If you get the technology wrong, you could end up with a solution that will not work in your deployed environment. If you do get it right, you can experience the geographic advantage many others have enjoyed over the past few decades.

Web-based service-oriented architecture solutions promise to move technology ahead at an even faster rate. The building of applications from Web services expands the benefits of a component development architecture and virtualization to a higher level—beyond vendor control and into the open marketplace. There will be a software performance cost in that too: more processing required to support the same information product. Nonetheless, many customers are choosing to take a chance once more, exchanging performance for a more adaptive application development environment.

Organizational GIS evolution

GIS's long history with users has informed its software development every step of the way. Organizations count among the user population, as you will see in this chapter. The more you know about how user needs have evolved, both on an individual and organizational level, the better you can prepare for the future and the opportunities it holds, even as you plan a system design to meet your current needs.

In today's world there are many configurations to choose from, thanks to the rapidly expanding GIS evolution. GIS implementations grew in size and complexity throughout the 1990s, and its evolution continues to flourish throughout the world today. It's still the case that many organizations begin with a single ArcGIS Desktop user and over time expand the use of GIS technology to provide department-level operations. In turn, as multiple departments use GIS in their work, data sharing between departments expands GIS to even newer levels. Such enterprise GIS operations offer more opportunities for data sharing and for leveraging geospatial resources to meet cross-department business needs. Mature GIS operations are expanding to integrate on a community and national level, and even globally.

Department-level GIS

As we've seen, GIS technology was initially introduced as a desktop application (ARC/INFO, in 1982), and geographic analysis and mapping work moved from data represented on paper and Mylar (clear plastic) sheets to digital datasets run by computer technology. As multiple GIS desktop users began working together, they found that sharing digital data allowed them to be more productive and provide better GIS information products. A department's local area network (LAN) was used to share data between desktops, as department managers saw the advantage in maintaining shared data on department file servers, with common user access to the shared data files. A majority of the GIS community today is still at the department-level. Figure 2-2 provides a simple overview of department-level GIS architecture.

In a typical scenario, as GIS data resources grow and mature at the department level, more users throughout the organization are able to take advantage of GIS technology. Many departments develop their own local GIS operations that benefit from data resources

developed and maintained by other departments. Then the organization's wide area network (WAN) becomes a way of sharing data between department servers. Data standardization and concurrency issues can become a challenge as data is developed and managed by different department-level sources, and organizations soon recognize a need for enterprise-level management of the shared spatial data resources.

The initial Spatial Database Engine release in the mid-1990s enabled the creation of this early enterprise-level GIS management in the form of data warehouse operations. Many organizations began sharing GIS data resources from a central ArcSDE data warehouse. IT departments hired GIS managers or geographic information officers to integrate shared enterprise GIS data resources and establish common data standards, to benefit operations throughout all departments and support enterprise management needs. The enterprise data warehouse provided a reliable, shared GIS data source for departments throughout the organization. This was a very common GIS technology evolution experienced by many local governments, and even today many smaller communities are growing together in the same way.

Enterprise-level GIS

Figure 2-3 provides an overview of the organization-level GIS architecture alternatives. Departments initially share data across the WAN until an enterprise-level data warehouse can be established to maintain and share common GIS data resources. The central IT GIS team can leverage the centralized GIS database by publishing a variety of GIS Web services. These services can support users throughout the organization, and provide a platform for using GIS technology to support enterprise-level operations.

In the early 1990s, when the electric and gas utilities began using company-wide GIS for work order management of power distribution facilities, most implementations were served by a central database. Remote users had terminal access to application compute servers (terminal servers) located in the central computer facility with the GIS database.

Many organizations today are moving their geospatial data to a common, central, enterprise geodatabase like this, shifting from a department-file-based GIS in favor of one that provides terminal client access to centrally managed server environments. In this scenario, departments retain responsibility for their data resources, updating and maintaining data through terminal access to central ArcGIS Desktop applications. The central IT computer center handles the general administration tasks of data backups, operating system upgrades, platform administration, and so forth. Users throughout the organization enjoy browser access to published Web services over the intranet.

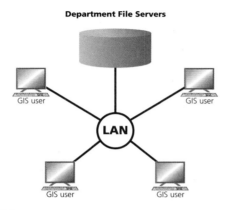

Department File Servers

GIS user GIS user

LAN

GIS user GIS user

Figure 2-2

Departmental GIS architecture.

Organizational GIS

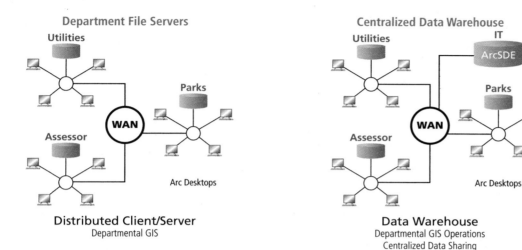

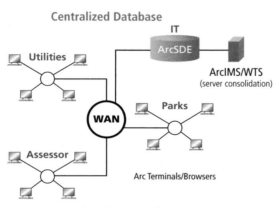

Figure 2-3

Architecture alternatives for using GIS at the organizational level.

The complexity and sophistication of the geodatabase, made possible by such central administration, has rendered centralized servers a more productive alternative for most organizations, from local municipalities to national-level government agencies. Most enterprise GIS operations today are supported by systems (application and data servers) maintained in a centralized data center.

Community GIS evolution

By the beginning of this century, a growing Internet was demonstrating the tremendous value of sharing information between communities, states, and organizations around the world. Internet access was extended from the workplace to the home, rapidly expanding the GIS user community. On the next page, figure 2-4 shows how communities and companies developed and deployed services to users and customers over the Internet. Because of the Internet, organizations could share data and exchange services while users, in turn, gained access to data and services from a multitude of organizations.

ESRI introduced the Geography Network, providing a metadata search engine that published information about GIS data services and provided direct Internet links between the ArcGIS Desktop application and the data or service provider. ArcIMS introduced a way for organizations throughout the world to share GIS data and services. The Geography Network established a framework to bring GIS data and services together, which helped foster a rapidly expanding infrastructure of communities everywhere sharing information worldwide. Promotion of data standards and improved data collection technologies continue to unlock enormous possibilities for sharing geospatial information, which should help us better understand and improve our world.

Community GIS

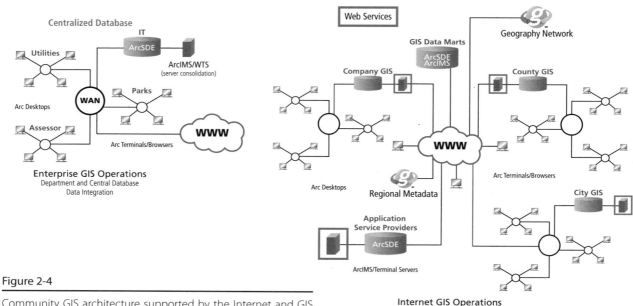

Figure 2-4

Community GIS architecture supported by the Internet and GIS services.

GIS data resources are expanding exponentially. In the 1990s, GIS data servers seldom held (or had a need for) a database that was more than 25 to 50 gigabytes (GB) in size. Today it is common for organizations to operate geodatabase servers handling several terabytes to petabytes of GIS data (one petabyte is equal to eight quadrillion bits of data). Community-level datamarts are being established to consolidate GIS data resources and provide Internet data sharing to organizations throughout county and state regional areas. State and national agencies are consolidating data and sharing it with each other and with communities and municipalities everywhere.

Many organizations are outsourcing their IT operations to commercial Internet service providers (ISPs). Application service providers (ASPs) support organizations with IT administration, providing opportunities for smaller organizations to take advantage of high-end GIS database and application solutions to serve their business needs. State governments are hosting applications and data for smaller municipalities throughout their state so that the smaller communities can take advantage of GIS technology to help their local operations.

Regional geography network (g.net) sites allow for the sharing of data throughout the region and within large state and federal agencies. Web portal software provides a metadata search engine that can be used by organizations to share their data and bolster their community operations. Cities can establish metadata

sites to promote local commercial and public interests. States can consolidate metadata search engines for sharing data and services with municipalities throughout the state. Law enforcement can establish search engines to feed and use national datasets. Businesses can establish metadata search engines to support distributed operational environments. Web services enable community data sharing and integrated workflows.

Deployment of ArcGIS Server with ArcGIS 9 expanded Web services technology to include geoprocessing and a broad range of service-oriented Web operations. GIS technology, in conjunction with Web standards and open systems architecture, has opened new opportunities for improving business operations. GIS software and computer infrastructure technology continue to expand GIS capabilities and introduce new business opportunities. Improved availability and capacity of wireless technology enables mobile communication connectivity for a growing number of GIS users.

The growth of both the technology and the sharing of it through data and services is establishing GIS as an integral part of community. The technology provides real-time access to geographic information, which means it's rapidly becoming a primary technology for understanding not only business opportunities but the world itself. With technology changing more rapidly today than ever before, national and global geospatial initiatives are bringing communities and people together to understand and solve the world's problems in ways we only dreamed possible just 10 years ago.

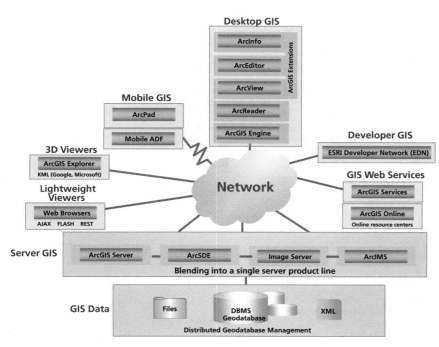

Figure 2-5

ESRI product family.

English is becoming a common second language for many people throughout the world, allowing many to share technology and work together like never before. Geography provides a common language for understanding our world, and brings people of different nations and cultures together to understand and solve common problems.

In my work I have had the pleasure to travel throughout the world and meet people from many different nationalities and backgrounds. My ability to communicate was limited by language barriers in the 1970s, 1980s, 1990s—I would communicate through interpreters, facial expressions, and sign language. It was a very big world, and there was much we did not understand. Today I share my understanding of GIS technology with our international distributors, business partners, and GIS customers all over the globe. We communicate ideas, problems, and solutions through a shared language, which is some combination of computerese and our common interest in geography. The language barrier is falling, perhaps because of some of the things we're doing with GIS:

- Working with global energy companies to better manage their geographic information resources
- Working with scientists throughout Europe to build a common geospatial data infrastructure for the European community
- Training United Nations staff responsible for peacekeeping operations, disaster relief support, and support for troubled nations throughout the world
- Working with people in national parks in Canada and the United States using GIS to better manage our natural resources

- Working with customers in Hong Kong who use GIS to better manage land resources and support services for one of the world's most complex metropolitan environments
- Sharing GIS technology with our international distributors in Munich, Paris, Rome, Rotterdam, London, Dubai, Costa Rica, São Paulo, Finland, Dublin, Stockholm, Hong Kong, Singapore, and Australia.

In other words, GIS is working; the technology is making a difference in our world, by employing geography to bring nations together in a very special way to better understand and solve the problems we share. From this perspective, the GIS community is as extensive as the world, and sometimes the software product options to choose from can seem just as wide-ranging. All this functionality is within reach, however; it's just a matter of assessing what you need to serve your own organization's mission. All the ESRI products available now have been developed over the years in response to user need. So looking more closely at what the software does, in the section that follows, becomes something of a journey through not only what is possible for you to do with it but also what your fellow users may already be doing.

ESRI product family

The ESRI product family, illustrated in figure 2-5, includes a mix of software developed to fulfill a wide range of GIS user requirements. ArcGIS software supports desktop, server, and mobile applications. Data management solutions support a variety of file-,

Scalable Architecture

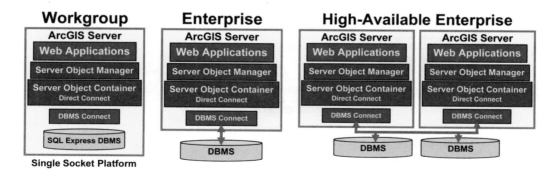

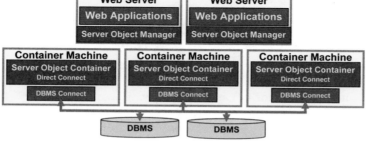

Figure 2-6

ArcGIS Server scalable architecture.

geodatabase-, and Extensible Markup Language (XML)-based formats.

GIS Web services support a variety of managed, hosted, and shared GIS Internet services. ArcGIS Server provides technology for publishing GIS services that can be consumed by ArcGIS Desktop, mobile GIS, and standard Web browsers. ESRI Developer Network (EDN) offers a range of technical services to the ESRI developer community through a bundled, low-cost developer software license.

Desktop GIS: ArcGIS Desktop software is focused on providing the professional GIS user with direct access to all GIS data resources, all published geodata services, and tools for geographic analysis and visualization that will lead to future advancement of GIS technology. Desktop GIS is divided into four licensed solutions based on user functional needs. These include ArcGIS Desktop (ArcInfo, ArcEditor, ArcView) and ArcGIS Engine, a desktop development environment that provides a complete set of ArcGIS ArcObjects components for custom desktop application development.

ArcGIS Desktop software is licensed at different software levels based on user needs. ArcReader is free desktop software for viewing and sharing a variety of dynamic geographic data. ArcView includes all the functionality of ArcReader and adds geographic data visualization, query, analysis, and integration capabilities. ArcEditor includes all the functionality of ArcView

and adds the power to create and edit data in an ArcSDE geodatabase. ArcInfo is the complete GIS data creation, update, query, mapping, and analysis system.

A range of desktop extension licenses are available that provide enhanced functionality for supporting more focused GIS operations. Desktop extensions operate with the foundation ArcGIS Desktop license and expand functions for geospatial analysis, productivity, solution-based, Web services, and a range of no-cost add-ons to address a variety of focused user needs. (For more detailed information on this and all ESRI software products, see www.esri.com.)

Server GIS: Server GIS is used for a variety of centrally hosted GIS services. Use of server-based GIS technology is expanding rapidly as more and more customers leverage Web-based enterprise technology. GIS technology can be managed on central application servers to deliver GIS capabilities to large numbers of users over local and wide area networks. Enterprise GIS users connect to central GIS servers using traditional desktop GIS as well as Web browsers, mobile computing devices, and digital appliances.

ArcGIS Server software (figure 2-6) is divided into three licensed solutions based on available functionality and system capacity. ArcGIS Server Basic includes geodatabase management (ArcSDE technology), geodatabase check-in/checkout, and geodatabase replication services. ArcGIS Server Standard includes all the functionality of ArcGIS Server Basic plus standard map publishing, ArcGlobe services (ArcGIS Explorer), and standard geoprocessing. ArcGIS Explorer is a free, lightweight, ArcGIS Server desktop client that can be used to access, integrate, and use GIS services, geographic content, and other Web services. ArcGIS Server Advanced includes all the functionality of ArcGIS Server Standard plus Web editing, mobile client application development framework (mobile ADF), advanced geoprocessing, and support for ArcGIS Server extensions.

Developer GIS: EDN is an annual subscription-based program designed to provide developers with comprehensive tools that increase productivity and reduce the cost of GIS development. EDN provides a comprehensive library of developer software, a documentation library, and a collaborative online Web site that offers an easy way to share up-to-the-minute information.

GIS Web services: GIS Web services offer a cost-effective way to access up-to-date GIS content and capabilities on demand. With ArcWeb Services, data storage, maintenance, and updates are handled by ESRI, eliminating the need for users to purchase and maintain the data. Users can access data and GIS capabilities directly using ArcGIS Desktop or use ArcWeb Services to build unique Web-based applications. An ArcWeb Services subscription provides instant and reliable access to terabytes (TB) of data, including street maps, live weather and traffic information, extensive demographic data, topographic maps, and high-resolution imagery from an extensive list of world-class data providers.

ArcGIS Online provides access to terabytes of cached map services. With ArcGIS Online you can access 2D maps, 3D globes, reference layers, and functional tasks via the Web to support your GIS work. You can contribute your own data for publishing through ArcGIS Online and make it broadly available to other users. You can also purchase data you see in ArcGIS Online and publish it on your own server.

Software vendors do modify pricing strategies when necessary to account for technology change. Prices normally go down over time, so building a solution from current technology and establishing your project budget based on current pricing models is a conservative management strategy. In any case, it is important to understand pricing and factor this in to your technology decision. (For example, the price of Web services has changed from ArcIMS to ArcGIS Server, which has affected how we design these systems.)

Customers are often aware of the basic core software, but many times they do not see what is really under the covers—what will make a difference in user performance and supportability. ArcGIS software includes a variety of ways to deploy GIS. Several technology options are available for ArcGIS Desktop users, Web users, and for Mobile users. There is an optimum architecture solution for each of the different types of user workflows. Understanding the choices available is a first step in finding it and putting the right pieces together for your environment.

Desktop operations: Figure 2-7 provides an overview of the various ArcGIS Desktop client choices you can make for your environment. Applications can use different levels of software (ArcInfo, ArcEditor, ArcView), any of the ArcGIS Desktop extensions, or custom ArcGIS Engine clients.

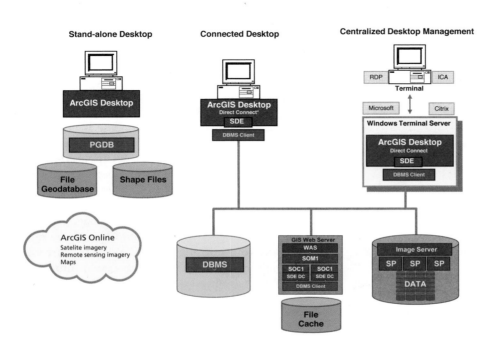

Figure 2-7

Desktop operations.

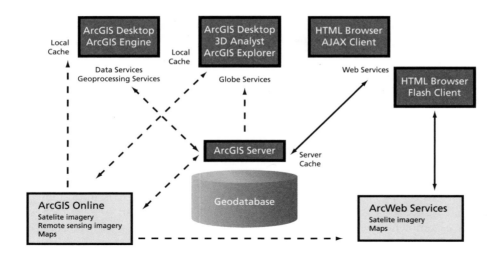

Figure 2-8

Web operations.

Stand-alone desktop applications can use a Microsoft SQL Server Express personal geodatabase (PGDB), providing up to 4 GB (per database server) of capacity for local user editing and viewing operations. The file geodatabase (FGDB), which was introduced with ArcGIS 9.2, supports 1 TB of geospatial data per table in a file format (can be configured to 256 TB). FGDB can be used as reference data or a single-user editing environment. Standard shapefiles can also be used for a local data source.

The same ArcGIS Desktop applications can be deployed in a connected LAN environment (with access to network data sources and Web services). Or, they can be deployed from a central data center on Windows Terminal Server platforms, which support remote WAN clients over lower bandwidth WAN environments, and still maintain full access to local LAN network data sources.

All of the ArcGIS Desktop applications can take advantage of the high performance of cached data sources, such as those provided over the Internet from ArcGIS Online. ArcGIS Online data is streamed to the client and cached locally, providing high performance reference data for local client applications.

Web operations: Figure 2-8 provides an overview of the various ArcGIS Server Web client choices you can make for your environment. Potential Web client candidates include ArcGIS Desktop, ArcGIS Engine, and ArcGIS Explorer. Also, standard Web browser applications can be client applications using a variety of Web applications and services.

ArcGIS Server can provide Simple Object Access Protocol (SOAP)/XML-based data services (published reference images) and geoprocessing services to ArcGIS Desktop and ArcGIS Engine client appli-

cations. It also provides a 3D globe, cached file data source for ArcGIS 3D Analyst and ArcGIS Explorer clients. ArcGIS Server also hosts a full range of map view and edit applications for Web HTML browser clients supported by out-of-the-box .NET and Java Web map and editor Server development kit components.

ArcIMS is a popular solution for delivering dynamic maps and GIS data and services via the Web. It provides a highly scalable framework for GIS Web publishing that meets the needs of corporate intranets and the demands of worldwide Internet access. ArcIMS customers are rapidly moving to ArcGIS Server software to leverage the rich functionality available with the new ArcGIS Server software release.

ArcGIS Image Server changes how imagery is managed, processed, and distributed. Image Server provides fast access and visualization of large quantities of file-based imagery, processed on the fly and on demand. It provides rapid display of imagery for a number of users working simultaneously, without the need to preprocess the data and load it into a database management system (DBMS). ArcGIS Image Server can be used as a data source for ArcGIS Desktop, ArcGIS Server, and ArcIMS operations. Additional support is provided for AutoCAD and MicroStation CAD clients. On-the-fly processing can include image enhancement, orthorectification, pan sharpening, and complex image mosaicking. Understanding these options and how they might best support your GIS needs is an important first step in selecting the right solution for your environment.

Mobile GIS: Mobile GIS supports a range of mobile systems from lightweight devices to PDAs, laptops, and Tablet PCs. ArcPad is software for mobile GIS and field-mapping applications. All ArcGIS Desktop products—

ArcReader, ArcView, ArcEditor, and ArcInfo—and custom applications can be used on high-end mobile systems, such as laptops and Tablet PCs. Figure 2-9 provides an overview of the primary connected mobile workflow alternatives.

The ArcGIS Server 9.2 Basic license supports distributed geodatabase replication. Geodatabase replication provides loosely connected synchronization services for distributed geodatabase versions maintained in supported database platforms. Web-based disconnected check-in and checkout services are also provided. Distributed geodatabase replication is discussed later in chapter 6.

ArcGIS 9.2 also provides geodatabase support in a Microsoft SQL Server Express personal geodatabase. Microsoft SQL Server Express is bundled with each ArcGIS Desktop software license. ArcGIS Desktop clients (including custom ArcGIS Engine runtime deployments) can support a distributed geodatabase client replica and synchronize changes with the central parent geodatabase. The SQL Server Express database has a data capacity of 4 GB.

ArcGIS 9.2 also supports a file-based geodatabase. ArcGIS Server 9.3 enables Web-based data checkout/check-in and one-way geodatabase replication to distributed file geodatabase clients.

ArcGIS Server 9.2 Advanced license comes with the ArcGIS mobile software development kit. ArcGIS Mobile lets developers create centrally managed, high-performance, GIS-focused applications for mobile clients. Mobile applications powered by ArcGIS Server

contribute to increased field productivity and more informed personnel.

Expanding GIS technology trends

Advances in GIS software and computer infrastructure technology fuel a continuous expansion in GIS capabilities and new business opportunities. At the same time, the increasing availability and capacity of wireless technology is providing improvements in mobile communication connectivity for a growing number of GIS users. Wireless technology is changing both how we work now and the types of architecture strategies available for generations to come.

Mobile devices are available for work when and where you need them, and now it's possible to *manage* their work wirelessly, too. Integrating loosely coupled mobile workers into enterprise workflows, mobile technology has become one of the most appealing ways to improve business operations. Traditionally, mobile data collection and management were set apart from the internal business workflow, but now they're embedded.

This is just one example of how enterprise architecture strategies are changing. The list of common ArcGIS deployment alternatives is growing. It started with traditional GIS workstations, expanded to the more centralized, enterprise GIS option, and now includes the newly emerging federated and service-oriented architectures. In looking for ways to improve access and data sharing with other organizations, traditional department-level

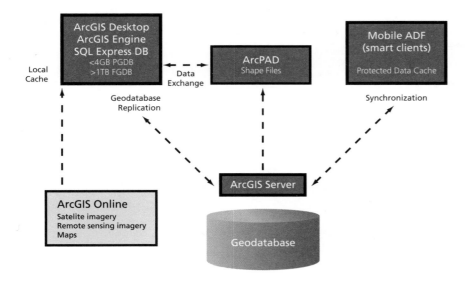

Figure 2-9

Mobile operations.

GIS client/server operations are shifting to a federated GIS architecture. In seeking ways to integrate GIS with other centrally managed business operations, some traditional enterprise GIS organizations are finding the integrated business solutions they're looking for within a service-oriented architecture (SOA).

Federated GIS technology

Database and Web technology standards provide new opportunities to better manage and support user access to a rapidly growing volume of geospatial data resources. Web services and rich XML communication protocols enable efficient data migration between distributed databases and centralized storage locations. Web search engines and standard Web mapping services provide a means to discover and consume integrated geospatial information products published from a common portal environment with data provided from a variety of distributed service locations. Federated architectures identified in figure 2-10 promote better data management by integrating community and national GIS operations. Geodatabase replication services and managed extract, transform, and load (ETL) processes support loosely coupled distributed geodatabase environments.

Federated systems are composed of parties that share networked applications and data resources. GIS is all about sharing. Many local government, state, and federal agencies share GIS data to support community operations. Data resources owned by each party are brought together to provide community-level information products. Data maintenance responsibilities are distributed between different groups, with databases

configured to share GIS resources between the different sites on a scheduled basis. Web portals provide applications supported by a variety of Web services and act as a broker to connect users with published community-wide data services.

Service-oriented architecture

Service-oriented architecture is the next computer frontier, if indeed Web services is our future. It's possible that at some point we may no longer have applications on our desktop, and IT expenses (which consume as much as half of the operational budget today) may go away. Changes of that magnitude have come to seem like a regular feature of this age of technology in which we live. Moving to a component-based software architecture in the late 1990s accelerated the growth of software vendor technology—software upgrades that happened once a year in the 1990s are released as quarterly service packs with today's software technology. Software upgrades that would happen once a year are now streamed to your computer for download and install on a daily basis. One can only imagine what might happen if these functional application components were replaced by Web services. A good guess is that the acceleration of technology change is just getting started—we've seen nothing yet. Yet we've seen enough to compel organizations to search for more effective ways to manage technology change.

Obviously, business environments are influenced by the rate of technology change and because change introduces risk, contributing to business success or failure. Selecting the right technology investment is

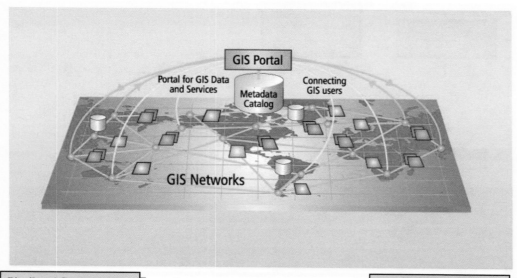

Distributed Geodatabase	**Spatial Data Infrastructure**	Portal Toolkit	Figure 2-10
- ETL Data Flows	**Heterogeneous GIS Integration**	- Map Viewer	
- Geodatabase Replication		- Metadata Catalog	Federated architecture for GIS technology.

critical. Service-oriented architecture deployment strategies reduce business risk through diversification and reduced vendor dependence. Open standards reduce the time and effort involved in developing integrated business systems, providing integrated information products (common operating picture) that support more informed business decisions. Advantages of a service-oriented architecture are highlighted in figure 2-11.

even more course-grained or at a higher level of function). SOA services have well-defined interfaces (service contracts) with the software technology providing the service abstracted from the external accessible service interface. The trend is to support SOA through Web services based on SOAP, XML, and KML protocols.

Common Web protocols and network connectivity are essential to support this type of architecture. The SOA infrastructure may be implemented using a variety

Advantages of an SOA
- Technology change ➤ Component architecture
- Business continuance ➤ Reduce vendor dependence
- Leverage investments ➤ Reuseable components
- Customer flexibility ➤ More vendor choices
- Business integration ➤ Open system communications

- Building applications from component Web services will accelerate technology change.
- Supporting business functions from multiple vendor sources can reduce vendor dependence and improve options for business continuance when others fail.
- Service architectures can support reusable components—the same published services can be used by several business applications.
- Web services can be abstracted from the vendor technology providing the service—this opens the door for more choices among the vendor products that can support the same business application.
- Business systems can interface through a services architecture; services from each business function can be integrated by an enterprise application.

Figure 2-11

Advantages of service-oriented architecture.

Basically, SOA is an approach for building distributed computing systems based on encapsulating business functions as services, services that can be easily accessed in a loosely coupled fashion. Business functions are encapsulated as Web services that can be consumed by Web clients and desktop applications. The core components that make up a service-oriented architecture include service providers, service consumers, and implementation of a service directory. The SOA infrastructure connects service consumers with service providers and may be used to communicate with service directories.

From a business perspective, new business functions are provided as Web services, which are IT assets that correspond to real-world business activities or recognizable business functions; accessibility is based on service policies established to support enterprise operations (loosely coupled to the business applications).

From a technical perspective with SOA, services are coarse-grained, reusable IT assets. (Coarse-grained indicates that the components provide a complete service rather than bits and pieces—functions—pulled together to support a service. The ArcGIS Server ADF components are course-grained reusable IT assets, for example. A service published by the ADF would be

of technologies, of which ESRI's software, with its support of open standards, can be a part. ESRI embraced open standards during the 1990s and has actively participated in the Open GIS Consortium and a variety of other standards bodies in an effort to promote open GIS technology. The initial ArcIMS Web services, Geography Network metadata search engines, Geospatial One-Stop, and the ESRI Portal Toolkit technology are all examples of the service-oriented solutions characterizing ESRI's current customer implementations. Figure 2-12 provides a view of how current ESRI software supports the evolving SOA enterprise by fitting in with its standard IT infrastructure.

The service-oriented architecture (SOA) framework includes multiple access layers connecting producers and consumers, based on current client/software technology and incorporating Web application and service communication tiers. Consumers connect to producers through a variety of communication paths. This framework supports a presentation tier of viewers with access to available published services, a serving/publishing tier of services, and an authoring tier of professional ArcGIS Desktop users. This framework supports current client/server connections (client applications),

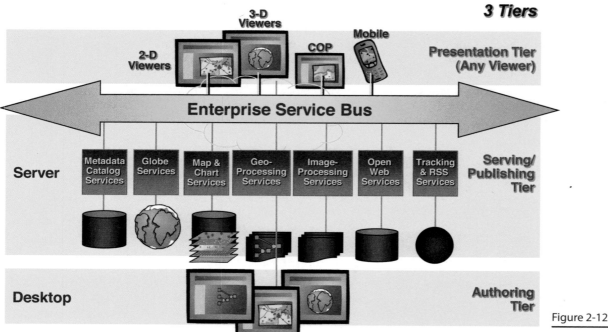

Figure 2-12

ESRI fits into SOA.

Web applications, and Web services—all available today with current technology. Future vendor compliance and maturity of Web interface standards are expected to gradually migrate business applications from tightly coupled proprietary client/server environments to a more loosely coupled SOA. The ideal environment would separate business services and workflows from the underlying software technology, providing an adaptive business environment in which to effectively manage and take advantage of rapid technology change.

GIS is by nature a service-oriented technology with inherent fundamental characteristics that bring diverse information systems together to support real-world decisions. GIS technology flourishes in a data-rich environment, and ArcGIS technology can help ease the transition from existing "stovepipe" GIS environments. The geodatabase technology provides a spatial framework for establishing and managing integrated business operations. Many spatial data resources are available to support organizations as they shift their operations to take advantage of GIS technology.

Migration to a service-oriented architecture is more a change in attitude than a change in technology. Moving a business from high-risk, tightly coupled, monolithic stovepipe operations to a more integrated, responsive SOA will take time. Figure 2-13 provides some basic guidelines for moving existing systems to a more dynamic and supportable SOA environment.

Understanding SOA and how it enables business process integration and helps control and manage tech-

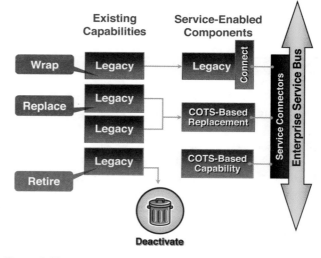

Figure 2-13

Migrating to an SOA.

nology change is important. Organizations must build an infrastructure that can effectively take advantage of new technology to stay competitive and productive in today's rapidly changing environment.

With this overview of how GIS-related technology has evolved over the last 20 years, it becomes apparent that geography is making a difference in how people think. GIS software provides tools with which to understand the world better. Selecting the right software and configuring it properly can make a big difference in how well you are able to support your needs.

GIS technology today

Current GIS technology is available to support a rapidly expanding spectrum of GIS user needs as depicted in figure 2-14. Solutions are supported by ESRI products integrated with a variety of vendor technologies. In fact, this integration or "playing well together" is more important today than it ever was, as users and their organizations are becoming more and more focused on integrating commercial-off-the-shelf solutions than on building customized applications. Of course, there will always be a need for the latter, but in the past, *most* applications had to be developed from scratch. Yet these days, with open standards becoming the norm, technology changing so rapidly, and solutions becoming more available at less cost, you could almost say that we are becoming more shoppers than developers.

Data storage and data management technologies are growing in importance as organizations continue to develop and maintain larger volumes of GIS data. Individual server storage solutions are being replaced by more adaptive storage area networks (SANs), enhancing

the IT's ability to respond to changing data storage needs and providing options for efficiently managing large volumes of data.

GIS data sources include file servers, geodatabase servers, and a variety of business database solutions. Desktop ArcGIS applications can be supported on local workstation clients or on centrally managed Windows Terminal Server farms.

ArcGIS Server (and legacy ArcIMS mapping services) provide Web services to Web browser clients throughout the organization and the community. ArcGIS clients are able to connect to ArcGIS Server Web products as intelligent browser clients, enabling connection to unlimited data resources through the ESRI Geography Network, and to organization resources served through a variety of ESRI customer portals. Users can access applications from the Internet or through intranet communication channels. Mobile ArcGIS users can be integrated into central workflow environments to support seamless integrated operations over wireless or remote connected communication. ArcGIS Desktop applications can include Web services as data sources integrated with local geodatabase or file data sources, expanding desktop operations to include available Internet data sources.

GIS enterprise architecture is typically supported by a combination of ArcGIS Desktop, ArcGIS Server, and geodatabase software technology. Selecting the right

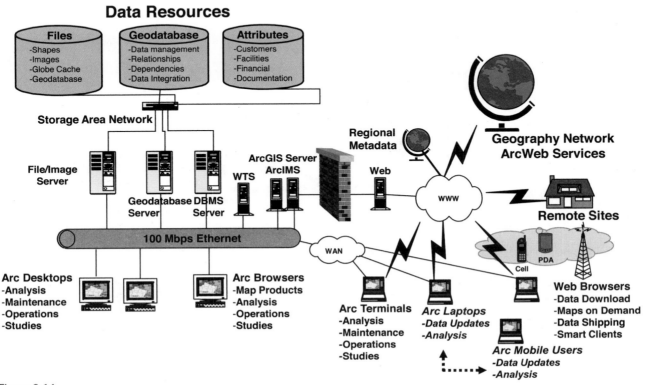

Figure 2-14

ESRI core GIS technology.

combination of technology will make a big difference in the level of support for user operational needs and business productivity.

GIS software selection

Selecting the right software and the most effective deployment architecture is very important. ArcGIS technology provides many alternative architecture solutions and a wide variety of software, all designed to satisfy specific user workflow needs. We looked at the ESRI products earlier. Now figure 2-15 shows these GIS software technology alternatives in terms of their everyday configuration. What is the best data source? What user workflows should be supported by GIS desktop applications? What can be supported by cost-effective Web services? What business functions would be best supported by network services? Where will mobile applications improve business operations? Understanding the available technology alternatives and how each will perform and scale within the available user environment can provide the information needed to make the right technology decisions.

GIS data source: Data can be accessed from local disks, shared file servers, geodatabase servers, or Web data sources. Local data sources support high-performance

productivity requirements with minimum network latency, while remote Web services allow connection to a variety of published data sources, with the drawback of potential bandwidth congestion and slow performance. There are, however, other more loosely connected architecture solutions that reduce potential network performance latency and support distributed data integration.

Desktop applications: The highest level of functionality and productivity is supported with the ArcGIS Desktop applications. Most professional GIS users and GIS power users will be more productive with the ArcGIS Desktop software. These applications can be supported on the user workstation or through terminal access to software executed on centralized Windows Terminal Server farms. Some of the more powerful ArcGIS Desktop software extensions perform best on the user workstation with a local data source, while most ArcGIS Desktop workflows can be supported efficiently on a terminal server farm. Selecting the appropriate application deployment strategy can have a significant impact on user performance, administrative support, and infrastructure implementations.

Web services: The ArcIMS and ArcGIS Server technologies provide efficient support for a wide variety of more focused GIS user workflows. Web services also provide a very efficient way to share data for

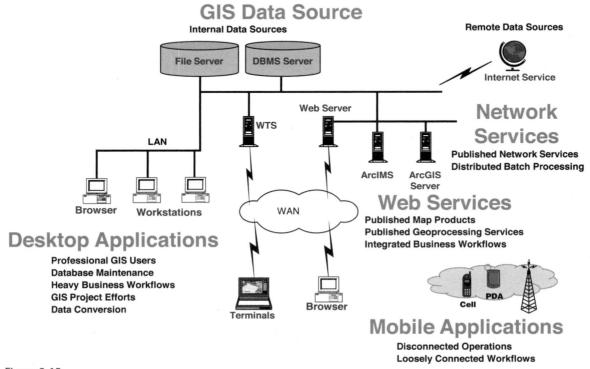

Figure 2-15

GIS software technology selection.

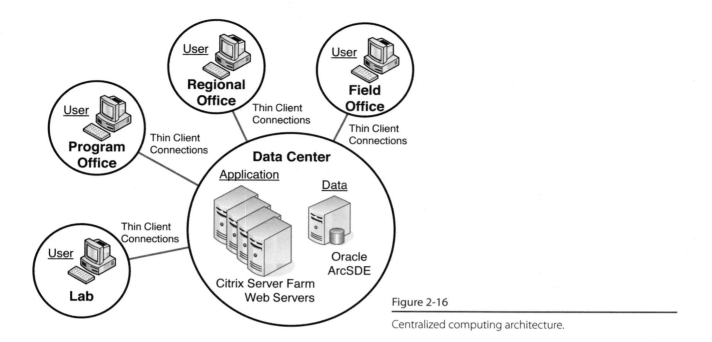

Figure 2-16

Centralized computing architecture.

remote client workflows. ArcIMS provides an efficient way to publish standard map information products, while ArcGIS Server provides enhanced functionality to support more advanced user workflows and services. Web services are a cost-effective way to leverage GIS resources to support users throughout the organization and associated user communities.

Network services: Intranet applications can access services provided by ArcGIS Server connecting directly through the server object manager. Network services can be used to support a variety of Web and network applications.

Mobile applications: A growing number of GIS operations are including more loosely connected mobile GIS solutions. ArcGIS technology enables continuous workflow operations that include disconnected editing and remote wireless operations. A disconnected architecture solution can significantly reduce infrastructure costs and improve user productivity for some operational workflows. Leveraging mobile services can provide alternative solutions to support a variety of user workflow environments.

Selecting the proper software and architecture deployment strategy can have a significant impact on user workflow performance, system administration, user support, and infrastructure requirements.

GIS architecture selection

As we've said, GIS environments commonly begin with single-user workstations at a department level. Many organizations start with one GIS team and evolve from the department level to an enterprise operation. (This was common through the early 1990s, as many organizations worked to establish digital representation of their spatial data.) Once this data is available, organizations frequently expand their GIS operations to support enterprise business needs. GIS is a very compute-intensive and data-rich technology. A typical GIS workflow can generate a remote user desktop display every 6–10 seconds; every time it does requires hundreds of sequential data requests to be sent to a shared central data server, to render each desktop display. As a result, GIS workflows can place high processing demands on central servers and generate a relatively high volume of network traffic. Selecting the right configuration strategy can make a significant impact on user productivity.

Data can be shared between users in a variety of ways. Most organizations today have user workstations connected to local area network (LAN) environments and locate shared spatial data on dedicated server platforms. However, there are configuration alternatives, as described below.

Centralized computing architecture
The simplest system architecture uses a single, central GIS database with one copy of the production database environment, minimizing administrative management requirements and ensuring data integrity. Standard enterprise data backup and recovery solutions can be used to manage the GIS resources.

GIS desktop applications can be supported on user workstations located on the central LAN, each with

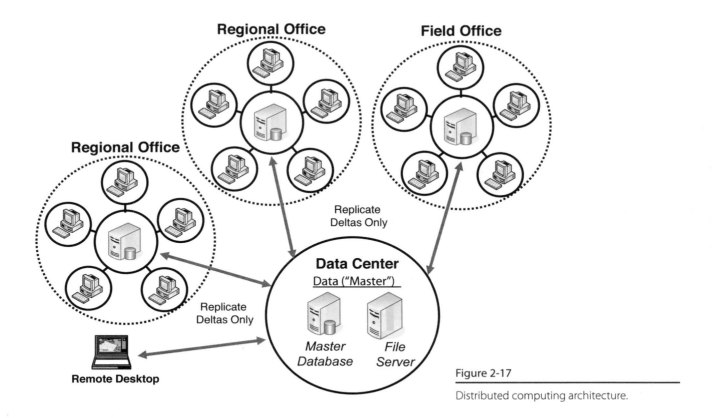

Figure 2-17

Distributed computing architecture.

access to central GIS data sources. Data sources can include GIS file servers, geodatabase servers, and related attribute data sources as shown in figure 2-16.

Remote user access to central data sources can be supported by central Windows Terminal Server (WTS) farms, providing low-bandwidth display and control of central application environments.

Centralized application farms minimize administration requirements and simplify application deployment and support throughout the organization. Source data is retained within the central computer facility, improving security and simplifying backup requirements.

A variety of Web mapping services can be provided to publish data to standard browser clients throughout the organization. Web mapping services allow low-bandwidth access to published GIS information products and services.

Today, centralized computing technology can support consolidated architectures at a much lower risk and cost than similar distributed environments. For this reason, many organizations are in the process of consolidating their data and server resources. GIS can benefit from consolidation for many of the same reasons experienced by other enterprise business solutions: reduced network traffic, improved security and data access, and less cost for hardware and administration. Centralized GIS architectures are generally easier to deploy, manage, and support than distributed

architectures and provide the same user performance and functionality.

Distributed computing architecture

Distributed solutions are characterized by replicated copies of the data at remote locations, establishing local processing nodes that must be maintained to be consistent with the central database environment, as shown in figure 2-17. Data integrity is critical in this type of architecture, requiring controlled procedures with appropriate commit logic to ensure changes are replicated to the associated data servers.

Distributed database environments generally increase initial system complexity and cost (more hardware and database software requirements) and demand additional ongoing system administration and maintenance. There can be increased network traffic with distributed data solutions, along with somewhat higher implementation risk. Distributed solutions are provided to support specific user needs.

In many cases, standard database solutions do not allow for replication of spatial data. GIS users with distributed database requirements must modify their data models and establish procedures to administratively support data replication. The complexity of current geodatabase environments has complicated the implementation of an efficient commercial spatial replication solution. Many GIS users are interested in replicating regional or selected versions of a geodatabase, which is not understood by commercial replication technologies. ArcGIS software functions are available to support custom geodatabase replication solutions. Arc-GIS 9.2 provides support for distributed geodatabase replication, providing alternative options for supporting distributed operational requirements. The ArcGIS 9.3 release adds one-way replication to the FGDB and PGDB data sources.

Selecting the right technical solution

Understanding the available software technology choices is critical in building effective GIS enterprise solutions. Technology is changing faster every day, and the rate of technology change has altered the way people build and maintain effective enterprise solutions. In the 1990s, many organizations hired their own programmers and developed their own technology solutions. Building enterprise GIS operations would take many years and involve hundreds of hours of planning, data collection and conversion, application development, and incremental system deployment. Today, these same systems can be deployed within months. Custom application development today focuses on the user interface, data model, and system integration— building systems from current technology components. The right technology selection and integrated solution architecture is the key to getting ahead.

Selecting the right solution starts with a clear understanding of your business needs. You are building a solution from commercial off-the-shelf technology— software that must be adaptive and play well together. Choosing the right technology at the right time can save hundreds of hours of grief—which quickly translates into success at the bottom line. Much has been learned over the last 20 years of software development: what works, what doesn't, and what you need to be successful. Learn from the experience of others, and keep abreast of technology trends. Understand why the technology is moving in the direction that it is, and keep your eyes open to new ideas that might signal a change in direction. Software and hardware technology trends play together to expose future business opportunities. Take advantage of every chance you get to learn about both.

The next chapter holds what you need to know about network communications. For many organizations, network communications provide the infrastructure for building and supporting enterprise GIS operations. You need to understand your network limitations to make the proper technology selections.

3 Network communications

Network communications provide the infrastructure to connect GIS operations. Networks connect user applications with shared data resources, remote office workflows with corporate data centers, and enable GIS users throughout the community and nation to share GIS data and services. Many geographic information systems today are globally connected, offering real-time information products to serve users around the world.

In providing an interactive model of our world, GIS uses a variety of data types (such as satellite imagery and aerial photography) to identify points, polygons, and lines that can be displayed on maps to represent spatial relationships. Often, these various data types must be brought together from a variety of data sources distributed in multiple locations. Nonetheless, the map display is rendered within seconds, with hundreds of spatial features converging. The GIS user may view the display for a few seconds, only to request a new map display (a different place, another resolution, etc.) again and again. All of this real-time (dynamic) map production—back and forth between the data source and the client application—can generate a considerable amount of traffic over the network. Therefore, understanding how much traffic your workflows will generate at peak times—and how much the network will bear—becomes a critical part of the system architecture design process.

Everyone can benefit from a fundamental understanding of network communications, but for decision makers that benefit resides on the bottom line. Although this may change in the future, the electronic products that compose this communications infrastructure can represent a major portion of the system budget. For that sort of investment, you want to be sure the network meets organizational needs, and that requires understanding an organization's workflows. Performance and scalability are the two signs of a computer system's efficiency, but both can be hamstrung by an existing network that can't stand up to how much you're asking it to do. GIS introduces a lot more data to move. Yet the answer is not as simple as "bigger is better." The bigger the network bandwidth, the higher the cost; and why pay for size you don't yet need? The answer lies in knowing your peak traffic loads, the network bandwidth that can handle those loads, and doubling the bandwidth to ensure optimum traffic flow.

While serving as an introduction to network technology in general, this chapter identifies the network protocols used with GIS specifically. These include standard IP disk-mounting protocols for accessing file data sources, message protocols for communicating between the GIS application and database data sources, and Web protocols for connecting intranet and Internet data sources and services. Design standards are provided for each supported communication protocol, along with general design guidelines for network bandwidth sizing. You will also find some standard network design planning factors. These factors, expressed in terms of traffic per map display, will be the numbers used for capacity planning purposes in later chapters.

The fundamentals

For GIS, a network is like a transportation system for data, and there are two classes of the technology, the first less expensive than the second: local area network (LAN) and wide area network (WAN). (Actually, the cost for LAN equipment is relatively inexpensive in general, compared to other hardware costs within the system environment.) A LAN can handle heavy traffic over a short distance, as on a college campus or in a building. The Web is an example of a WAN, which supports communication or data transport over long distances. An organization's WAN can also become a way of sharing data between various departments as well as a way to leverage service-oriented Web operations through the latest integrative technology (i.e., ArcGIS Server 9.3).

For both a LAN and WAN, the volume of data (measured in bits) that can be transported per second is referred to as the capacity or transport rate of a particular network segment. This capacity is called *network bandwidth* and is typically measured in millions of bits (megabits or Mb) or billions of bits (gigabits or Gb) per second. Bandwidth specifications (the rate of moving data) provide a simple way for GIS design architects to calculate and talk about an organization's network capacity needs.

Those needs are best expressed in terms of what bandwidth is necessary to handle the network traffic during the hours when the most work is done. If you ever doubt the importance of considering network traffic in your design analysis, remember the analogy of an old transportation system that hasn't been upgraded to accommodate population growth: too many cars, coming from all directions, funneling down to one lane. During peak work periods, operational workflow performance can slow to a crawl similar to what is experienced driving onto major highway arteries during rush hour—it seems as if it takes forever to get anywhere or you just can't get anywhere at all. In fact, sometimes it is the latter and networks do "crash," temporarily refusing to function at all. Insufficient bandwidth capacity is at the root of many remote client performance problems. Data has to move at a rate appropriate to supporting user productivity during peak business hours. Therefore, calculating what will be a sufficient bandwidth is a critical part of the network component of system design.

To help you do that, best practices in terms of GIS network traffic transport times are identified later in this chapter (figure 3-5) for each of the configuration alternatives currently available for client/server communications. You will see in the chart that desktop client/server applications perform best in a LAN environment. Remote desktop users should be supported from a Windows Terminal Server or distributed desktop environment: for example, the application runs in the computer room with local data access, and persistent remote client access is provided for active session display and control. Web services, on the other hand, work fine over widely distributed WAN and Internet environments.

Eventually, you can use the capacity planning methodology described in this book to identify the bandwidth your GIS operations require. But clearly, capacity is not the only factor to consider in planning network communications. The network's configuration must also be a fit; it, too, influences how fast data appears on your screen. Where are your data sources in relation to the desktop users? How much data will your applications require from one source as opposed to another? Also, the appropriate conventions (protocols) must be in place, to allow the network to interface with the various computer products and multiple data formats encountered among its members.

Smaller factors within bandwidth, configuration, and protocols play a role in data transport and therefore in how fast and reliably data traffic moves through your network. In turn, how well the network performs is a major factor in overall system performance. Cumulatively, every factor down to the smallest either detracts from or contributes to creating and maintaining optimal system performance. Right now, during initial system design, you simply need to size and configure a network that will support the system during implementation. You can test and tune it later to verify your applications perform as expected. Nonetheless, understanding network components and processes and their interrelationships well enough to model them is prerequisite to creating a system that works at all, and this chapter will help you in this. Math and physics provide wonderful tools to model our world down to the smallest interaction. Over the years, we have used these tools to model the world of network communications. As a result, we can offer network design planning guidelines and best practices in this chapter that can help you get it right. But first, you need to understand that world: the physical components and processes of a network communication infrastructure for GIS. Then you can model the interactions between them (using the capacity planning models and workflow performance targets introduced in this book) according to the realities of your situation. Doing so can get you within range of a network bandwidth and configuration adequate for your needs. And that's where you need to start—with sufficient bandwidth to support your peak traffic needs.

Network components and GIS operations

GIS is different than other information systems, and its data traffic may put a bigger load on the network than it's accustomed to. GIS operations rate among the heaviest data movers (joined by document management and video conferencing enterprise solutions). Geographical information systems are data heavy because geography is data rich, and the beauty of GIS analysis is in its ability to examine high volumes of data quickly and turn it into information that is useful to you (an information product). What used to take hours and days of research and analysis, GIS can do in moments. With information layered so that you can readily see its relationship to place (and time) and other information, GIS information products are typically represented in a user-friendly map display. What goes on behind the scenes to create this user-friendly experience—transforming complexity into simplicity—is the stuff of system design.

LANs and WANs

The fundamentals of network technology set the scene. For many years (1970s–1990s), network technology remained a relatively static environment, while computer performance increased at an accelerating rate. Recent advances in communication technology, however, have enabled a dramatic shift in network solutions—and your system design options. Worldwide communications over the Internet bring information from millions of sources directly to the desktop in real time. Wireless communications is fast becoming mainstream; now data can virtually be transmitted from any place at any time.

There are also a variety of physical media used to transport data, some of which may be among your network segments. Data is typically transported from one server to another over physical networks made of copper wire or glass fiber (LANs and WANs). Other types of transport media include microwave, radio wave, and satellite digital transmissions. Wireless radio frequency bands and laser beams are also used as a communication medium.

Normally, it's the transmission medium that limits how fast the data can be transmitted. This rate of data transport is identified by the specifications applied to its communication procedure, called the network *protocol*. Data and applications may reside at many sites throughout an organization. Allowing for the fact that data may also exist in different formats at these various sites, a protocol is a set of conventions that governs how the network treats the data. For example, protocols may specify different levels of compression that can reduce the volume of data, thereby increasing transport efficiency. Today, network products are made to encourage a stable and dependable environment overall for data transport, whatever communication methods are used among the variety of protocols that allow applications and data resources to be useful and shared wherever they are located.

Network transport solutions can be grouped into two general technology classes. Figure 3-1 illustrates these two types of networks and some of the fundamental terms associated with the technology of each.

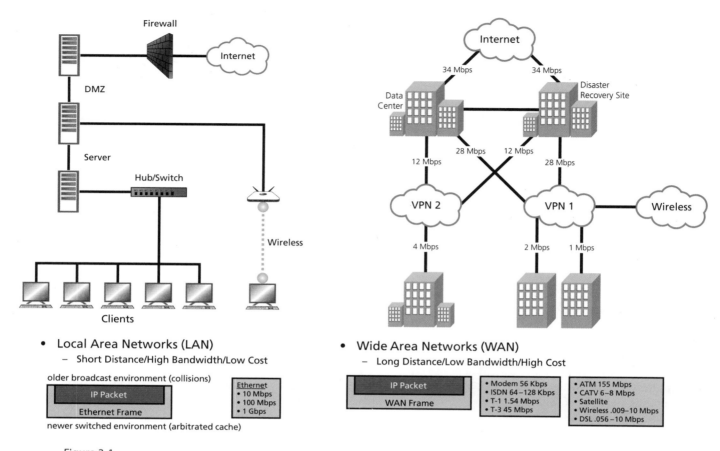

- **Local Area Networks (LAN)**
  - Short Distance/High Bandwidth/Low Cost

older broadcast environment (collisions)

IP Packet
Ethernet Frame

Ethernet
• 10 Mbps
• 100 Mbps
• 1 Gbps

newer switched environment (arbitrated cache)

- **Wide Area Networks (WAN)**
  - Long Distance/Low Bandwidth/High Cost

IP Packet
WAN Frame

• Modem 56 Kbps
• ISDN 64–128 Kbps
• T-1 1.54 Mbps
• T-3 45 Mbps

• ATM 155 Mbps
• CATV 6–8 Mbps
• Satellite
• Wireless .009–10 Mbps
• DSL .056–10 Mbps

Figure 3-1

Types of networks.

Local area networks (LAN)

A LAN supports high-bandwidth communication over short distances. Data transport over a single technology like this is single threaded, which means only one data transmission can be supported on a single LAN segment at any time.

The local operating environment is much more efficient now than it was when local area network technology began. LAN protocols were initially introduced in the 1970s to provide electronic communications between central compute-server environments. By the early 1990s, several LAN protocols were in use, and protocol exchange devices (bridges) were needed to connect the different LAN environments (Microsoft Windows operating system, AppleTalk network system, UNIX operating system, Novell's network software, etc.). By the late 1990s, Ethernet technology became the LAN standard, and today Ethernet bandwidths provide the capacity to transport volumes of data per second; measured in megabits and gigabits as 10 Mbps, 100 Mbps, 1 Gbps, and 10 Gbps. The bigger bandwidths are used in computer rooms and between buildings over campus-type environments. User desktops are connected by dedicated network cable to a central wiring closet where the network switches are located. Higher-bandwidth shared connections are provided between closets to the central computer facility.

Wide area networks (WAN)

A WAN supports long distance communications between remote locations. WAN protocols typically provide much lower bandwidth than what is available for the LAN, but the WAN is a data transport environment nonetheless: the source data is packaged in a series of packets and transported as a stream of packages (data packets) along the transmission medium. As mentioned earlier, the cost for WAN connections is relatively high compared to LAN environments.

The number of WAN protocols has expanded rapidly since the introduction of the Internet, to allow for the expanding communication medium, including telephone lines, cable TV lines, satellite communications, infrared transmitters, and wireless radio frequency bands, to name just a few. Data is transferred by light, electricity, radio frequencies, and laser beams—and these are only the common examples we see today.

Data

As you either know or have surmised by now, in GIS terms and in computer terms in general, data is essentially a collection of digital computer information stored in media that have the capability to record and retain the data structure. This data is represented by little pieces of information called *bits*. Each bit takes up the same space on storage or transmission medium. For convenience, these little bits can be grouped into *bytes* of data, with each byte containing eight bits. Data can be transported from one location to another within packets that protect the integrity of the data.

GIS users identify data capacity in terms of megabytes or gigabytes in reference to data stored on a computer disk. Network administrators, on the other hand, tend to think of data in terms of bandwidth specifications, identifying data in megabits or gigabits as the peak traffic volume a network can support per second. Megabyte is abbreviated using a capital "B," while megabit is abbreviated using a lowercase "b." These subtle differences can cause confusion when GIS specialists and systems administrators work together to design a system for GIS. So remember, 1MB = 8Mb when converting data volume from disk storage to data as traffic, and be very sensitive about using the proper abbreviation.

Another thing to remember: Network traffic includes some protocol overhead. The amount of overhead depends on the protocol packaging (size of network packets and the data volume). A simple guideline: translate 1MB of data to about 10 Mb of traffic.

What is data really?

The dictionary defines how we use the term in our language, and in today's world *data* is most commonly associated with computer technology and science. This tells us *how we use* the word, but does not tell us *what it is*. In a computer, it is represented in terms of switches (as a pattern of ones and zeros). In a communication medium, data is represented in frequency patterns, transmission patterns, etc. Data really is a pattern that represents a thought, an idea, or something we see or imagine (for example, a picture). That's not the whole story, but we're getting closer.

Where did it come from?

One of the first recorded datasets represented the names of animals or plants. We find pictures in ancient caves that tell us stories. Computers exist because someone thought of a way to represent our thoughts and

observations by patterns that could be manipulated by computer processors and stored on disk (on and off switches). Yet, all this and we still can find no way to define data except in terms of representations that have to be used or processed to make them meaningful.

Does it actually exist?

The physicist's answer is no, data does not exist in the usual way we think of as "being." In itself—stripped of all means of conveyance—it is definitely not something we can touch, smell, or lift. We can observe how it is depicted, and we can translate this impostor from one medium into another. But there is no physical requirement for data itself to comply with the laws of physics—it has no weight or mass that must be moved when transmitted from one location to another. The limitations exist only by way of the medium in which data is represented. In other words, in system design, the medium helps us understand the message.

No substance, no weight, no mass, just a pattern representing a thought. This suggests there may be no physical limitations in how we move data—nothing really to limit data communication or what we think of as bandwidth capacity anyway. So then, might we not expect current infrastructure limitations to be resolved, over time, as we find more efficient ways to record (store) and communicate (transport) our ideas and observations (data)? It's just a thought. And one not half as unlikely as "Beam me up, Scotty," although it does have something in common with that Star Trek phenomenon: The only requirement (*and this is not small*) is that the pattern that arrives at the client processor is exactly the same as the pattern sent by the server processor.

Communication protocols

For arriving in the same shape it was sent, data owes its thanks to protocols, (or procedures) that make sure it does. Applications move data over the network through proprietary client/server communication protocols, which work this way: Communication processes located on the client and server platforms define the communication format and address information. Data is packaged in communication packets, which contain the communication control information required to transport data from its source client process to the destination server process while maintaining the data's original structure.

Communication packet structure

So many bits, so little time; the possibility for corruption of the data while on the road would be great were it not for the packet structure in which it travels. Data is transmitted within packets that protect the integrity

Figure 3-2

Communication packet structure for data delivery across a network.

of the data. Within these packets is information that allows for its delivery across the network medium. The basic Internet protocol (IP) packet structure, as shown in figure 3-2, includes destination and source addresses and a series of control information, in addition to the data structure itself. Multiple packets are used to support a single data transfer.

Network transport protocol

The framework for client-server communications over a network—let's say, host-to-host over the Internet—is a series of step-by-step processes, with each protocol like a gateway to the next step. The terms "communication packet," "packet structure," and "data frame" are often used interchangeably; we'll simply use "packet" here, because "data frame" means something else in ESRI software terminology. But in describing how network transmission operates, let's be precise: the communication that goes across the network from host to host includes everything the packet structure is composed of (which has to be built every time). Just as we think of data in terms of the medium that moves it, we can think of that medium—in this case, the packet—in terms of the process that constructs it.

The packet is constructed at different layers during the transmission process. Data starts out in a stream to become a framework acceptable for travel across the network only after going through a layered experience. Using the standard way network administrators view the protocol stack (application layer, transport layer,

Internet layer, and network access layer), figure 3-3 shows how a data stream from the host A application is sent through the protocol layers to establish a packet (data frame) with access to network transmission:

- The transmission control protocol (TCP) header packages the data at the transport layer.
- The Internet protocol (IP) header is added at the Internet layer.
- The medium access control (MAC) address information is included at the physical network layer.

The packet is then transmitted across the network to the host B side of the pendulum where the process—in reverse—moves the data to the host B application. A single data transfer can include several communications back and forth between the host applications, each time through these sequences.

GIS communication protocols

Figure 3-4 on the opposite page shows the primary communication protocols that GIS applications use for network data transfer. Component processes of client and server both take part in implementing each protocol, enabling delivery in this way: the client process prepares the data for transmission, and the server process delivers the data to the application environment where analysis and display take place.

Network file services and common Internet file system protocols

All GIS applications are able to access a variety of file formats from a local disk. Shared data can be provided over the network on a file share. The server platform operating system includes a remote disk-mounting protocol enabling the client application to access data from a distributed server platform. UNIX and Windows each offer their own network-mounting protocols. UNIX

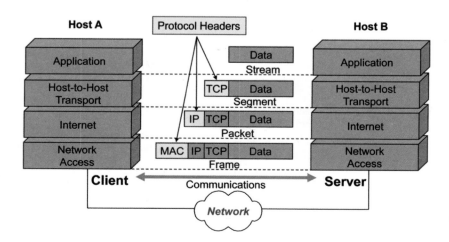

Figure 3-3

Network transport protocol.

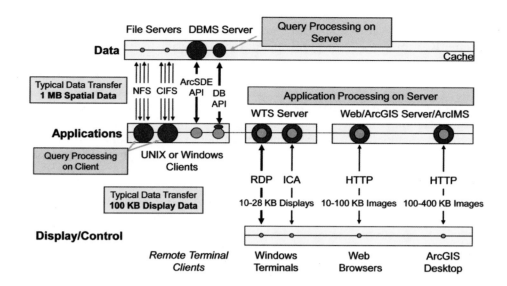

Figure 3-4

GIS communication protocols.

provides a network file services (NFS) protocol and Windows provides a common Internet file system (CIFS) protocol, each allowing the client application to access the remote file share as if it were a local drive. The client and server platform must be configured with the same protocol stack to support remote file access.

When accessing data from a file server, all program-specific executables reside on the client platform. They provide direction to the server operating system, through the connection protocol for access to data located on the server platform. Each piece of data must be transferred to the client application to enable query, analysis, and display.

Many GIS and image file formats are optimized to minimize the volume of data that has to be transferred. The file (which might be quite large) can include an index, which the client application can use to identify the specific segment of the data file required to respond to the query. The client can then request solely the portion of the file needed for the display. These types of data structures improve input/output (I/O) performance when accessing large files.

Much more data must be managed in client memory when working with the file data source as a whole, as opposed to a selected segment of the file, so client memory requirements are much higher when accessing file data sources.

Database access protocols
ArcSDE (Spatial Database Engine) technology provides a schema and communication protocol that enables spatial data to be centrally managed within a commercial database management system. The geodatabase schema and the open application program interface (API) support six DBMS platforms: Oracle, Microsoft SQL

Server, SQL Server Express, IBM DB2, PostgreSQL, and Informix. The DBMS server platform includes executables that enable query processing while spatial data is compressed within the geodatabase (roughly 50 percent compression). The data remains compressed during network transfer and is uncompressed by the client application, where it must go for analysis and display.

All ESRI client applications include the ArcSDE executables as part of a direct connect API that communicates directly with the DBMS network client software. (An SDE executable is a middleware translator, enabling ArcObjects to query the supported databases and manage the geodatabase schema.) These ArcSDE middleware functions are embedded within the client applications. The DBMS network client transmits data to the DBMS server, where the query processing takes place.

ArcSDE executables can be installed on a remote server or on the DBMS server platform. If installed on a remote server, the ArcSDE application server would use a DBMS connection to a local DBMS network client to enable communication to the DBMS platform. When the ArcSDE application server is installed on the same platform as the DBMS, it will use the server client libraries and spawn an ArcSDE client connection that will communicate directly with the DBMS server client connections.

Terminal client access protocols
ESRI customers use Windows Terminal Server to deploy centrally managed ArcGIS Desktop sessions for display and control by remote terminal clients. Two terminal client protocols are available to support the remote client desktop environment. Microsoft

provides a remote desktop protocol (RDP) that works with Windows client platforms, and Citrix provides an independent computing architecture (ICA) protocol through implementation of its software. Both protocols implement compressed data transmissions that perform well over limited bandwidth connections.

The Citrix XenApp server (formerly Presentation Server) provides advanced server administration, security, and a variety of supported terminal client platforms beyond what is available with the Microsoft RDP client. Most ESRI customers include Citrix XenApp to provide access to Desktop GIS applications.

Web communication protocols

Many people have grown accustomed to easy access over the Web with HTTP (hypertext transfer protocol), a standard Web transmission protocol. It allows for publishing Web applications for "thin" browser clients (the curious and casual) and Web services for heavier desktop applications. In this transaction-based environment, the browser or the desktop client controls the application service selection and display. A Web application service provider publishes a map for browser clients to see. Browsers can zoom in on it and query it, but because the map is published based on the Web application, the size of the display traffic can be optimized for fast delivery. The ArcGIS Desktop display environment, however, is controlled by the client application. Traffic for the ArcGIS Desktop is higher because of the larger image transfers. Image size is proportional to the physical screen display size; thus, larger image displays can result in higher display traffic.

Network communications performance

Network communications can affect how the user experiences computer performance in several different ways. The primary and most obvious performance impact is data transport time: the time it takes to transfer the data for the client display. A typical GIS application requires up to 1 MB of data to generate each map display, and that translates to about 10 Mb of traffic (uncompressed) for the data alone. (For the entire network traffic per display, you must also account for the extra traffic required to process the data transmission.)

In figure 3-5, a simple traffic display analysis shows the minimum transport times over the network for the typical GIS communication protocols. Network traffic transport times are computed for a 56 Kbps dial-up connection, 1.54 Mbps T-1 WAN, and for 10 Mbps, 100 Mbps, and 1 Gbps local network connections. Tightly coupled client/server workflows are listed first on the left, followed by Windows Terminal Server and Web Services workflows. The chart demonstrates the importance of considering network infrastructure capacity in the overall system design: you want to be safe within the shaded area for best practices and reasonable transport times.

You can use the data transfer volume (traffic per display) and available network bandwidth to calculate the minimum network transport times for a single map display transaction, for various workflow configurations, as shown in the chart. For the client/server configuration at the top (file server to workstation client), each application display needs 1 MB of data, which is the equivalent of 10 Mb of traffic. Up to 40 Mb of additional traffic is generated for the file server access, bringing the total to 50 Mb of traffic per display. To come to a best-case estimate of the data transport time, you simply divide the total traffic required (50 Mb) by the network bandwidth available. In the case of our client/server configuration, you would need a 100 Mbps bandwidth to transfer display traffic in less than one second.

It is easy to see that client access to a file data source makes sense only in a LAN environment. (There, workstation connections are 100 Mbps these days to

Client/Server Communications Configurations	Network Traffic Transport Time (Seconds)				
	Wide Area Network (WAN)		Local Area Network (LAN)		
	56 Kbps	1.54 Kbps	10 Mbps	100 Mbps	1 Gbps
File Server to Workstation client (CIFS)					
1 MB => 10 Mb + 40 Mb = 50 Mb	893	32	5	0.5	0.05
Geodatabase to Workstation Client					
1 MB => 10 Mb >> 5 Mb	89	3.2	0.5	0.05	0.005
				Best Practices	
Windows Terminal Server to Terminal Client (ICA)					
Vector 100 KB => 1 Mb >> 280 Kb	5	0.18	0.03	0.003	0.0003
Raster 100 KB => 1 Mb	18	0.6	0.1	0.01	0.001
Web Server to Browser Client (HTTP)					
Light 100 KB => 1 Mb	18	0.6	0.1	0.01	0.001
Standard 200 KB => 2 Mb	36	1.2	0.2	0.02	0.002
Web Server to ArcGIS Desktop Client (HTTP)					
Light 200 KB => 2 Mb	36	1.2	0.2	0.02	0.002
Standard 400 KB => 4 Mb	72	2.4	0.4	0.04	0.004

Figure 3-5

GIS network transport time for display traffic.

optimize user productivity in a file-based environment.) Geodatabase access is about 10 times better than file access. Access over a dial-up connection still does not make sense; transport times are far too slow for most user workflows. Some customers use one or two geodatabase connections over a T-1 connection, although this is not recommended.

Terminal and browser clients access only the display environment over the network. Windows Terminal Server displays require about 100 KB of data, or about 1 Mb of traffic. Vector traffic is compressed to less than 280 Kb per display—image traffic does not compress as well. Windows Terminal Server clients provide the best user performance over limited bandwidth environments.

By using published Web applications to provide displays, you generate less traffic, thereby optimizing network performance. To do this, you need to be sensitive to the traffic required to provide each display. Consider that standard ArcIMS client displays tend to be around 100 KB in size. Most of the newer ArcGIS Server published applications and services are more like 200 KB per display. ArcGIS Desktop data services can be much higher, since they usually request an image service in order to produce a full, high-resolution ArcMap desktop display.

If possible, the most frequently accessed displays should be very simple, without lots of pictures or maps that require much network traffic. Opting for small and simple map displays reduces traffic transfer time, but using cached maps is even better. If real-time data is not involved and you can get by with displays or maps that are prerendered, caching is one of the best ways to optimize performance. As an example of displays, we've been talking mostly about dynamic maps here, ones produced on demand or "on the fly." But what if you didn't have to transfer all that data back and forth every time you wanted a map? For desktop clients, data makes one trip to the client and then is available in a local cache for follow-on displays. Cached maps and cached data make sense if the maps and data don't change too often; even caching a basemap containing only the static data will save network traffic and transport time.

As opposed to the typical remote client displays requiring 100 KB of data, the richer ArcGIS Server environments require more like 200 KB of data to provide a dynamic display, but the 9.2 and 9.3 versions provide a variety of data caching options to compensate for that and optimize performance. Static data can be preprocessed and provided as a client basemap file-based image service. The image files are delivered to a client cache during the initial display, so that follow-on displays can use the data now localized in a cache. Implementation of client data cache architecture can reduce the dynamic layer display requirements and

network display traffic, using less server and network resources and improving client performance.

It is important to consider network traffic in your design analysis. During peak work periods, workflows can slow to a crawl similar to what is experienced in big city driving on major highway arteries during rush hour. Many remote client performance problems stem from just such a traffic jam scenario on the network. Sufficient bandwidth is critical to enabling user productivity during peak business hours.

Display traffic transfer times are only part of the overall network performance challenge. Other traffic on the network will also reduce available bandwidth. ArcGIS Desktop DBMS session-based workflows are tightly coupled, and unstable network connections can cause the client application to get out of sync with the DBMS server session. When this happens, the database should recover to the last consistent state (last save). This can be a very frustrating experience for database maintenance workflow operations or even GIS analysis or project work. Network latency is also a consideration, since geodatabase queries can require hundreds of sequential server communications to support a single map display.

You have seen the best practices identified for each of the available configuration alternatives. Distributed client/server applications perform best in a LAN environment. Remote desktop users should be supported from a Windows Terminal Server or distributed desktop environment (application runs in the computer room with local data access, and persistent remote client access is provided for active session display and control). Web services work fine over widely distributed WAN and Internet environments. Be sensitive to the published display traffic and clients will be more productive.

It may be necessary to upgrade the existing bandwidth. I was in Atlanta this past year to teach a class, and was staying at a hotel only five miles from the classroom. It took over an hour to drive to class during the morning rush hour—it was a parking lot the whole way. People like housing developments with pretty parks, swimming pools, and schools for the kids. Cities like to build technology centers for company offices, hotels, and restaurants away from family and children. Atlanta has both—but the roads connecting the housing community with the business park were never upgraded to support the traffic. Transportation bandwidth is an important design consideration.

Network latency

Several key performance factors contribute to overall display response time (chapter 7). These factors include display, network, and database processing time plus any queue time (processing delays) and network latency (medium travel time). Queue time is the duration when

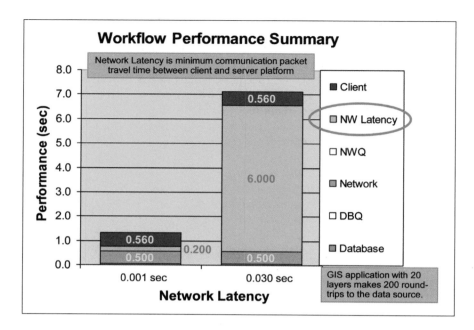

Figure 3-6

Network latency.

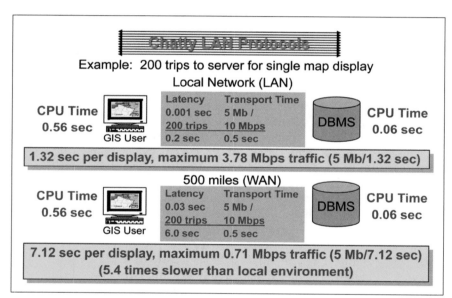

Figure 3-7

Network latency considerations.

program instructions are waiting in line to be processed; these processing delays are caused by random arrival times. (We begin to experience processing delays when traffic reaches about 50 percent of network capacity, and the delay time continues to increase as network traffic increases beyond 50 percent capacity.) Network latency is the time it takes for a network packet to travel from client to server (travel time), while network transport time represents the processing time incurred at the network connection—time required to get the data on the network medium (based on bandwidth capacity). Network latency can be measured with a simple ping utility ("tracert" <trace route> DOS command is one example of a latency measurement), which identifies the

hardware components (routers) in the transmission path and the travel time between each network connection.

Client/server protocols that make several trips back and forth between the client and server to complete processing for one display are called "chatty" transactions. Each round-trip will incur the network latency (packet travel time) delays. Figures 3-6 and 3-7 provide an example of network latency for a low latency and high latency connection, where bandwidth capacity over the network connection are both 10 Mbps. The complete client map display processing time is approximately 0.56 seconds. Network capacity is less than 50 percent, so there is no queue time in this example.

Database access protocols are "chatty": a typical database query requires a large number of trips to and from the server to complete the client display transaction. The total number of trips to and from the server (query transactions) will vary based on the data model complexity (primarily based on the number of feature classes or layers in the display). Figure 3-7 demonstrates how network latency can make a difference even when there is plenty of network bandwidth.

What affects display response time the most? For LAN environments, network latency is very low (typically less than 0.001 milliseconds per trip to the server). Even numerous trips between the server and client don't limit performance very much. Primarily, what determines how long you wait for the display are the client and server processing and network data transport times. In the example, computer processing time is 0.62 (0.56 + 0.06) seconds; network transport time is 0.5 (5 Mb/10 Mbps) seconds; and network latency is 0.2 (0.001 x 200) seconds. Total display response time is 1.32 seconds. Average network traffic is 3.78 (5 Mb/1.32) seconds. This traffic is well below 50 percent of the 10 Mbps network bandwidth capacity, so we would not expect queue time delays.

For WAN distances, which are longer and normally include multiple router hops, there can be measurable network latency delay. Network latency can have a considerable performance impact when using chatty database protocols. In the figure 3-7 example, the total transaction time over the WAN (including cumulative network latency) is 7.12 seconds. The maximum bandwidth used by a single user on this WAN connection is 0.71 Mbps, so you can see that user performance has been limited by network latency, not by WAN bandwidth. Many global WAN connections these days include satellite communication links. The fastest packet travel time is limited by the speed of light, which for very long distances (satellite connections) can result in network latency that is unacceptable. Good performance over WAN environments can be achieved from protocols that minimize sequential round-trips to the server (communication chatter). Latency is an important consideration when selecting remote client software solutions—good remote client solutions make use of Citrix Windows Terminal Server and Web software technology.

Shared network capacity

The total number of clients that can be supported on a single network segment (network backbone, server network interface card—NIC, campus network between buildings, etc.) is a function of network traffic transport time (amount of data traffic divided by network bandwidth) and the total number of concurrent clients.

Only one client packet can be transmitted over a shared network segment at any time.

With older switch technology, multiple transmissions on the same Ethernet network segment would result in collisions. Recovery from a collision would require each client to send a repeat transmission to complete their packet delivery. Ethernet segments became quickly saturated when multiple collisions occurred because of the rapidly increasing number of transmissions. Ethernet switches today include a cache buffer (concurrent transactions wait in a cache until they can be transmitted over the network segment), which when configured properly can avoid network collisions and improve transmission efficiency.

Figure 3-8 shows multiple client sessions sharing the same network segment; each data exchange is represented by the small boxes. Only one data exchange can be supported at one time on the same network segment.

A GIS application can require 1 MB of spatial data, or up to 10 Mb of network traffic, to enable each map display. A 1 MB map display is illustrated in figure 3-9 (1:2,400 scale [feet], average features = 250). Applications can be tuned to prevent display of specific layers when the map extent exceeds defined thresholds. Reducing the number of layers underpinning the display improves performance. Only required data layers should

Figure 3-8

Shared network capacity.

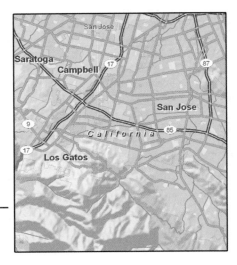

Figure 3-9

Typical 1 MB map display.

Source: Data from ESRI Data & Maps, StreetMaps USA.

be displayed for each map extent (e.g., you would not display individual parcel layers in a map showing the full extent of Santa Clara County, California). Proper tuning of the application can reduce network traffic and speed up performance.

Network configuration guidelines

Standard published guidelines are used for configuring network communication environments. These standards are GIS application–specific and based on typical user environment needs. Communication environments are statistical in nature, since only a percentage of user processing time requires transmission over the network. Network design standards represent typical GIS user workloads and address statistical application use, thereby providing the basis for the initial system design—general guidelines for establishing network requirements, as shown in figure 3-10.

The Capacity Planning Tool (chapter 10) will provide a flow analysis based on specific workflow traffic loads that address communication bandwidth requirements. Network data transfer time is a small fraction of the total display response time (on properly configured networks). Network data transfer can be the largest factor contributing to display response time when bandwidth is too small or when too many clients are on the same shared network segment.

The network must be designed to support peak traffic demands. The amount of traffic varies based on the different types of applications and user work patterns.

Standard guidelines provide a place to start configuring a network environment. Once the network is operational, network management becomes an ongoing traffic management task, strongly affected by the work environment and changes in computer technology. Network traffic demands should be monitored and necessary adjustments made to support peak user requirements.

Network design standards
Figure 3-10 provides recommended design guidelines to show you how big a bandwidth you will need to handle the network traffic generated by all the workflows you expect might be operating at once. These guidelines for network bandwidths establish preliminary guidelines for configuring distributed LAN and WAN workflows. Four separate GIS communication environments are included for each network bandwidth. The number of recommended clients is based on experience with actual system implementations and may not represent your worst-case scenarios. Networks should be configured with flexibility to provide special support to power users whose data transfer needs exceed typical GIS user bandwidth requirements.

Web services configuration guidelines
Implementation of Web mapping services places additional demands on the network infrastructure. The amount of system impact is related to the complexity of the published mapping services: Map services with small (less than 10 KB) or a limited number of complex images will have little effect on network traffic. Large

Local Area Networks	Concurrent Client Loads			
Bandwidth	File Servers	SDE Servers	Windows Terminals	Web Products
10 Mbps LAN	2-4	10-20	350-700	150-300
16 Mbps LAN	3-6	16-32	550-1100	250-500
100 Mbps LAN	20-40	100-200	3,500-7,000	1,500-3,000
1 Gbps LAN	200-400	1,000-2,000	35,000-70,000	15,000-30,000
Wide Area Networks	Concurrent Client Loads			
Bandwidth	File Servers	SDE Servers	Windows Terminals	Web Products
56 Kbps Modem	NR	NR	2-4	1-2
128 Kbps ISDN	NR	NR	5-10	2-4
256 Kbps DSL	NR	NR	10-20	5-10
512 Kbps	NR	NR	20-40	10-20
1.54 Mbps T-1	NR	NR	50-100	25-50
2 Mbps E-1	NR	1-2	75-150	40-80
6.16 Mbps T-2	1-2	1-3	200-400	100-200
45 Mbps T-3	10-20	6-12	1,500-3,000	700-1500
155 Mbps ATM	30-60	50-100	5,000-10,000	2,500-5,000

Figure 3-10

Network design guidelines: Staying within the green is recommended.

Wide Area Network	Peak Web Map Requests/Hour (based on Average Image Size)						
Bandwidth	10 KB	30 KB	50 KB	75 KB	100 KB	200 KB	400 KB
56 Kbps Modem	2,016	672	403	269	202	101	50
1.54 Mbps T-1	55,440	18,480	11,088	7,392	5,544	2,772	1,386
6.16 Mbps T-2	221,760	73,920	44,352	29,568	22,176	11,088	5,544
45 Mbps T-3	1,620,000	540,000	324,000	216,000	162,000	81,000	40,500
155 Mbps ATM	5,580,000	1,860,000	1,116,000	744,000	558,000	279,000	139,500

Note: 1 KB = 10 Kb HTTP traffic

Wide Area Network	Image Transfer Time (sec) based on Average Image Size						
Bandwidth	10 KB	30 KB	50 KB	75 KB	100 KB	200 KB	400 KB
19 Kbps Modem	5	16	26	39	53	105	211
28 Kbps Modem	4	11	18	27	36	71	143
56 Kbps Modem	2	5	9	13	18	36	71
256 Kbps	0.4	1	2	3	4	8	16
512 Kbps	0.2	1	1	1	2	4	8
1.54 Mbps T-1	0.1	0.2	0.3	0.5	1	1	3
6.16 Mbps T-2	0.02	0.05	0.1	0.1	0.2	0.3	1
45 Mbps T-3	0.002	0.01	0.01	0.02	0.02	0.04	0.09
155 Mbps ATM	0.001	0.002	0.003	0.005	0.006	0.01	0.03

Figure 3-11

Web services network performance.

images (greater than 100 KB) can significantly impact network performance.

In figure 3-11, you'll see an overview of network performance characteristics you should consider when deploying a Web mapping solution. The top portion of the chart shows the maximum number of requests per hour that can be handled over various WAN bandwidths, based on average map image size. The bottom portion of the chart shows the optimum transmission time for various map images. You must design Web information products according to user performance needs. What's the available network bandwidth? That may be your primary performance consideration. Simple, high-performance map services can make do just fine with map images from 50 KB to 100 KB in size. Less traffic per display will minimize network transport time. (Consider that a 100 KB image uses up more than 36 seconds of network transport time for 28 Kbps clients, a response time most users would consider unreasonably slow.) Many Web developers overlook the fact that peak site capacity with an average of 100 KB display traffic means a maximum of 5,544 requests per hour over a single T-1 Internet service provider connection. You should ask the question, do we have enough bandwidth to support our peak transaction loads? A single entry-level ArcGIS Server platform can support up to 25,000 map requests per hour, requiring more bandwidth than most customer sites have for Internet access services. More complex ArcGIS Server map services generate 100 KB to 200 KB traffic per display. ArcGIS Desktop users may request image services from 200 KB to 400 KB in size (image size varies with user display

size and resolution). Users generally demand reasonable performance or they will not be pleased with the service. Adequate network bandwidth capacity and careful information-product design are primary considerations when developing popular high-performance Web applications.

ArcGIS Server can make high-performance, complex services possible by using a preprocessed data cache. The more intelligent clients (ArcGIS Desktop, ArcGIS Explorer, Web applications with Adobe Flash clients, etc.) are able to overlay Web-based vector and image services on top of a high-performance, local, cache layer. The local cache data can be sent from the server once and used by the client many times, since the images are stored on the local client machine. Client data caching can reduce network transport times and increase display performance, when configured and used properly.

You can get data delivery as a service from the Web, downloading data to clients over the Internet. The data delivery service extracts data layers from the geodatabase based on identified extent, then zips the data into a compressed file, and downloads the data to the client. Standard Web-server, file-transfer applications can do similar things. Figure 3-12 identifies minimum download times based on available bandwidth and the size of compressed data transfers. Data downloads should be restricted to protect Web service bandwidth, as they can very easily consume all available bandwidth and slow down performance for other Web mapping clients. You need to control or limit the amount of traffic capacity that anyone can use for downloading data.

Wide Area Network Bandwidth	Peak FTP Downloads/Hour (based on Average File Size)				
	1 MB	5 MB	10 MB	20 MB	50 MB
56 Kbps Modem	17	3	2	1	0
1.54 Mbps T-1	462	92	46	23	9
6.16 Mbps T-2	1,848	370	185	92	37
45 Mbps T-3	13,500	2,700	1,350	675	270
155 Mbps ATM	46,500	9,300	4,650	2,325	930

Note: 1 KB = 10 Kb FTP traffic

Wide Area Network Bandwidth	File Transfer Time (sec) based on Average File Size				
	1 MB	5 MB	10 MB	20 MB	50 MB
19 Kbps Modem	526	2,632	5,263	10,526	26,316
28 Kbps Modem	357	1,786	3,571	7,143	17,857
56 Kbps Modem	179	893	1,786	3,571	8,929
128 Kbps	78	391	781	1,563	3,906
256 Kbps	39	195	391	781	1,953
1.54 Mbps T-1	6	32	65	130	325
6.16 Mbps T-2	2	8	16	32	81
45 Mbps T-3	0.2	1	2	4	11
155 Mbps ATM	0.1	0.3	1	1	3

Figure 3-12

Data download performance.

Client Platform	Data per display		Traffic per display		Kbps Traffic per user	
	KBpd	Adj KBpd	Kbpd	Mbpd	6 dpm	10 dpm
File Server Client	1,000	5,000	50,000	50	5,000	8,333
Geodatabase Client	1,000	500	5,000	5	500	833
Terminal Client (vector)	100	28	280	0.28	28	47
Terminal Client (raster)	100	28	280	0.28	28	47
Web Browser Client (light)	100	100	1,000	2	100	167
Web Browser Client (standard)	200	200	2,000	2	200	333
Web Desktop Client (light)	200	200	2,000	4	200	333
Web Desktop Client (standard)	200	200	2,000	4	200	333

Figure 3-13

Network design planning factors. The yellow-highlighted traffic per display (Mbpd) figures above are used as network load factors in the Capacity Planning Tool in chapter 10.

Network planning factors

Many network administrators establish and maintain metrics on network use. These metrics help them estimate increased network demands as they plan for future user deployments. Figure 3-13 identifies standard network design planning factors for typical GIS clients (highlighted in yellow), based on their target data source. You will be seeing these numbers again in later chapters as you learn to project network bandwidths adequate to GIS user needs in planned GIS deployments.

Network traffic delays can have a significant effect on user response times. Ever wonder what happens when you click and then . . . nothing? It might be all those intricate and beautiful maps—higher display traffic requirements—that are consuming all available network bandwidth resources. Delays begin to occur once traffic exceeds 50 percent of the bandwidth capacity, determined by the weakest network connection. Network latency can be a major concern with chatty client/server protocols (desktop applications accessing database or file data sources).

Network protocols enable distributed software communications. The next chapter will take a look at all the different ways GIS software can be distributed to support your enterprise operations. Now that you know what these communication protocols are, and what they mean in regard to network traffic and performance, you are prepared to review these communication options and select the best platform architecture to support your GIS needs.

4 GIS product architecture

A geographical information system is composed of multiple parts that can be configured in various ways. All these components and the architecture they describe together have a great deal to do with how the system as a whole performs. We've just looked at the performance factors affecting network capacity (traffic and bandwidth). We have identified a conversion factor (figure 3-13) between data and network traffic, which is a function of the network protocol. Now we look at the component architecture of a GIS, as well as the matter of its configuration (how the software processing loads are to be distributed on the hardware platforms).

GIS software and data resources are supported on computers in a distributed network configuration. This chapter identifies the software components for each of the standard GIS workflow solutions, the protocols used to enable communication between the software components, and the platform installation requirements that allow for scalable production operations.

You can look at each standard GIS workflow solution in terms of the software components installed on the selected platform environments. Some of these software components are included with the core GIS software, but you may need additional ones from third party vendors to complete the GIS workflow solution. The complete set of software components required to support each workflow solution is identified here.

For each standard GIS software workflow, you'll get a look at the possible platform configurations you might choose to support it. A series of configuration options are provided for each workflow, one for each type of data source and network connection protocol. General workflow solutions include ArcGIS Desktop (workstation and Windows Terminal Server) and GIS Web Services (ArcIMS and ArcGIS Server). File servers, geodatabase servers, and the ArcGIS Image Server number among the data sources for the included configurations. You will also see both standard and high-available server configurations for clustered-database and load-balanced Web application servers.

The software architecture for GIS and its multiple configuration options play significant roles in your selection of the proper hardware. How much processing must be done and where it will happen are primary factors to consider when putting together the system design puzzle. Software is executed on the installed server platform. The host platform processing load is determined by the installed software. In later chapters, you will find the standard processing loads for the software components of the workflows identified here. Then you will be able to take the platform processing requirements that match your situation and use them in a capacity planning tool to select a platform that will meet your needs.

What are the configuration options available for the core ArcGIS Desktop and ArcGIS Server technology? Many more than there used to be. Early in my ESRI consulting career, ESRI software was delivered as a software package (ARC/INFO, ArcView, Librarian spatial database, facilities tabular database, customer tabular database, etc.). At that time, there were no descriptions of how these software packages could work together at the system level. Customers were interested in an integrated system solution, one that did not require a team of system architects and installers to put together. Customers wanted a simple description of how the software and hardware components fit together to support a system solution. This chapter provides it.

This chapter describes the software components and platform configuration options available to support distributed GIS operations. Selecting the right system architecture depends on the user requirements and the existing infrastructure. Software must be configured properly to provide optimum user performance and peak system capacity. An understanding of application architecture alternatives and associated configuration strategies gives you the necessary foundation for an integrated GIS systems architecture design.

Enterprise-level GIS operations provide applications for a variety of users throughout an organization, all requiring access to shared spatial and attribute data sources. These distributed GIS applications, and the system hardware and software environments that enable them, are buttressed by a multitier client/server or Web services architecture. The primary architecture components, which are arranged in tiers in figure 4-1, are explained below.

Data server tier: Shared spatial and tabular database management systems provide central data repositories for shared geographic data. These database management systems can be located on separate data servers or on the same central server platform (these are the platforms represented as cylinders in figure 4-1).

Application server tier: Herein lie the hardware platforms that support GIS applications within the integrated system solution. In a centralized solution, application server platforms are used as terminal server platforms in Windows Terminal Server farms and for Web Application Server platforms in a Web server configuration. Windows Terminal Server and Web application server platforms can provide host compute services to a large number of concurrent GIS clients. Windows Terminal Servers host GIS desktop applications on centrally managed server farms allowing remote terminal clients to display and control applications executed on the terminal server platforms. Web application servers provide a variety of Web applications and services accessed by standard browser clients or other desktop applications.

Desktop workstations: Display and execution of application processes are functions performed by desktop workstations which, in many cases, are desktops or laptops that also can function as Windows terminal clients or Web browser clients. In many GIS environments, the user workstation supports execution and display of the desktop applications (in other words, the processing for both analysis and display happen right there on the desktop).

ArcGIS system software architecture

As described in chapter 3, ArcGIS is an integrated collection of software products for building a complete

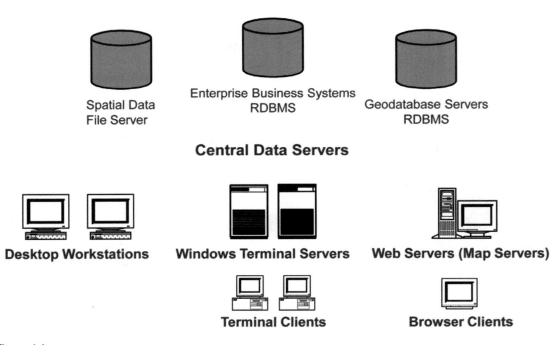

Figure 4-1

GIS multitier architecture.

geographic information system. The ArcGIS family of software products is used to deploy GIS functionality and business logic where needed—in desktops, servers, custom applications, Web services, and mobile devices. The ArcGIS applications are compiled from a common set of ArcObjects components developed using Microsoft Component Object Model (COM) programming technology. Figure 4-2 provides an overview of the ArcGIS system environment. In this diagram, the database tier would include file servers, geodatabase servers, and ArcGIS Image Server, and could also be configured on the application tier. Terminal services are executed on Windows Terminal Server farms and Web services are deployed on Web application servers, both types of services provided from an application tier. The client display tier includes the full range of desktop hardware, including simple-terminal or Web-browser-application display to the more complex mobile clients and ArcGIS Desktop application clients. (The latter two types are versatile enough for local disconnected application processing as well as client display.)

How spatial data is maintained and published within the organization is a huge factor in system performance and scalability, and therefore a major consideration in the system design. The volume of spatial data used to support a GIS has grown exponentially over the last 10 years, primarily due to improved technology for collecting and distributing spatial data types. Many GIS environments maintain several terabytes of active GIS data resources; megabytes of that data must be reviewed and processed within a typical user display session. All this data management needs to be organized in such a way as to promote effective and efficient GIS operations.

For years now, ArcGIS technology has included a spatial database engine (ArcSDE) for just such a purpose.

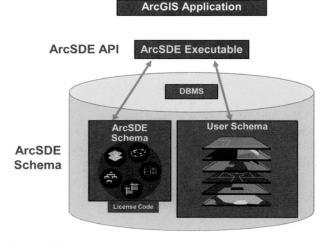

Figure 4-3

ArcSDE technology components.

ArcSDE technology provides a communication interface between GIS applications and several standard database management systems (DBMS). The ArcSDE components that enable DBMS applications to manage and publish GIS data are illustrated in figure 4-3. (Note that ArcSDE executables are included in the direct-connect API.)

Every ESRI software product includes an ArcSDE communication client. ESRI publishes standard data models that provide samples of user schema for a variety of standard GIS operations. Geodatabase is a term used to describe spatial data resources stored in an ArcSDE user schema. ESRI provides a published API supporting an open development interface to the ArcSDE schema.

The rest of this chapter describes the standard ArcGIS desktop and Web services configuration strategies applied to integrated enterprise GIS solutions, as outlined in figure 4-4. ArcGIS Desktop configurations include distributed workstation clients and centralized Windows Terminal Server farms. Web Services configurations include ArcIMS and ArcGIS Server single-, two-, and three-tier standard and high-available Web services implementations. Configurations are supported by file server, geodatabase server, and image server data sources.

ArcGIS Desktop can be deployed on client workstations or hosted by a Windows Terminal Server. Custom ArcGIS Engine applications include the same ArcObjects components underpinning the ArcGIS Desktop commercial software, so they share common configuration strategies. Different configuration alternatives are available to enable communications between the client application and the GIS data source.

GIS applications support open-standards-based systems architecture protocols. In other words, ESRI

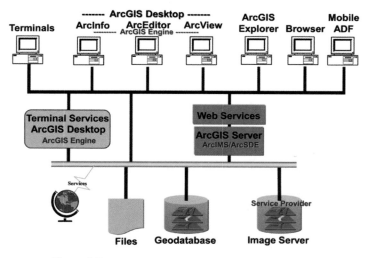

Figure 4-2

ESRI ArcGIS system environment.

ArcGIS Desktop client/server configurations
- Distributed workstation architecture
- Centralized Windows Terminal Server architecture

ArcIMS/ArcGIS Server Web services architecture
- ArcIMS component architecture
- ArcGIS Server component architecture
- Web platform configuration strategies

Figure 4-4

ESRI software environments.

products "play well with others": GIS enterprise architecture combines a variety of closely integrated commercial products to establish a fully supported system solution. All commercial software products must be maintained to accommodate evolving communication interface standards. Selecting well-established (popular) software architecture—solutions that support open-standard design protocols—is crucial: all parts of the distributed configuration are critical to the whole and must work together to ensure communication interfaces are properly maintained and supported.

ArcGIS Desktop client/server configurations

ArcGIS Desktop software is supported on Microsoft Windows desktop and terminal services operating system environments. ArcGIS Engine is a software development kit providing ArcObjects components for custom desktop application development. Visual Studio is used as the primary programming framework for ArcGIS Desktop applications. Programming options include the Visual C++ and .NET languages.

ESRI recommendations for deploying GIS desktop applications stem from many years of implementation experience. We've conducted careful monitoring and testing over the years, to identify any changes introduced with evolving technology trends. Many of the customer performance problems we've found over time have been the result of infrastructure limitations; a better understanding of technology fundamentals has led to resolving many performance issues.

Here are some of these fundamentals around how ArcGIS Desktop is deployed in an open-standard, client/server architecture. The client applications are tightly coupled with the GIS data source, with hundreds of sequential data requests exchanged to complete each user transaction. A typical map display, for example, is refreshed in less than a second, requiring a very chatty protocol exchange with the connected data source. This degree of communication dependency between software components (which "tightly coupled" refers to) allows no wiggle room: tightly coupled applications will fail when client/server communication is interrupted. Therefore, communications between the ArcGIS Desktop application and the GIS data source should be provided over stable, high-bandwidth, local network environments with minimum communication latency. Remote clients should be bolstered by terminal access to a central Windows Terminal Server located with the GIS data source. ("Terminal access" here means client terminals with software providing user display and control access to a terminal server compute environment that is centrally managed. For example, Citrix offers an ICA client that provides terminal access to a centrally managed Windows Terminal Server environment.)

The primary software components that enable the ArcGIS Desktop application workflow are pictured in figure 4-5. (Again, note that ArcSDE executables are included in the direct-connect API.) The tightly coupled executions in the center represent hundreds of

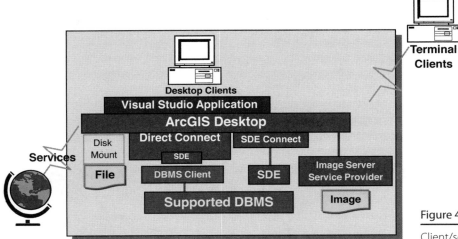

Figure 4-5

Client/server software architecture.

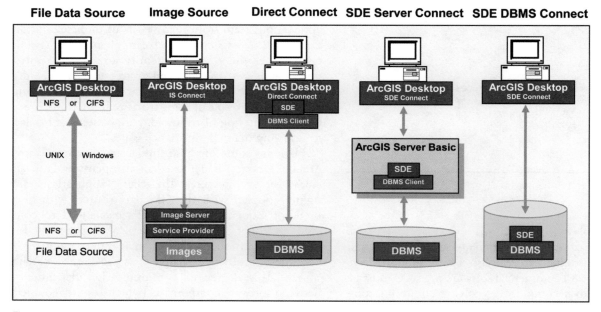

Figure 4-6

Distributed ArcGIS Desktop client.

sequential database transactions per map display, while the loosely coupled ones at the top right can accomplish displays simply through data streaming.

ArcGIS Desktop software can make connections to local file data sources, image services, DBMS data sources through an ArcSDE interface (direct connect or application server connect), and published Web data services. Web data services can be integrated with local data in a standard GIS map display.

Distributed workstation architecture

Figure 4-6 shows five distributed configuration alternatives using ArcGIS Desktop as the client. What distinguishes these configurations one from the other is the type of data source and connection protocol. From left to right, the five configurations include those with access to the following:

- A network file data source
- An ArcGIS image server data source
- An ArcSDE direct-connect access to a DBMS data source
- An ArcSDE connect access through an ArcGIS Server Basic remote server to a DBMS data source
- An ArcSDE connect to ArcSDE installed on the DBMS server.

ArcGIS Desktop software can access GIS data and files located on a local disk (i.e., native file access). GIS applications can access a remote file data source by using Microsoft's CIFS or similar UNIX NFS disk-mount protocol. When mounting a remote disk located on the file data source, the remote file appears as a local file share to the desktop application—you can see the file on your local desktop Explorer located on the remote server file share. (File server folders must be shared for network access before clients can connect to these file shares.) Query processing for a file data source is performed by the ArcGIS Desktop application; same processing required when accessing a local file. Where the processing happens makes a big difference on network traffic and platform processing loads (see NFS/ CIFS protocols discussed in chapter 3).

ArcGIS Desktop software provides two options for geodatabase access. The direct-connect option uses the ArcSDE executables (included in the direct-connect API) to connect to a local database client. The database client passes the network communications to the parent database server. The ArcSDE connect option offers network communication with a remote geodatabase server (ArcSDE can be installed as an ArcGIS Server geodatabase service—ArcGIS Server Basic— or installed on the DBMS server). Query requests received by the ArcSDE Application Server are sent to the data server and processed by the supported DBMS software. All data is stored and maintained in the DBMS repository.

Centralized Windows Terminal Server architecture

The Microsoft Windows Terminal Server product establishes a multihost environment on a Windows server. (*Multihost* is an IT term used to signal when mul-

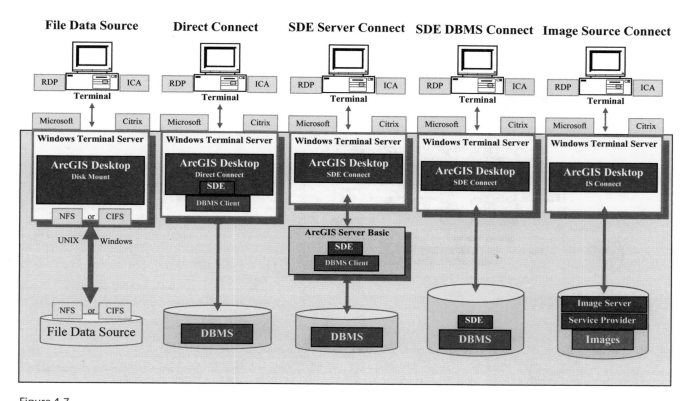

Figure 4-7

Centralized ArcGIS Desktop client.

tiple clients can be supported on a single "host" server.) A Windows terminal client provides display and control of applications executed on the Windows Terminal Server. Microsoft uses a standard remote desktop protocol (RDP) to allow for communication between the terminal server and the Windows client.

The Citrix XenApp server software extends system administration functions and improves client performance by using a proprietary communication protocol between the terminal server and client. Citrix calls this protocol an "independent computing architecture." The Citrix ICA protocol includes data compression technology that reduces network traffic to less than 280 Kb per display (rendering vector data information products). Traffic can increase to 1 Mb per display when accessing a map display that includes an image data source. XenApp server includes terminal client software for Windows, UNIX, Macintosh, and embedded Web client environments.

Five distributed ArcGIS Desktop configuration alternatives are identified in figure 4-7. All configurations provide remote terminal client access to ArcGIS Desktop applications hosted on a central Windows Terminal Server. The ArcGIS Desktop Windows Terminal Server configurations include the following:

- Connections to a network file data source
- ArcSDE direct-connect access to a geodatabase
- Two ArcSDE connect options to a geodatabase (one connecting through an ArcGIS Server middle tier and the other connecting to ArcSDE installed on the database server)
- An ArcGIS Image Server data source

The Windows terminal client communicates with the Windows Terminal Server through a compressed message-oriented communication protocol. Terminal clients have a persistent connection with the Windows Terminal Server ArcGIS Desktop session; lost connections are reinstated without losing the session. The application display is provided over the network to the terminal client, requiring much less data transfer than the spatial data query chatter between the ArcGIS Desktop application and the data source. The terminal client display traffic requirements are very small, supporting good single-user application performance over 28 Kbps modem dial-up connections (displays with an image backdrop may require more bandwidth).

Each ArcGIS Desktop user session hosted on the Windows Terminal Server connects to each data source the same way it would if executed on a client workstation, as explained in the previous section. Most current

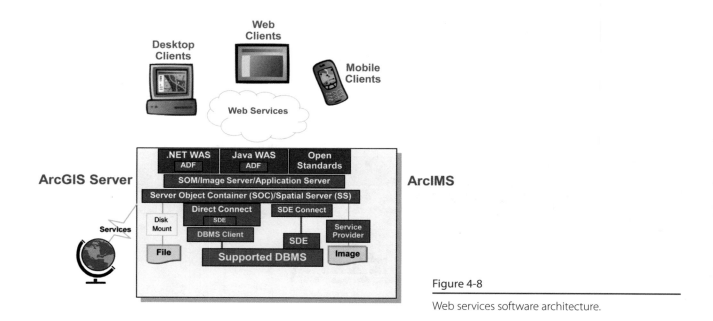

Figure 4-8

Web services software architecture.

ESRI customers using Windows Terminal Server also use the Citrix XenApp Server software. The Windows Terminal Server farm hardware are commodity Windows server platforms (Intel or AMD). Client session load balancing across the terminal server farm is managed by the Citrix software. Client profiles and security options provide an optimum GIS user display experience.

Web services architecture

Web mapping services offer an efficient and effective approach to publishing map products and services over the Internet. The Web services architecture is unlike the ArcGIS Desktop workstation architecture presented earlier in that it is not composed of tightly coupled client/server processes that demand stable high-bandwidth communications over relatively short distances. Rather, Web client communications use a transaction-based hypertext transfer protocol (HTTP); this allows for optimum communications over long distances and the ability to perform over less stable communication environments. Web applications share many of the same benefits achieved when using Windows Terminal Server for remote ArcGIS Desktop clients: persistent display (performs well over unstable remote networks), lower bandwidth (requires less network traffic), and minimum communication chatter (works well over high-latency network connections).

Figure 4-8 shows the software components associated with the Web services communication architecture. Above the tightly coupled (chattier) executions inside the box are the loosely coupled ones that require only one server request per display. (Also, loosely coupled

applications do not fail if the connection is lost—they can repeat the request or save requests for a later time when communication is reinstated.)

Web services are published on a Web server. When accessing the Web site, clients of that server are presented with a catalog of those published services and applications. Web applications consume map services and render a client presentation layer to create a published application workflow. (In other words, published services are used as component functions brought together in a Web application process flow. The client and Web servers are loosely coupled, with each client communication representing a complete transaction.)

Looking at exactly what a transaction is—in terms of how much and where on the system such activity happens is a relevant piece of the puzzle the system architect is engaged in figuring out. A GIS Web transaction is defined as the complete Web server response to a client request, which includes all associated updates to the client display. Transactions are processed by the appropriate Web-based GIS server and returned to the client.

ESRI Web GIS services are provided by ArcIMS and ArcGIS Server software. ArcIMS was the first ESRI Web mapping technology. Initially released in 1997, it provides several service engines to efficiently publish map services on the Web. ArcGIS Server technology, initially released in 2000, is ESRI's latest Web services technology. ArcGIS Server provides a rich development environment for server-based GIS applications and services. It can be deployed as Web-based or LAN/WAN local network-based services, enabling desktop application clients to use local GIS services in addition to Web-based services. ArcGIS Server comes with a mobile application development framework

ID	ArcIMS	ArcGIS Server	Image Server
WA	Web Server	Web App Server	
SM	App Server	Server Obj Manager	Image Server
SS	Spatial Server	Server Obj Container	
SDE/SP	ArcSDE	ArcSDE	Service Provider
DS	Data Server	Data Server	File Server

Figure 4-9

Web software nomenclature.

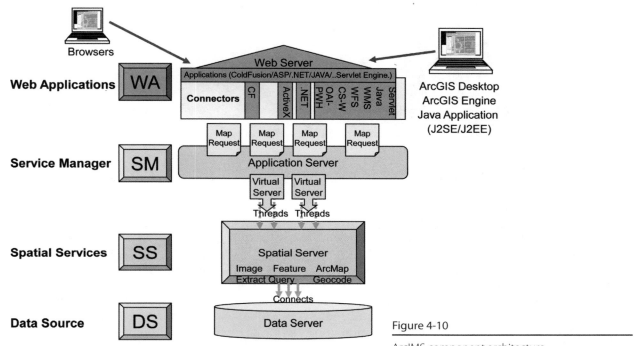

Figure 4-10

ArcIMS component architecture.

(ADF), which can support loosely coupled, lightweight, handheld or desktop computers with persistent data cache. The local data cache and lightweight application can be used for both connected and disconnected GIS client operations: ArcGIS Server parent services manages client application deployment and data synchronization.

The software architecture components for ArcIMS and ArcGIS Server have different names and functionalities, although in a logical sense they perform similar functional roles and are based on common platform configuration strategies. Both software solutions include a Web application (WA), service manager (SM), spatial server (SS), and data server (DS)—software that can be deployed on different platform combinations. Both types of software can support scalable architecture and system availability needs. Figure 4-9 sets forth the Web software nomenclature used in this book, which you'll see again here and there throughout, for ease of reference.

Both the location of the various software components and the software configuration that you select will directly affect system capacity, service reliability, and overall output performance. Keep this in mind as we run through the two types of Web service architectures and the various configuration strategies associated with them.

ArcIMS component architecture

ArcIMS delivers dynamic maps and GIS data and services via the Web. The standard product includes an ArcIMS manager and map development wizards that allow the design and authoring of most standard map products without special programming. ArcIMS services can be used by a wide range of clients, including custom Web applications, ArcGIS Desktop, and mobile and wireless devices.

Figure 4-10 provides an overview of the ArcIMS component architecture and the associated software configuration groups. The architecture includes four software configuration groups identified as WA, SM, SS, and DS (defined in figure 4-9).

These configuration groupings indicate how the software processing loads will be distributed over the

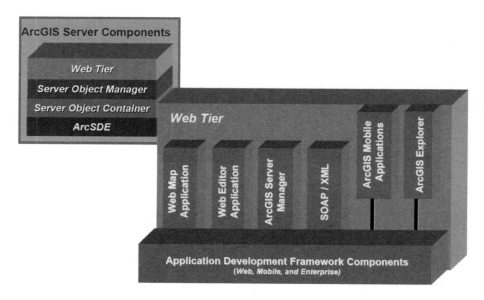

Figure 4-11

ArcGIS Server software components.

hardware platforms, determined by how these software components are installed on the system. The ArcIMS software configuration groups are described here:

ArcIMS Web applications: The Web applications software includes the Web HTTP server and the Web applications. The HTTP server enables communication between ArcIMS map services and the Web client. Web applications can be involved to manage and enhance user workflow and client display presentation. Connectors on the Web server translate Web HTTP traffic and/or Web application program calls to communication ArcIMS Web services can understand.

ArcIMS service manager: The ArcIMS application server software manages the inbound map-service request queues (virtual servers) and connects to the ArcIMS public service engines (Image, Feature, Extract, Query, ArcMap Image, Geocode, Route, Metadata). Inbound requests are routed to available service instances for processing. A relatively small amount of processing is required to support the application server functions.

ArcIMS spatial services: The ArcIMS spatial servers are installed on a map server platform. The spatial servers include the ArcIMS service engines (Image, ArcMap Image, Feature, etc.) that service the map requests. (*ArcIMS Monitor* is another term used for an ArcIMS spatial server.)

Data source: The data server (GIS data source) is where the GIS data is stored. An ArcSDE data source provides the query processing functions. Standard GIS image or file data sources are also represented at this level.

ArcGIS Server component architecture

ArcGIS Server is a set of processes: objects, applications, and services that make it possible to run ArcObjects components on a server platform environment. Server objects are managed and run within the GIS server. Server applications make use of server objects and may also use other ArcObjects components that are installed on the GIS server.

The Web server hosts server applications and Web services developed using the ArcGIS Server application programming interface. These Web services and Web applications can be developed using the ArcGIS Server ADF, which is available for both .NET and Java developers and supported within the associated Web application server development environments.

Figure 4-11 provides a high-level overview of the ArcGIS Server component architecture.

Figure 4-12 provides an overview of the Arc-GIS Server component architecture with its associated software-function locations. The ArcGIS Server architecture includes four configuration groups of software—Web applications, service manager, spatial services, and data source—which are explained below.

ArcGIS Server Web applications: The ArcGIS Server WA components include a commercial Web HTTP server enabling communications between the Web clients and a Web application server development environment for .NET or Java Web applications or Web service catalogs. ArcGIS Server enables a Web tier, with its .NET and Java ADF.

ArcGIS Server service manager: A server object manager (SOM) controls service object deployment and initial assignment of the client application requests to available server object containers. The SOM performs as a parent process, providing adaptive service deployment and container machine load balancing based on concurrent service demands.

ArcGIS Server spatial services: The "container machines" (one or more depending on peak transaction requirements) are so named because they host the server object containers (SOCs), which are managed by the SOM. Each service configuration has its own dedicated SOCs. The server objects hosted within each SOC are compiled ArcObjects components installed on the container machine.

Data source: The data server (GIS data source) is where the GIS data is stored. An ArcSDE data source connects the GIS applications with the DBMS query processing functions. Standard GIS image or file data sources are also represented at this level.

Services available on ArcGIS Server can be published for use by intranet LAN or WAN applications. Local GIS applications can access published services with direct access to the SOM without using the Web server interface. The SOM will assign SOC instances to support the service connection.

Web platform configuration strategies

A Web site can be supported with as few as one platform or as many as six or even more, depending on site capac-

ity and availability requirements. ArcIMS and ArcGIS Server platform configuration options are similar, and criteria for establishing the proper configuration are common to both technologies. Here, we'll address these component configurations strategies using the ArcGIS Server nomenclature. (ArcGIS Server and ArcIMS capacity planning guidelines related to these configurations will be provided in later chapters.)

The Web system architecture design alternatives are grouped as single-tier, two-tier, and three-tier configurations. Simple configurations are easier to maintain, while more complex configurations satisfy scalable-capacity and system-availability requirements. Initial prototype operations are often published on a single platform (identified as a standard configuration), while most GIS production operations are ordinarily supported with high-availability configurations (configurations that will continue providing services regardless of any single platform failure).

ArcIMS and ArcGIS Server are designed to foster a scalable Web architecture. Optimum platform environments are configured using single- or multiple-core

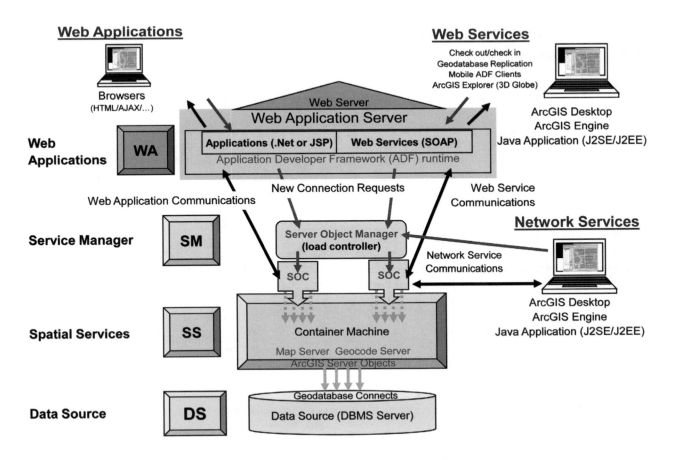

Figure 4-12

ArcGIS Server component architecture.

Standard Configuration

High-Availability Configuration

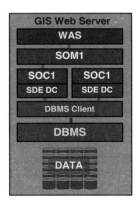

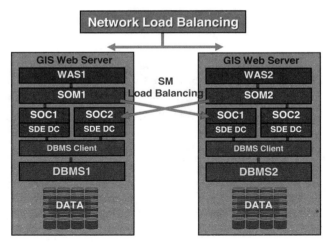

Figure 4-13

Single-tier platform configurations: Commonly used for prototype GIS deployment.

server platform technology. Over the next few pages, let's take a look at the options you have among the platform configurations recommended for supporting primary GIS services.

Single-tier platform configuration

Single-tier platform configurations, shown in figure 4-13, use one or two platforms to handle all Web service components. Most initial customer prototype deployments—using a small relatively static database— work just fine in a single-tier architecture, in either a standard or high-availability configuration. (For the latter, the data can be replicated to take advantage of two servers to provide access to the common dataset.)

Standard configuration: A complete Web site can run on a single hardware platform. This standard configuration is appropriate for Web service development and testing, for sites with a limited number of service requests, and for initial prototype deployments (see chapter 12). Multiple SOC instances can be deployed to support optimum capacity requirements.

High-availability configuration: In the event of a single platform failure, it makes sense to have another one to fall back on. Most GIS operations do require redundant servers, configured so that the site remains operational. The two-platform configuration is your insurance that production will continue during periods when single platforms are down for maintenance or upgrading or very busy configuring and publishing new services. This high-availability configuration

requires the following assets: *Network load balancing* routes the traffic to each of the servers during normal operations, but only to the active server if one server fails. *Service manager (SM) load balancing* distributes the processing load between the two platforms to prevent requests from backing up on one server when extra processing resources are available on the other server. Separate SOCs are required on each platform to enable the latter. Of course, two data servers require a complete copy of the data, which requires duplicating the data. The SOC1 instances are deployed by SOM1 and the SOC2 instances are deployed by SOM2. Multiple SOC instances can be deployed to support optimum capacity requirements.

Two-tier platform configuration

A two-tier architecture provides the best solution for sites that have a separate database server. The two-tier, high-availability is the most popular and practical configuration for most ArcGIS Server deployments.

The two-tier architecture in figure 4-14 (an enterprise GIS with supported database) includes GIS server and data server platforms. The Web server and GIS server components are located on the GIS server platform, while the data server is on a separate data server platform. This is a popular configuration for sites with large volumes of data resources or existing data servers. A single copy of the data can support multiple server components in conjunction with other GIS data clients in the enterprise.

Standard Configuration   **High-Availability Configuration**

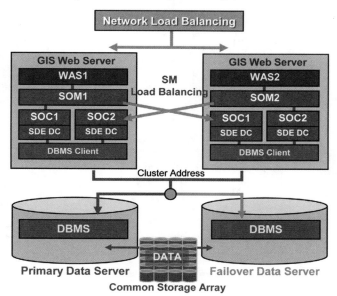

Enterprise Server With Supported Database

Figure 4-14

Two-tier platform configurations (separate data servers):
Commonly used for production GIS deployments.

Standard configuration: The standard, two-tier configuration is composed of one GIS Web server platform and a separate single data server platform. The Web server is installed on the GIS Web server platform. Multiple SOC instances can be deployed to support optimum capacity requirements (see chapter 8).

High-availability configuration: High-availability operations require redundant server solutions, configured so the site remains operational in the event of any single platform failure. This configuration includes the following: network load balancing to route the traffic to each of the GIS Web servers during normal operations but only to the active GIS Web server if one of the servers fails; SOM load balancing to distribute the SOC processing load between the two GIS Web server platforms, and thus to avoid having requests back up on one server while extra processing resources are available on the other server (multiple SOC container groups are required on each GIS Web server platform to support this configuration); and two data servers that are clustered and connected to a common storage array data source. The primary data server handles geodatabase services during normal operations, and the secondary data server takes over geodatabase services when the primary server fails. Data server clustering is not required if availability requirements are satisfied with a single data server. The SOC1 instances are deployed by SOM1 and the SOC2 instances are deployed by SOM2. Multiple SOC instances can be deployed to support optimum capacity requirements.

Three-tier platform configuration

Three-tier configurations include Web server, container machine, and data server tiers. Two configuration options are the choices, based on the location of the server object manager (SOM).

Figure 4-15 shows a three-tier configuration with the SOM located on the Web server tier. This configuration provides the simplest three-tier architecture, and is the most popular solution. Within it, network load balancing handles the Web tier failover. The three-tier configuration provides a scalable architecture, wherein the middle tier can support two or more platforms as required to handle peak workflow loads.

Standard configuration: The standard three-tier configuration includes a single Web server with a separate map server/container machine layer and a separate data server. The map server/container machine layer can be a single platform or can be expanded to include several platforms, depending on the required site capacity. SOM load balancing is provided by the GIS Web

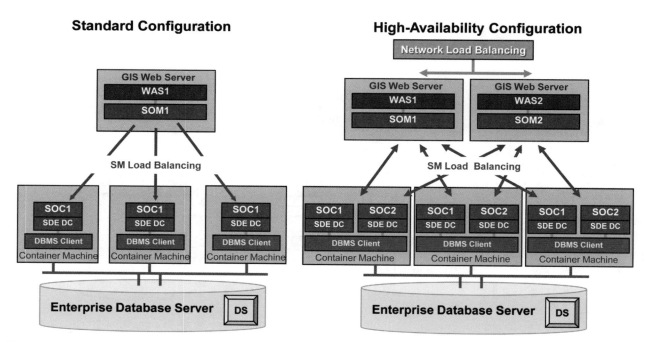

Figure 4-15

Three-tier platform configurations with SOM on Web tier
(separate container machine tier) provide the simplest three-tier
high-capacity architecture.

server service manager. Multiple SOC instances can be deployed to support optimum capacity requirements.

High-availability configuration: High-availability operations require redundant server solutions, configured so the site remains operational in the event of any single platform failure. This configuration includes the following: network load balancing to route the traffic to each of the GIS Web servers during normal operations and only to the active GIS Web server if one of the servers fails; SOM load balancing to distribute the SOC processing load between the two container machine platforms to avoid having requests back up on one server when extra processing resources are available on the other server (separate SOC containers are required on each container machine platform to support this configuration); and a data server configuration that satisfies your enterprise availability requirements. The SOC1 instances are deployed by SOM1 and the SOC2 instances are deployed by SOM2. Multiple SOC instances can be deployed to support optimum capacity requirements.

Figure 4-16 shows a three-tier configuration with the service manager located on the map server/container machine tier. The Web server and spatial server connectors are located on the Web server platform, and the SOM and SOC components are located on the container machine platforms. This could be a preferred configuration when supporting Java applications

on Linux-based Web servers. In this configuration, all the COM-based software is located on the container machine tier. This is the preferred architecture when the Web server is not a Windows platform.

The failover scenarios are more complicated when the SOM is supported on a separate platform tier from the Web applications. You will still need to configure SOM load balancing for optimum capacity during peak loads. You will also need to configure the Web application to support failover to the remaining SOM when one of the SOM platforms fails. For example, if the SOM1 platform fails, WAS1 will need to be configured to send map requests to SOM2 for processing. The SOM2 output file will need to be shared with the WAS1 server for return to the client display.

Standard configuration: The standard for this configuration includes a single Web server with a separate map server/container machine layer. The map server/container machine layer can be a single platform or can be expanded to support several platforms, depending on the required site capacity. Web application traffic balancing is supported by the GIS Web server connectors for the ArcIMS implementation. The ArcGIS Server implementation can be configured in a failover mode (SOM2 would be activated only if SOM1 fails). SOM load balancing is provided by the server manager components (preferably no more than two SOM components on the container machine tier).

Separate SOC executables (separate group of SOC instances deployed by each SOM) must be supported on all container machines to support load balancing across multiple platforms. A separate data server is provided as a common data source. Administration of this architecture can become increasingly complex as additional container machines are added. Multiple SOC instances can be added to support optimum capacity requirements.

High-availability operations: High-availability operations require redundant server solutions, configured so the site remains operational in the event of any single platform failure. This configuration includes the following: network load balancing to route the traffic to each of the GIS Web servers during normal operations and only to the active GIS Web server if one of the servers fails; Web application traffic load balancing to distribute inbound load between the two SOM located on the container machine tier; SOM load balancing to distribute SOC processing loads across the container machine platforms to avoid having requests back up on one server when extra processing resources are available on the other server (two SOC container groups

are required on each container machine platform to support this configuration, with each SOC assigned to a parent SOM); and data server configuration supporting enterprise requirements. Administration of this architecture becomes increasingly complex as additional container machines are deployed. The SOC1 instances are deployed by SOM1 and the SOC2 instances are deployed by SOM2. Multiple SOC instances can be deployed to support optimum capacity requirements.

Three-tier service-oriented platform configuration
ArcGIS Server Web applications can be developed and deployed using data services provided entirely by remote Web service providers. It is also possible to provide these HTTP SOAP-based services from a separate local ArcGIS Server Web site. Figure 4-17 provides an example of a Web application supported entirely by HTTP SOAP services. This ArcGIS Server Web configuration supports an enterprise-services architecture with the Web application communicating with a published Web service across a firewall.

The internal GIS Web Servers are configured the same as the high-availability example seen earlier in figure

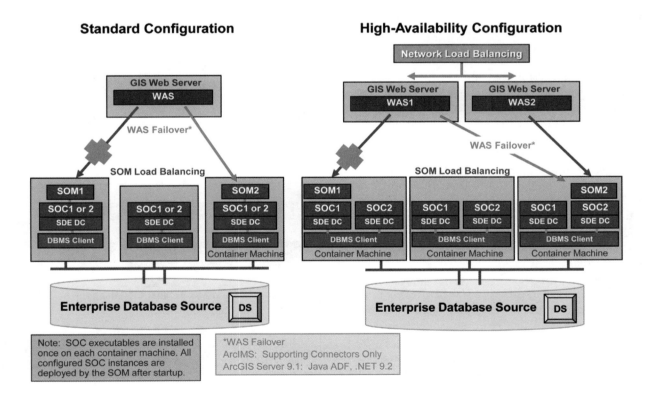

Figure 4-16

Three-tier platform configuration with SOM on map server/SOC tier (separate Web servers) is an alternative architecture when the Web server is not a Windows platform.

High-Availability Configuration

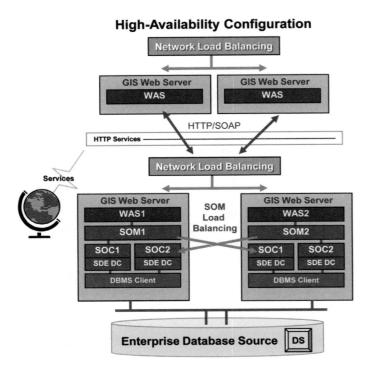

Figure 4-17

Three-tier platform configuration: Web services architecture.

4-14. Web services can be passed through to the external Web applications using standard HTTP SOAP and XML service protocols. This is the preferred way to support ArcGIS Server Web applications that need to connect to Web services provided over a firewall connection. (Firewall configurations are discussed in chapter 5.)

Many of the more powerful ArcGIS Server applications benefit from a more tightly coupled DCOM communications. Each application is directly coupled to an assigned SOC to support each transaction. Results from these applications can be provided as services to more loosely coupled enterprise applications supported in a separate security zone by using standard HTTP SOAP protocols.

ArcGIS Server provides a broad range of functionality that can be used to support standard Web mapping or for more complex geospatial workflows that previously were not available through open standard Web protocols. Use of preprocessed cached data layers available with ArcGIS Server can improve user performance and expand server capacity beyond what we saw with ArcIMS technology. Client caching can reduce network traffic requirements. Careful attention to user requirements and proper deployment of

the technology can make a big difference in system capacity and user experience.

ArcGIS Image Server

ArcGIS Image Server provides fast access to large quantities of file-based imagery (image source does not require a database structure). Imagery can be maintained in its native format and accessed by ArcGIS Desktop and Server applications on the fly and on demand. ArcGIS Image Server can also perform dynamic imagery processing; providing better image visualization, while at the same time reducing storage costs, data processing, and maintenance.

In the overview of its software architecture (figure 4-18) you can see that Image Server consists of two primary software components, the image server (IS) and the service providers (SP). The IS publishes available services and connects service requests to the SP instances. The SP instances process the service request. The processed image is returned to the requesting application.

Client communications are loosely coupled, supporting communications over widely dispersed networks. Communications between the service providers and

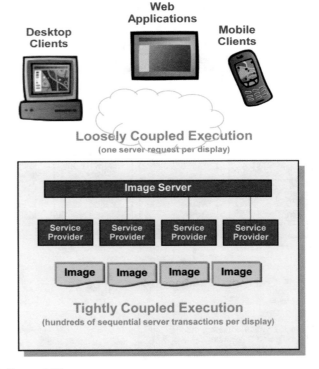

Figure 4-18

ArcGIS Image Server software architecture.

Standard Configuration   **High-Availability Configuration**

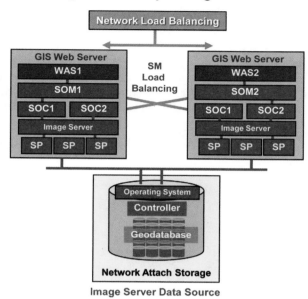

Figure 4-19

ArcGIS Image Server as a direct image data source.

the image files is tightly coupled. Regardless of the fact that these are tightly coupled executions, ordinarily requiring hundreds of sequential server transactions per display, here between the IS and SP image processing is very fast, since only the extent of the requested application display is processed. The service providers can be configured to carry on a variety of image processing services. Any ArcGIS application can use ArcGIS Image Server as a primary image data source for a standard map display, as described in the next example.

Figure 4-19 shows a two-tier Web configuration using ArcGIS Image Server as an image data source. ArcGIS Server can extend the image services to serve a full range of Web clients. ArcGIS Desktop applications can also access Image Server as a direct image data source.

Standard configuration: ArcGIS Image Server can be configured as a stand-alone server, with the IS and SP components all supported on the same platform along with the imagery. ArcGIS Server can access Image Server as an image data source. ArcGIS Desktop clients can access the same image server data source.

High-availability configuration: High-availability operations require redundant server solutions, configured so the site remains operational in the event of any single platform failure. The ArcGIS Image Server components can be configured alone or installed with

ArcGIS Server to provide a variety of image services. This configuration includes the following: network load balancing to route the traffic to each of the GIS Web servers during normal operations and only to the active GIS Web server if one of the servers fails; SOM load balancing to distribute the SOC processing load between the two GIS Web server platforms to avoid having requests back up on one server when extra processing resources are available on the other server (two SOC container groups are required on each GIS Web server platform to support this configuration); and a mirror set of image server service providers on each platform. A single image data source can be provided by a high-available NAS (network attached storage) solution or clustered file servers.

Selecting the right architecture

These are the configuration options available for the core ArcGIS Desktop and ArcGIS Server technology. All the architecture options identified in this chapter are available for the full range of business operations developed on top of the core ESRI software technology. What you select depends on what you need. To deploy ArcIMS and ArcGIS Server Web applications, most

ESRI customers have selected among the Web platform configurations just described. For the more complex, heavy ArcGIS Desktop workflows (utility work orders, county infrastructure maintenance, etc.), you would choose among the ArcGIS Desktop workstation and Windows Terminal Server configurations described earlier in the chapter. Selecting the right architecture does depend on the software technology and the architecture options incumbent with it. But there's a lot more to consider: user locations, available network communi- cations, security and availability requirements (see next chapter), and performance and scalability of the selected technology solution. Yet in looking at all these considerations, as you are doing with this book, don't forget the fundamental consideration underlying it all: your user requirements. There's no substitute for understanding your needs and how the selected technology is going to meet them. This is your foundation for choosing the right architecture solution.

5 | Enterprise security

Security comes at a price; yet not having it is even more costly. Too many security controls can reduce productivity and increase cost. Too little control can result in loss of property and the ability to perform. Finding the right balance is the answer, even though looking for it is like aiming at a moving target. There is no single solution for security, but you can manage operational risk. By understanding security risk, you can apply the appropriate security controls, and at the same time, control security costs.

Security is about protecting our ability to do work. Business operations have become so dependent on computer technology that losing the use of our computer or access to the network can shut us down. Our dependence on network communications, both within the workplace and for access to Internet resources, has changed the way we work—and how much we need to protect our ability to do work. There is a price to pay if we don't protect our business environment. It's not unheard of for computer viruses or breaches in security to restrict or completely shut down commercial and government operations for periods of time. Yet the solution is not about locking up our resources so tight that nobody can use them. The answer is in security and control measures that both enable and protect required business communications.

All must be weighed, then (threats and vulnerabilities as well as safeguards and controls), in the context of these three primary goals of information security:
1. Confidentiality—prevent intentional or unintentional disclosure
2. Integrity—prevent unauthorized data modifications
3. Availability—ensure reliable and timely access to data

The third goal goes right to the heart of system design's objective overall, which is a scalable system that performs optimally. Enterprise security can be a challenge for IT architects and security specialists. Frequently, until the last few years, entire IT systems were designed around a single objective and "community of interest," served by physically isolated systems, each with its own data stores and applications. With the new emerging standards, however, business operations include organization-wide enterprise solutions that are interactive and comprehensive. All this is enabled by more mature communication environments, more intelligent operating systems, and a variety of standard integration protocols.

For most organizations, the sense of the importance of security is growing, as is the range of focused technology intended to secure information system environments. The security measures an organization requires to provide secure operations—and how they affect building a GIS—will depend on the level of threat associated with each business operation. Of course, information security is a primary design consideration for system deployments that connect with World Wide Web.

Recent industry advancements, especially in the areas of Web service standards and service-oriented architectures, are enabling organizations to more effectively secure the enterprise. Figure 5-1 shows how security can be integrated throughout the system design to support an in-depth security solution; note how no single element of the system remains uninvolved. In the same way, in any matter of security, all of us who participate in an organization must contribute to the solution, too.

ESRI's commitment to standards continues, and along with providing highly interoperable software components, offers security architects a high level of flexibility (see the white paper at http://www.esri.com/library/whitepapers/pdfs/arcgis-security.pdf). This chapter introduces the broad range of technical controls implemented with ArcGIS Desktop applications, Web applications, and Web Services; also described are standard practices for configuring ArcIMS and ArcGIS Server components across a firewall. By the end of the chapter, you'll be able to identify the security control types, along with some examples of technical controls that can be implemented with ArcGIS enterprise systems to provide secure system environments.

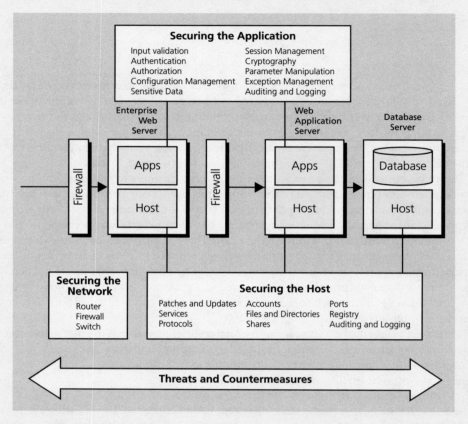

Figure 5-1

Security overview.

Security solutions are unique to each client situation. The right security solution depends on your enterprise risks and your selection of enterprise controls. The challenge is to implement reasonable and appropriate security controls. Therefore, you must assess regularly and keep these security risk assessments current. After establishing security guidelines and controls, you should perform ongoing security audits to ensure that objectives are being maintained.

Selecting the right security solution

To develop and support a security risk management program, you can use standard risk management frameworks, offered by trusted industry experts. These include *Gartner's Simple Enterprise Risk Management Framework, Microsoft Risk Assessment Guide* for Microsoft-centric customers, and material from the National Institute for Standards and Technology, providing a baseline for security certification and accreditation of federal government sites.

Figure 5-2 identifies the most common security risks according to the dollar amount of losses they actually caused in 2006. More than 30 percent of these realized threats came from virus contamination, followed by more than 20 percent from unauthorized access to information. After a dramatic rise the year before, unauthorized access continued to increase in 2006. Laptop theft also increased, making it to the number three spot that year.

Figure 5-3 highlights the most common security controls for that year, with firewall technology holding a slight lead over antivirus software. A diverse range of controls are available to address security concerns, including antispyware technology, which continues to be used increasingly.

The top two security threats, virus contamination and unauthorized access to information, are on the rise because of our expanding use of distributed network solutions.

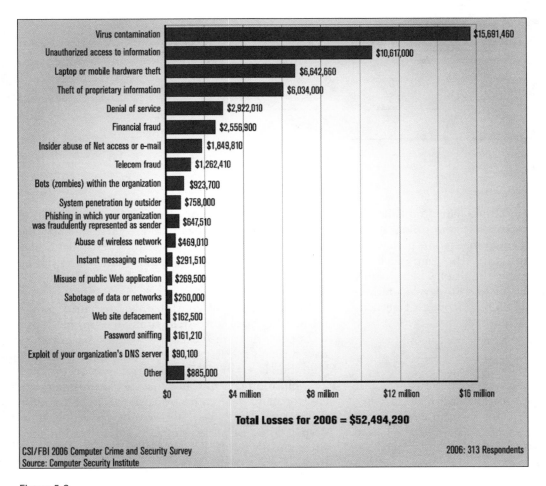

Figure 5-2

Dollar amount losses by threat.

Courtesy of Computer Security Institute, from CSI Computer Crime and Security Survey, 2007.

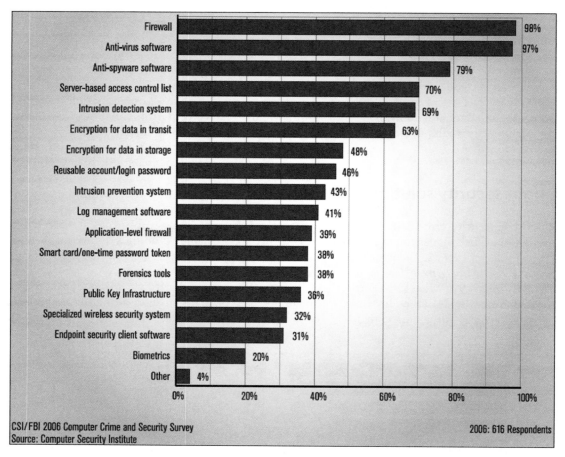

Figure 5-3

Security technologies used.

Courtesy of Computer Security Institute, from CSI Computer Crime and Security Survey, 2007.

Traditionally, many organizations have implemented systems that allow all network traffic unless it has been specifically blocked, a configuration policy that opens the door for compromise. There are much better ways of securing a network, all of which require a deeper understanding of day-to-day operations in particular and of the fundamentals of security controls in general.

Security and control

Enterprise protection is provided through multiple levels of security controls. This is because no security measure is infallible, so protection can only be achieved through a layered defense. Three types of security controls compose these levels of defense—physical, administrative, and technical controls—which work together to provide a secure environment. The three basic levels of security are the purview of many who work together to assure coordination of the different types of controls

and the ways the technical controls will work best within a particular system configuration. Figure 5-4 shows the types of controls at the three security levels (left), illustrating in detail (right) what defensive measures can be implemented at each layer (middle) of the technical control level.

Multiple types of control provide security at the technical control (or system configuration) level, protecting information at each layer where data and assets pass through the system as it's configured. These configuration layers or technical control types are grouped as application controls, host/device controls, network controls, and data controls, as shown in the middle column of figure 5-4. Within this layered defense, deeper layers of authentication and validation measures actually do the detail work of securing the enterprise. These security measures are listed to the right of their corresponding technical control type in the illustration.

Deeper still at the layer of application controls, there are many commercial off-the-shelf (COTS) functionalities

Security Control Types

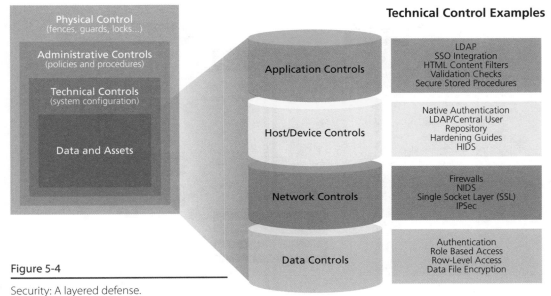

Figure 5-4

Security: A layered defense.

available to further secure the environment. Refer to the security definitions in appendix B for specific measures and procedures that can be included to address enterprise security needs.

Enterprise security strategies

Business operations today are exposed to a variety of information security threats. These threats can be generated by friendly and unfriendly sources and may include both internal and external users. Whether intentional or inadvertent, these threats can result in loss of resources, compromise of critical information, or denial of service. Figure 5-5 (on the next page) provides an overview of security options available for client/server, Web application, and Web services architecture.

Client/server architecture

It's a combination of security techniques working together throughout the information flow that provides the highest level of protection: Desktop and network operating systems should require user identification and password based on defined system-access privileges. Networks can include firewalls that restrict and monitor content of communications, establishing different levels of access criteria for message traffic. Communication packets can be encrypted (Secure Sockets Layer [SSL]) to deny unauthorized information access, even if the data is captured or lost during transmission. Specific content exchange criteria can be established between servers (IPSec) to restrict communication flow and to validate traffic sources. Traffic activity can be monitored

(intrusion detection) to identify attempts to overcome security protection. Data can be protected on disk to avoid corruption or prevent access as appropriate (encryption). Database environments provide access control (privileges) and row-level security.

Web application architecture

Be aware that sometimes protection comes in exchange for performance, so you must understand your security needs, and provide only what is required to meet those needs. Standard firewall, SSL, IPSec, intrusion detection, data file encryption, and RDBMS security solutions continue to protect Web operations. Additional security can be implemented to protect and control HTTP communications; basic and digest authentication and implementation of digital certificate authentication (PKI) procedures promote secure communications and support restricted-user access to published Web applications. Secure HTTP protocols (HTTPS) encrypt data transfers, fostering a higher level of communication protection. Web applications can assume user rights for data access (impersonation), and there are options that both enhance security and control access to the data source, such as options for passing user authentication (SSO) for database access. Again, be selective. Security is not free—there is a price to pay—and this is often a cost in reduced performance and reduced user productivity.

Web services architecture

First over the Web, secure HTTP communications encrypt data transmissions and improve communication security. After that, it makes sense that the more

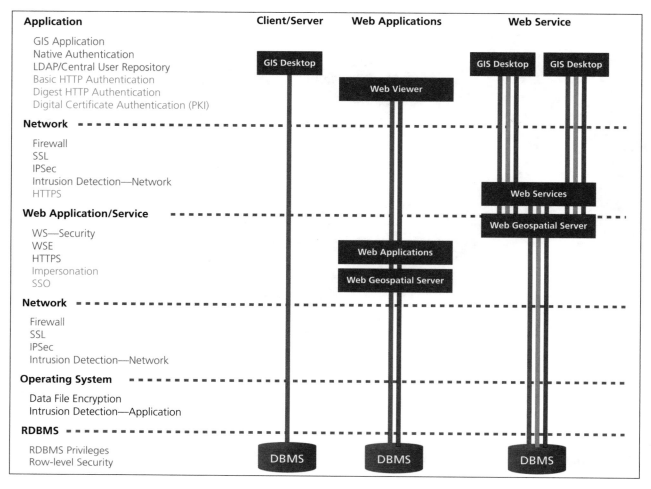

Figure 5-5

Security in depth (ArcGIS architecture).

wide-ranging the architecture, the more security options are available—to counteract the increased risk.

The most security controls are available with an enterprise service-oriented architecture (SOA). SOA includes third-party security solutions that enhance the protection provided by the Web application architecture. Additional options are available to enhance access controls. Other security features may be included in client applications to ensure their proper use and control. You can require user authentication and further restrict access to Web services through more Web services security solutions (WS-Security). Web server technology also provides specific Web services security implementations through Web services extensions (WSE).

Web firewall configuration alternatives

A firewall is a hardware or software device configured to permit, deny, or proxy communication between computer networks that have different levels of trust. Standard security practices dictate a "default-deny" firewall rule-set, in which the only network connections permitted are ones that have been explicitly allowed. An effective "default-deny" configuration requires a detailed understanding of the network applications and endpoint requirements necessary to support day-to-day operations. Many businesses lack such understanding, and therefore implement a "default-allow" rule-set, in which all traffic is allowed unless it has been specifically blocked. This

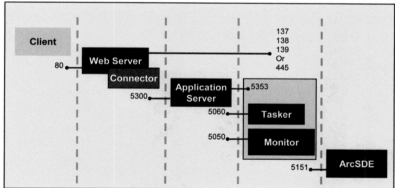

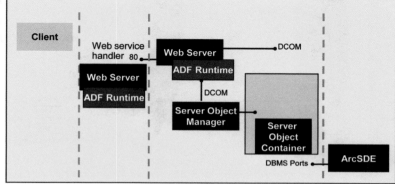

Figure 5-6

Firewall communications.

configuration makes inadvertent network connections and system compromise much more likely.

A number of firewall configuration options are identified to support ArcGIS Web communications. For example, see the options in the white paper at http://www .esri.com/library/whitepapers/pdfs/securityarcims.pdf.

Figure 5-6 provides an overview of default TCP ports used with ArcIMS and ArcGIS Server firewall configurations. ArcIMS Web mapping functions are well defined and controlled, and software components can be distributed as needed through shared firewall port connections.

You can see that ArcGIS Server provides a much richer programming environment, one that requires tighter communications between Web applications and the server object container executables. The server object container (SOC) exposes the ESRI ArcObject code, providing application access to a broad range of GIS functionality. ArcGIS Server communications

between Web applications and the SOC executables use distributed-component-object-model (DCOM) protocols. The use of DCOM involves dynamically assigning dedicated TCP/IP ports for communication between components; therefore, separation of these components over a firewall configuration is not recommended.

ArcGIS Server Web applications can communicate across a firewall using HTTP service protocols. Many enterprise-level Web applications can use data sources provided entirely by Web service transactions. Web applications requiring more tightly coupled communications can connect through a reverse proxy server to the internal Web application server environment.

The following sections address available ArcGIS firewall-configuration strategies, discussing the advantages and disadvantages of each configuration. Understanding available configuration options and associated implications can help the security architect select the best solution for supporting your enterprise security needs.

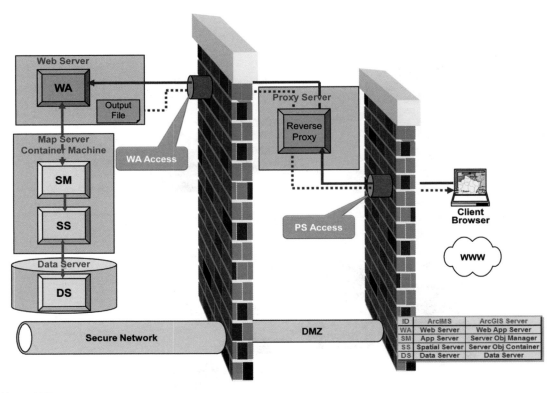

Figure 5-7

Web services with proxy server: The preferred firewall configuration for GIS Web services.

Web services with proxy server

Figure 5-7 shows the interface-with-intranet-Web-application configuration supported by a proxy server. This solution provides private network security through a reverse proxy server and supports the complete Web services configuration on the private network. Enabling full management of the Web site on the private network (through use of a reverse proxy architecture), this is the preferred firewall configuration for ArcGIS Server deployment.

A rapidly growing market of enterprise-quality edge servers, proxy servers, network traffic load balancing solutions, and Web traffic accelerators are introducing industry standards for Web-access management. These industry standards provide organizations with a variety of management tools to use to maintain and support enterprise security needs.

The proxy server provides virtual Internet access with all Web components supported on the internal private network. Ease of Web administration and support make this the preferred Web-access solution for most ESRI customers.

Web application server in the DMZ

Figure 5-8 shows the Web application server located in the demilitarized zone (DMZ), with the map server/

container machine and data server located on the secure network. (DMZ is a term for the interface zone between an internal private network and the public Internet, and its purpose is to add an additional layer of security to an organization's LAN.) The service manager and spatial services must be located on the internal network for this configuration to be acceptable. The output file, located on the Web server, must be shared with the map server. This disk mount will allow one-way access from the map server through the firewall to the Web server. This configuration is not recommended for ArcGIS Server.

Figure 5-8 became a popular solution for many ArcIMS implementations as organizations looked for ways to promote secure Web access. Connections between the Web applications and the ArcIMS Application Server were relatively stable, and only the Web traffic was sent across the firewall. Many security architects were concerned with supporting a disk mount across the firewall, so output streaming (available with the Java and .NET applications) was provided as an alternative solution.

ArcGIS Server implementations across the firewall are faced with additional security concerns. DCOM communications between the Web application and the server object containers would make Swiss cheese of the firewall (dedicated ports are required for each concurrent Web transaction). An internal Web application server can provide published HTTP services to support a Web application outside the firewall. Web applications that require DCOM communications would be supported on the internal Web server. Web client connections to the internal supported applications would be provided through reverse proxy communications.

A note about the Swiss cheese reference above: ArcGIS Server uses dedicated communication ports for each concurrent transaction; this is the foundation for recommended firewall deployment limitations. Each Web ADF application uses a dedicated DCOM port assigned by the SOM to enable execution of each Web transaction. Depending on the service profile, a standard ArcGIS Server platform can support up to five concurrent transactions per server core. A high-available ArcGIS Server platform configuration would include two or more 4-core server platforms, each platform able to handle up to 20 concurrent ArcGIS Server transactions. Two production platforms would require a minimum of 20 DCOM ports available to support concurrent transactions during peak service loads. A larger system capacity would require more DCOM ports,

which would make the firewall port lockdown look like Swiss cheese (the firewall would not be effective).

All Web services components in the DMZ
Of course, the most secure solution provides no physical connection between the secure network and the outside world. In some network environments, the optimum level of security is of such critical importance that an air gap is required to maintain secure communications on the internal network. Figure 5-9 shows the Web application, service manager, spatial services, and data source, all located outside the secure network firewall and within the Web access tier or DMZ. This configuration requires that a copy of the required GIS data be maintained in the DMZ. Data updates are provided from the internal GIS data server to the external data server to feed published Web services.

ArcGIS 9.2 software provides new technology advances that simplify management and support of a distributed geodatabase (see next chapter). ArcGIS Server geodatabase replication can provide incremental updates of a defined subset (version) of the internal parent geodatabase to the child geodatabase supporting the published Web services. Updates can be provided

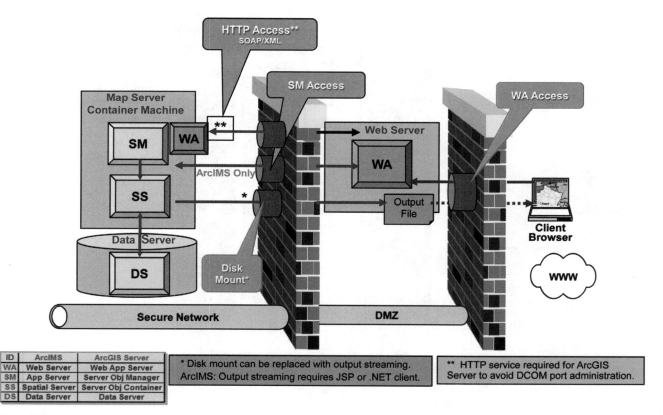

Figure 5-8

Web application in DMZ: Popular with ArcIMS but not recommended for ArcGIS Server.

through a temporary firewall connection or exported to an XML file for manual transfer to the external database. With the ArcGIS 9.3 release, one way geodatabase replication will support incremental updates of a child file geodatabase.

All Web services components in the DMZ except the data server

Figure 5-10 shows the Web application, service manager, and spatial services located in the DMZ, accessing the internal ArcSDE data server located on the secure network. In this configuration, the spatial services use a direct connection to the database client software, which in turn uses standard database port access through the secure firewall, allowing limited access to the internal geodatabase server. Because a high volume of traffic must pass between the spatial services and the data source, there can be problems with firewall disconnects. Any network disconnects with the data server generate delays (to allow for all publish-service connections to be reestablished).

Figure 5-10 was a popular solution for many ArcIMS implementations as organizations looked for ways to secure Web access. With Web services provided outside

the private network in the DMZ, virtual access to the central database management system (DBMS) is provided through standard database security protocols. Now this solution can be applied to both ArcIMS and ArcGIS Server.

Be aware of the challenge that network traffic and connection stability can pose within this configuration, however. Network traffic between the spatial server executables can be relatively high, since all required data must be transferred to the GIS executables to create the map service. Connection stability can be a problem if the firewall connection is not stable. If connections between the published Web services and the database are lost, it can take some time to reinstate these database connections and bring the services back online. So when implementing this solution, be sure there is adequate network bandwidth and a stable, keep-alive connection across the firewall.

Avoid this one

Figure 5-11 shows Port 80 Web access to the secure network, which is not a secure configuration at all: The Web application, map server/container machine, and data server components are all inside the firewall on the

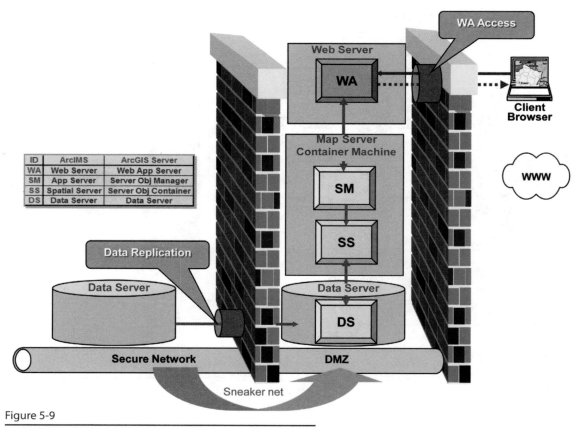

ID	ArcIMS	ArcGIS Server
WA	Web Server	Web App Server
SM	App Server	Server Obj Manager
SS	Spatial Server	Server Obj Container
DS	Data Server	Data Server

Figure 5-9

All Web services components in the DMZ: The most secure Web services architecture.

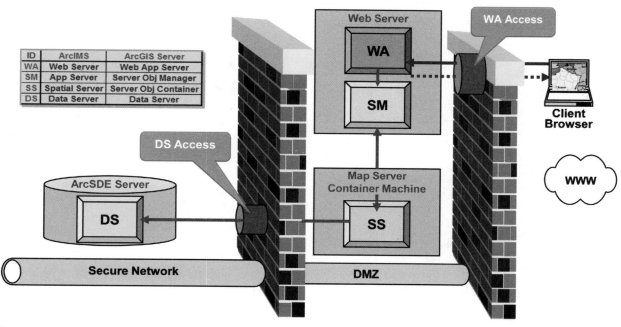

Figure 5-10

All Web services components in the DMZ except the data server:
Can be used for both ArcIMS and ArcGIS Server.

secure network. Port 80 provides open access to HTTP traffic through the firewall. Many organizations have demonstrated that this is not a secure solution. Adequate firewall protection must be in place to prevent general public access to the internal network.

By paying attention and making the right trade-offs, you can identify and implement the optimum security solution for your organization. Information in the next chapter, on data management, will also help you in this effort.

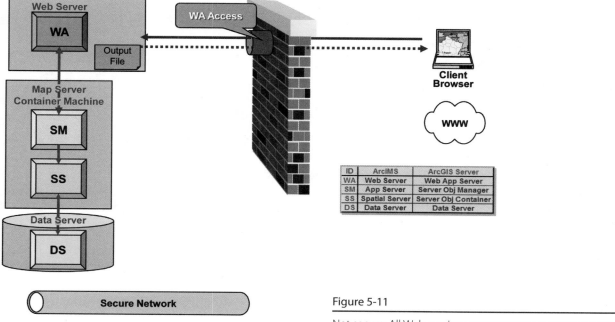

Figure 5-11

Not secure: All Web services components on secure network.

6 GIS data administration

OVERVIEW

Data management and administration is a primary consideration when selecting the right enterprise GIS architecture. Enterprise GIS often benefits from efforts to consolidate GIS data resources—combining and sharing data resources, regardless of whether they are centralized or distributed. Data consolidation improves user access to data resources, allows for better data protection, and enhances the quality of the data. In turn, the consolidation of the IT hardware and management resources needed to support the data reduces hardware costs and system administration requirements.

The simplest and most cost-effective way to manage data resources is to keep one primary copy of the data in a central data repository and provide user access to this data as is necessary for data maintenance and operational GIS query and analysis. This is not always practical, and many system solutions require that organizations maintain distributed copies of the data. This became easier with ArcGIS 9.2 geodatabase replication technology, which significantly reduced the complexity involved in maintaining distributed database architectures. Now it will be easier still, with one-way geodatabase replication to a file geodatabase implemented in the 9.3 release.

Geodatabase versioning, introduced with ArcSDE 8 technology, provides the foundation technology for distributed geodatabase management. Geodatabase replication functionality was introduced during the ArcGIS 8.3 release to provide disconnected editing for mobile ArcGIS Desktop clients and mobile database operations. Full distributed geodatabase management functions became available with the ArcGIS 9.2 release; these functions fully support remote versions of a single geodatabase located at regional locations. Changes are automatically synchronized with a central corporate database that integrates regional data through standard versioning, reconcile-and-post functionality. The capability to deploy and support distributed geodatabase operations comes with the core ArcGIS Server technology—functionality provided with the entry-level geodatabase license.

While software can help you manage data, there's no silver bullet that automatically accomplishes all that is involved in data administration. In the distributed geodatabase scenario above, for example, the data administrators must carefully plan and manage update and backup operations to ensure distributed database replicas are maintained in a consistent state with the central parent database. Depending upon the size of your GIS operation, various specialists/professionals might work together to ensure that the GIS architecture is compatible with the way the existing organization manages data and systems administration.

This chapter identifies how spatial data is stored, protected, backed up, and moved within standard IT data center environments. There are many ways to maintain and support spatial data, and you must choose the system architecture that satisfies your data management needs. These various ways are examined in more detail after the following brief summary of them.

Storing spatial data. Spatial data has been stored on most every type of storage media. Entry-level systems may rely on a local disk drive. Workgroup operations can be enhanced by maintaining shared spatial data on a local file server disk share. Higher performance, better data management, and more functionality are provided by a database management system, which can store data on internal storage, direct-attached storage, and storage area networks (SAN). Network-attached storage provides an appliance file-based solution as an alternative for file-server disk shares and for direct-attached storage. The best storage solution is likely to be the one currently supporting your existing business operations, allowing you to leverage how you protect, back up, and move data throughout the organization.

Protecting data. GIS data is key to a variety of critical business processes within an enterprise GIS, so it must be protected. The primary data protection strategy is provided by the storage architecture (usually certain types of disks organized in particular ways). Most storage vendors have standardized on RAID (redundant array of independent disks) as the primary way to organize storage solutions for better access performance and protected recovery from disk hardware failures. RAID combinations that best foster performance goals as well as data protection are the most desirable.

Backing up data. If the storage architecture is the first line of defense, then data backups provide the last line. Essentially, when it comes to backups there are two types of people: those who backup regularly, and those who wish they had! Squirreling away a current backup copy not only protects your data investment but also keeps your GIS running in emergencies. A strategy for backing up data should take into consideration many factors, but mainly costs in time and money. Though the size of databases has increased in recent years, the cost of disk storage has decreased, rendering disk copy solutions competitive in price with tape storage solutions. Using a backup disk copy source to simply restart the DBMS could make sense in terms of time as well, since recovery of large databases from tape backups is very slow. For network performance issues such as whether backups will be required during peak or off-peak hours, various servers can be configured to handle either or both, in order to avoid network bottlenecks.

Moving data. Backup and recovery methods are also involved when your focus is on moving data. Organizations move spatial data in various ways: by means of geodatabase transition functions in ArcGIS, by using database replication, and also by using replication specifically at the disk level. Traditional methods copy data on tape or disk and physically deliver this data to a remote location through standard transportation modes. Once at the remote site, data can be installed on a remote server environment. Technology has evolved to provide more efficient alternatives for maintaining distributed data sources.

Managing and accessing data. How to protect data while people are working on it simultaneously is a primary objective of data management and administration. ArcGIS technology allows for reliable ways to manage and access distributed spatial data, which are explored in the last half of the chapter.

Skilled data management and administration preserve and make the best of an organization's significant investment in data. Planning for both plays a role in any system architecture design, and success in both depends heavily on the administrative team that keeps the system operational. There are many ways to store and protect GIS data resources. GIS data is maintained and supported much like other enterprise data systems (accounting systems, facility management systems, customer service systems). Standard IT best practices are applied to store, back up, move, and protect GIS data resources. There are many similarities; however, there are also differences. This chapter focuses on the ways of supporting GIS data resources that might not be so familiar to you. The architecture should still be selected to leverage your existing enterprise data management practices, such as system backup, disaster recovery, and enterprise storage network technology. There is seldom a need to introduce new and different ways to manage your enterprise data resources.

As we've seen, most GIS implementations started with a few users at the working level and evolved over time into enterprise operations. Most GIS environments today are composed of a distributed array of departmental systems working to share available data resources and integrate operations. Most organizations are finding that centralized configurations that consolidate geodatabase resources while providing efficient access

to remote users provide the most adaptive, low-cost solutions. Distributed environments (where the data is not centralized but rather stored at various locations) may be needed to fulfill user needs. Hybrid solutions are now available that provide the efficiency of centralized integrated data management coupled with the high display performance of distributed systems.

The ways to store and protect GIS data resources are improving all the time. In this chapter, you'll find the various types of data storage solutions and RAID configuration strategies addressed, along with guidelines for protecting and moving data within storage networks. Database and storage replication services provide solutions for disaster recovery and moving complete database environments, using standard data center solutions.

Storage architecture strategies

Storage technology has evolved over the past 20 years to improve data access and provide better management of available storage resources. Understanding the advantages of each technical solution will help you select the storage architecture that best supports your needs. Figure 6.1 provides an overview of the technology evolution from internal workstation disk to the storage area network architecture.

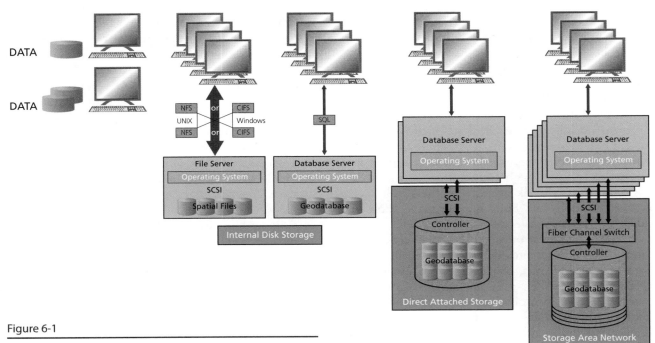

Figure 6-1

Advent of the storage area network.

Internal disk storage. The most elementary storage architecture puts the storage disk on the local machine. Most computer hardware today includes internal disk storage (meaning that a storage disk is included in the computer box). Workstations and servers can both be configured with internal disk storage. The fact that access to it is through the local workstation or server can be a significant limitation in a shared server environment: if the server operating system goes down, there is no way for other systems to access the internal data resources.

File server storage provides a network share that can be accessed by many client applications within the local network. Disk-mounting protocols—network file server (NFS) and common Internet file services (CIFS)—provide local application access over the network to the data on the file server platform. Query processing is provided by the client application (as opposed to applications located on the server platform), which can involve a high amount of chatty communications between the client and server network connection.

Database server storage provides query processing on the server platform, and significantly reduces the required network communication traffic. Database software improves data management and provides better administrative control of the integrity of the data.

Internal storage can include RAID mirror disk volumes that will preserve the data store in the event of a single disk failure. Many servers include bays that support multiple disk drives for configuring RAID 5 configurations and support high-capacity storage needs. Internal storage access is limited to the host server, however. Therefore, as many data center environments grew larger in the 1990s, organizations came to have many servers in their data center with too much disk (disk not being used) and other servers with too little disk. This made disk volume management a challenge, since data volumes could not be shared between server internal storage volumes. External storage architecture (direct-attached, storage area networks, and network-attached storage) is a way for organizations to break away from these "silo-based" storage solutions and build a more manageable and adaptive storage architecture.

Direct-attached storage (DAS). A direct-attached storage architecture provides the storage disk on an external storage array platform. Host bus adaptors (HBAs) connect the server operating system to the external storage controller using the same block-level protocols used for internal disk storage; so, from an application and server perspective, the direct-attached storage appears and functions the same as internal storage. The external storage arrays can be designed with fully redundant components (system would continue operations with any single component failure), so a single storage array can meet high-available storage requirements.

Direct-attached storage technology can provide several fiber channel connections between the storage controller and the server HBAs. For high availability purposes, it is standard practice to configure two HBA fiber channel connections for each server environment. Direct-attached storage solutions provide from four to eight fiber channel connections, so you can easily provide up to four servers with two redundant fiber channel connections from a single direct-connect storage array controller. The disk storage volumes are allocated to specific host servers, and the host servers control access to the assigned storage volumes. In a server failover scenario, the primary server disk volumes can be reassigned to the failover server. You will need to use external storage to support a high-available failover server solution (see chapter 4, figure 4-14).

Storage area networks (SAN). The difference between direct-attached storage and a storage area network is the introduction of a fiber channel switch to provide network connectivity between the servers and the external storage arrays. The SAN improves administrative flexibility for assigning and managing storage resources when you have a growing number of server environments. The server HBAs and the external storage array controllers are connected to the fiber channel switch, so any server can be assigned storage from any storage array located in the storage farm (connected through the same storage network). Storage protocols are still the same as with direct-attached or internal storage—so from a software perspective, these storage architecture solutions appear the same and are transparent to the application and data interface.

Network-attached storage (NAS). By the late 1990s, many data centers were using servers to provide client application access to shared file data sources. High-available environments require clustered file server storage, so if one of the servers fails, users would still have access to the file share. Hardware vendors now provide a hybrid appliance configuration to support network file shares, called *network-attached storage* or NAS. The NAS provides a file server and storage in a single high-available storage platform. The file server can be configured with a modified operating system that provides both NFS and CIFS disk-mount protocols, and a storage array with this modified file server network interface is deployed as a simple network-attached hardware appliance. The storage array includes a standard network interface card (NIC) interface to the local area network, and client applications can connect to the storage over standard disk-mount protocols. The network-attached storage provides a very simple way to deploy a shared storage for access by a large number of UNIX and Windows

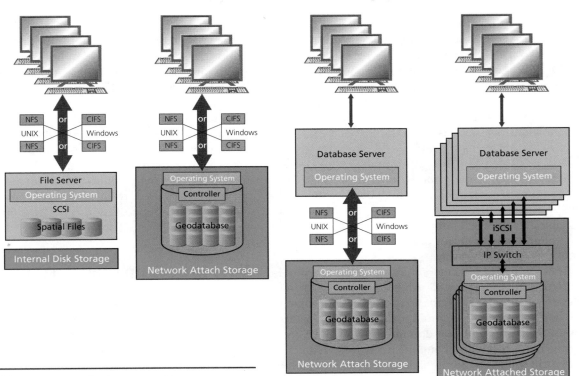

Figure 6-2

Evolution of the network-attached storage architecture.

network clients. Figure 6-2 shows the evolution of the NAS architecture.

Network-attached storage provides a very effective architecture alternative for supporting network file shares, and has become very popular among many GIS customers. As GIS data moves from early file-based data stores (coverages, Librarian, ArcStorm, shapefiles) to a more database-centric data management environment (geodatabase servers), the network-attached storage vendors suggest you use a network file share to support a database server storage. There are some limitations: It is important to assign dedicated data storage volumes controlled by the host database server to avoid data corruption. Other limitations include much slower database query performance over the chatty IP disk-mount protocols than with the traditional fiber channel SCSI protocols, and the bandwidth over the IP network is lower than the fiber channel switch environments (1 Gbps IP networks versus 2 Gbps fiber channel networks). Implementation of network-attached storage as an alternative to storage area networks is not an optimum storage architecture for geodatabase server environments. At the same time, it is an optimum architecture for file-based data sources and use of the NAS technology alternative continues to grow.

Because of the simple nature of network-attached storage solutions, you can use a standard local area network (LAN) switch to provide a network to connect your servers and storage solutions; this is a big selling point for the NAS proponents. There is quite a bit of competition between storage area networks and network-attached storage technology, particularly when supporting the more common database environments. The SAN community will claim their architecture is supported by higher bandwidth connections and the use of standard storage block protocols. The NAS community will claim they can support your storage network using standard LAN communication protocols and provide support for both database server and network file access clients from the same storage solution.

The NAS community provides a more efficient iSCSI communication protocol for its storage networks (basically SCSI storage protocols over IP networks). GIS architectures today include a growing number of file data sources (examples include ArcGIS Image Server imagery, ArcGIS Server preprocessed 2D and 3D file caches, and the file geodatabase). For many GIS operations, a mix of these storage technologies provides the optimum storage solution.

Ways to protect spatial data

Data is one of the most valuable resources of a GIS, and protecting data is fundamental to enabling critical business operations. Strategically, the storage architecture does provide the first line of defense. This brief overview of storage protection alternatives includes the following description of the most common configurations and a sidebar that describes other alternatives (top of page 87).

The two RAID configurations (RAID 1/0 and RAID 5) in figure 6-3 are the most commonly used for geodatabase (ArcSDE) storage, and they represent the RAID combinations that best support both data protection and performance goals. Additional hybrid RAID configurations (see sidebar) are available, but RAID 1/0 and RAID 5 are the most popular.

RAID 1/0: RAID 1/0 is a composite solution that includes RAID 0 striping and RAID 1 mirroring. This is the optimum combination for high performance and data protection. This is also the most expensive solution. Available data storage is limited to 50 percent of the total disk volume, since a mirror disk copy is maintained for every data disk in the array.

RAID 5: RAID 5 includes the striping and parity of the RAID 3 solution and distribution of the parity volumes for each stripe across the array to avoid disk contention bottlenecks. This improved parity solution provides optimum disk use and near optimum performance, supporting disk storage on all but one parity disk volume.

Hybrid solutions: Some vendors provide alternative proprietary RAID strategies to enhance their storage solution. New ways to store data on disk can improve performance and protection and may simplify other data management needs. Each hybrid solution should be evaluated to determine if and how it may support specific data storage needs.

ArcSDE data storage strategies are different based on the selected database environment.

SQL Server: Log files are located on a single RAID 1 mirror, and index and data tables are distributed on a RAID 5 disk volume. The number of disks required to support the RAID 5 volume depends on the user workflow. Heavier workflow loads can require a more distributed data volume.

Oracle, Informix, and DB2: Index tables and log files are located on a RAID 1/0 mirror, and data tables are distributed on RAID 5 disk volumes. The number of disks required to support the RAID 5 volume depends on the user workflow. Heavier workflow loads can require a more distributed data volume.

PostgreSQL: Support for this database is introduced with the ArcGIS 9.3 software.

Ways to back up spatial data

Data protection at the disk level minimizes the need for system recovery in the event of a single disk failure but will not protect against a variety of other data failure scenarios. For the sake of business continuance and disaster recovery, proactive planning is in order, and it should include remote backup: you should keep a current backup copy of critical data resources at a safe location away from the primary site.

Data backups provide the last line of defense for protecting data investments. Careful planning and attention to storage backup procedures are important factors in a successful backup strategy. Data loss can result from many types of situations, with some of the most probable

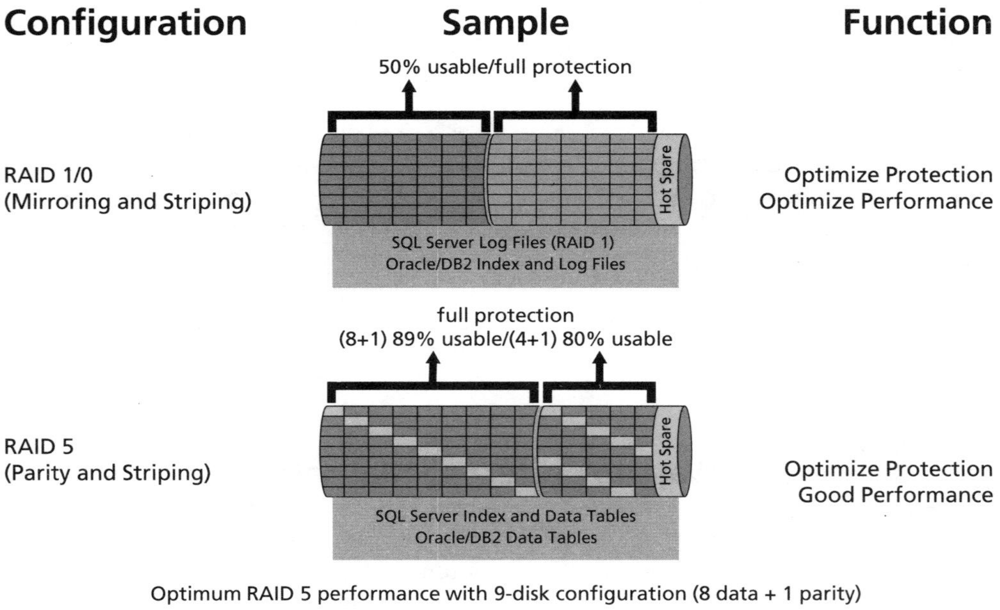

Configuration	Sample	Function
RAID 1/0 (Mirroring and Striping)	50% usable/full protection SQL Server Log Files (RAID 1) Oracle/DB2 Index and Log Files	Optimize Protection Optimize Performance
RAID 5 (Parity and Striping)	full protection (8+1) 89% usable/(4+1) 80% usable SQL Server Index and Data Tables Oracle/DB2 Data Tables	Optimize Protection Good Performance

Optimum RAID 5 performance with 9-disk configuration (8 data + 1 parity)

Figure 6-3

Standard RAID configurations: The most common geodatabase storage environment.

Other disk storage configurations

JBOD: A disk volume with no RAID protection is referred to as "just a bunch of disks," or JBOD. This represents a configuration of disks with no protection and no performance optimization.

RAID 0: A disk volume in a RAID 0 configuration provides striping (spreading) of data across several disks in the storage array. Striping supports parallel disk controller access to data across several disks, reducing the time required to locate and transfer the requested data. Data is transferred to the array cache once it is found on each disk. RAID 0 striping provides optimum data access performance with no data protection. One hundred percent of the disk volume is available for data storage.

RAID 1: A disk volume in a RAID 1 configuration provides mirror copies of the data on disk pairs within the array. If one disk in a pair fails, data can be accessed from the remaining disk copy. The failed disk can be replaced and data restored automatically from the mirror copy without bringing the storage array down for maintenance. RAID 1 provides optimum data protection with minimum performance gain. Available data storage is limited to 50 percent of the total disk volume, since a mirror disk copy is maintained for every data disk in the array.

RAID 3 and 4: A disk volume in a RAID 3 or RAID 4 configuration supports striping of data across all disks in the array except for one parity disk. A parity bit is calculated for each data stripe and stored on the parity disk. If one of the disks fails, the parity bit can be used to recalculate and restore the missing data. RAID 3 provides good protection of the data and allows optimum use of the storage volume. All but one parity disk can be used for data storage, optimizing use of the available disk volume for data storage capacity.

There are some technical differences between RAID 3 and RAID 4 that are beyond the scope of this discussion. Both of these storage configurations have potential performance disadvantages. The common parity disk must be accessed for each write, which can result in disk contention under heavy peak user loads. Performance may also suffer because of requirements to calculate and store the parity bit for each write. Write performance issues are normally resolved through array cache algorithms on most high-performance disk storage solutions.

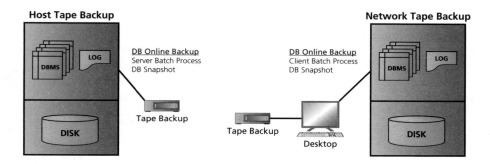

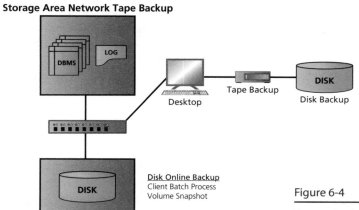

Figure 6-4

Ways to back up spatial data.

being administrative or procedural error. Figure 6-4 provides an overview of the different ways to back up spatial data (note that DB stands for database).

Host tape backup. Traditional server backup solutions use lower-cost tape storage for backup. Data must be converted to a tape storage format and stored in a linear tape medium. Backups can be a compute-intensive process taking considerable server processing resources (the backup process will consume a processor core) and requiring special data management procedures when supporting business operations.

For database environments, backup is provided based on a single point in time to maintain database continuity. Standard database backup procedures are employed to support a snapshot of the database. A copy of the protected snapshot data is retained in a snapshot table when changes are made to the database, supporting a point-in-time backup of the data and potential database recovery back to the time of the snapshot.

Host processors can be used to support backup operations during off-peak hours. If backups are required during peak-use periods, a network server can reduce data server loads.

Network client tape backup. The traditional online backup can often be supported over the LAN with the primary batch backup process running on a separate client platform. Client backup processes can contribute to potential network performance bottlenecks between the server and the client machine, however, because of the high data transfer rates during the backup process. So DBMS (database management system) snapshots may still be used to provide point-in-time backups for online database environments.

Storage area network client tape backup. Some backup solutions support direct disk storage access without affecting the host DBMS server environment. Storage backup is performed over the SAN or through a separate fiber channel access to the disk array with the batch process running on a separate client platform. A disk-level storage array snapshot is used to support point-in-time backups for online database environments. Host platform processing loads and LAN performance bottlenecks can be avoided with disk-level backup solutions.

Disk copy backup. The size of databases has increased dramatically in recent years, growing from tens of gigabytes to several terabytes of data. Recovery of large databases from tape backups is very slow, taking days to recover large spatial database environments. At the same time, the cost of disk storage has decreased dramatically, making disk copy solutions for large database environments competitive in price to tape storage solutions. A backup copy of the database on local disk, or a backup copy of the production disk volumes to a remote recovery site, can enable immediate restart of the DBMS following a storage failure.

Ways to move spatial data

Understanding the available options and risks involved in moving data is important in defining the optimum enterprise GIS architecture. Many enterprise GIS solutions require updates of distributed copies of the GIS data resources, replicated on a periodic schedule from a central GIS data repository or enterprise database environment. Organizations with a single enterprise database solution still have a need to protect data resources and update their business continuance plan, in the

event of an emergency such as fire, flood, accidents, or other natural disasters.

This section reviews the various ways organizations move spatial data. Traditional methods copy data on tape or disk and physically deliver this data to the remote site through standard transportation modes. Once at the remote site, data can be installed on the remote server. By now, technology has evolved to provide more efficient alternatives for maintaining distributed data sources.

Traditional data transfer methods

Figure 6-5 identifies traditional methods for moving a copy of data to a remote location.

Traditional methods include backup and recovery of data using standard tape or disk transfer media. Moving data using these methods is commonly called "sneaker net." These methods provide a way to transfer data without the support of a physical network.

Tape backup. Tape backup solutions can be used to move data to a separate server environment. Tape transfers are relatively slow. The reduced cost of disk storage has made disk copy a much more attractive option.

Disk copy. A replicated copy of the database on disk storage can support rapid recovery at a separate site. The database can be restarted with the new data copy and be online again within a short recovery period.

ArcGIS geodatabase transition

Moving subsets of a versioned geodatabase cannot be carried out easily using standard backup strategies. Data must be extracted from the primary geodatabase and imported into the remote geodatabase to enable the data transfer. Geodatabase transition can be supported using standard ArcGIS export/import functions. These tools can be used as a method of establishing and maintaining a copy of the geodatabase at a separate location. Figure 6-6 identifies the ways to move spatial data using ArcGIS geodatabase transition functions.

ArcSDE admin commands

Batch processes can be used with ArcSDE admin commands to support export and import of an ArcSDE simple feature database. Moving data using these commands is most practical when completely replacing the data layers. These commands are not optimum solutions when transferring data to a complex ArcSDE geodatabase environment.

ArcCatalog/ArcTools commands

ArcCatalog supports migration of data between ArcSDE geodatabase environments, extracting to a personal geodatabase or separate database instance, and importing to an ArcSDE environment.

Tape

Tape Backup

DBMS LOG

DISK

DB Online Backup
Batch Process
DB Snapshot

Tape Backup

Tape Restore

DB Restore
Batch Process
Slow

DBMS LOG

Tape Backup

DISK

Disk or CD

Disk Backup

DBMS LOG

DISK

DB Online Backup
Batch Process
DB Snapshot

Disk Copy

Disk Restore

DB Restore
Batch Process

Disk Copy

DBMS LOG

DISK

Figure 6-5

Tape backup/disk copy: Traditional methods for moving copied data to a remote location.

Extract / Transform / Load (ETL)

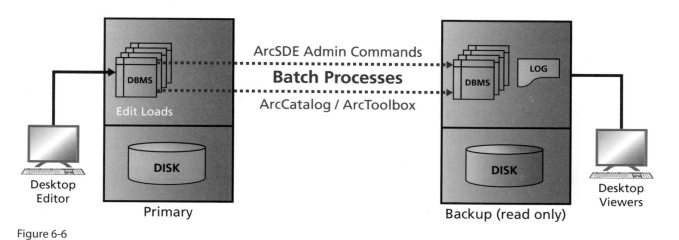

ArcSDE Admin Commands
Batch Processes
ArcCatalog / ArcToolbox

DBMS

Edit Loads

DISK

Desktop Editor

Primary

DBMS LOG

DISK

Desktop Viewers

Backup (read only)

Figure 6-6

Geodatabase transition: Moving spatial data using standard ArcGIS export/import functions.

ArcGIS Server geodatabase transition

ArcGIS ArcInfo 9.2 includes tools to create a data transformation between two different geodatabase schemas. This transformation can be published on ArcGIS Server to publish an automated extract/transform/load service.

Database replication

Customers have experienced a variety of technical challenges when configuring DBMS spatial data replication solutions. For example, ArcSDE data model modifications may be required to enable third-party DBMS replication. Heavy ETL processing would have to be imposed on both server environments, contributing to potential performance or server sizing impacts. Data changes must be transmitted over network connections between the two servers, causing potential communication bottlenecks. These challenges must be overcome to support a successful DBMS replication solution.

Customers have indicated that DBMS replication solutions can work but require a considerable amount of patience and implementation risk. Acceptable solutions are available through some DBMS vendors to support replication to a read-only backup database server. A dual master server configuration significantly increases the complexity of an already complex replication solution. Figure 6-7 presents an overview of ways to move spatial data using RDBMS-level replication. (Note that these ways are not appropriate when moving a portion of a versioned geodatabase.)

Synchronous replication

Real-time replication requires commitment of data transfer to the replicated server before releasing the client application on the primary server. Users can experience performance delays when carrying out edit operations with this configuration due to the heavy volume of spatial data traffic and required client interaction times. High-bandwidth fiber connectivity (1,000 Mbps bandwidth) is recommended between the primary server and the replicated backup server to minimize performance delays.

Asynchronous replication

Near-real-time database replication strategies decouple the primary server from the data transfer transaction to the secondary server environment. Asynchronous replication can be supported over lower bandwidth WAN connections, since transmission delays are isolated from primary server performance. Data transfers (updates) can be delayed to off-peak periods if WAN bandwidth limitations dictate, supporting periodic updates of the secondary server environment at a frequency based on practical operational needs.

Disk-level replication

Disk-level replication is a well-established technology, supporting global replication of critical data for many types of industry solutions. Spatial data is stored on disk sectors in the same way as other data, and does not require special attention beyond what might be required for other data types. Disk volume configurations (the data location on the disk and the volumes being transferred to the remote site) may be critical to ensure database integrity. Mirror copies are refreshed based on point-in-time snapshot functions supported by the storage vendor solution.

Disk-level replication provides transfer of block-level data changes on disk to a mirror disk volume located at a remote location. Transfer can be supported with active online transactions with minimum impact on DBMS server performance capacity. Secondary DBMS applications must be restarted to refresh the DBMS cache and processing environment to the point in time of the replicated disk volume.

Figure 6-8 presents different ways to move spatial data using disk-level replication. (Again, these ways are not appropriate for moving any portion of a versioned geodatabase.)

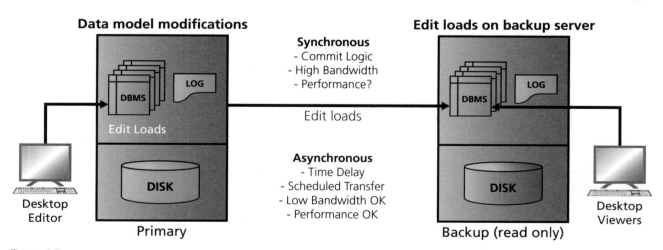

Figure 6-7

Database replication: Ways to move spatial data using RDBMS-level replication.

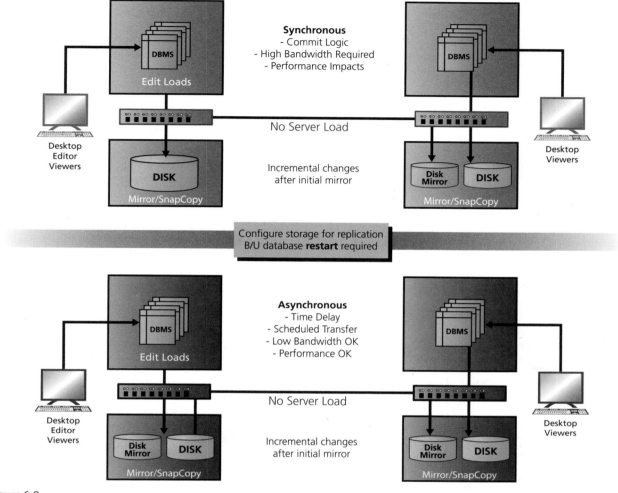

Figure 6-8

Using disk-level replication to move spatial data.

Synchronous replication

Real-time replication requires commitment of data transfer to the replicated storage array before releasing the DBMS application on the primary server. High-bandwidth fiber connectivity (1,000 Mbps bandwidth) is recommended between the primary server and the replicated backup server to avoid performance delays.

Asynchronous replication

Near-real-time disk-level replication strategies decouple the primary disk array from the transport of changes to the secondary storage array environment. Asynchronous replication can be supported over WAN connections, since the slow transmission times are isolated from primary DBMS server performance. Disk block changes can be stored and data transfers delayed to off-peak periods if WAN bandwidth limitations dictate, supporting

periodic updates of the secondary disk storage volumes to meet operational requirements.

Ways to manage and access spatial data

Release of the ArcGIS technology introduced the ArcSDE geodatabase, which provides a way to manage long transaction edit sessions within a single database instance. Single database instance is a single database environment executed on a single database platform. When designing and deploying a database, your data is supported in a single database instance. When you open an application on your PC, you are opening a single instance of that application in a window on your PC. ArcSDE supports long transactions using versions (different views) of the database. In other words, you can maintain the original version while one person or many people work on updating or otherwise changing the data. A geodatabase can support thousands of concurrent versions of the data within a single database instance. The default version represents the real world, and other named versions are proposed changes and database updates in progress.

Figure 6-9 shows a typical long-transaction workflow life cycle. The workflow represents the design and construction of a typical housing subdivision. Several design alternatives might initially be represented as separate named versions in the database to allow for planning a new subdivision. One of these designs (versions) is approved to continue on to the construction phase. After the construction phase is complete, the selected design (version) is modified to represent the as-built environment. Once the final design is validated, the as-built version can be reconciled with the geodatabase and posted to the default version to publish the new subdivision.

Geodatabase versioning

The simplest way to introduce the versioning concept behind what happens in the geodatabase is by using some logical flow diagrams. Figure 6-10 demonstrates the explicit-state model represented in the geodatabase. In an explicit-state model, each state represents an evolution of the data environment. (The explicit states are the numbered ovals.) The default version lineage is represented in the center of the diagram, and version states are added each time edits are posted to the default view. Each edit post represents a state change in the default view (accepted changes to the real-world view). There can be thousands of database changes (versions) at a time. As changes are completed, these versions are posted to the default lineage. Changes are recorded in add/delete transaction tables. States are modeled explicitly in the database using delta records. Explicit-state models allow multiple, parallel edit sessions with optimistic concurrency models for long transactions.

Long-transaction Workflow	QTR1												QTR2			
	Month1				Month2				Month3				Month1			
	1	2	3	4	1	2	3	4	1	2	3	4	1	2	3	4
Initiate Edit Session (version)	▲															
Develop Design	▬	▬														
Design Approved			▲													
Construction Phase					▬	▬	▬	▬	▬	▬	▬	▬				
Update As Build												▬	▬			
Reconcile and Post to Default														▲		

Figure 6-9

An example of the life cycle of a long-transaction workflow.

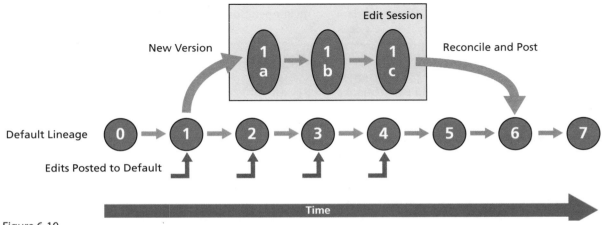

Figure 6-10

Explicit-state model: A representation of geodatabase versioning. In the diagram, the reconcile operation increases the default lineage by 1 state. Technically, however, the reconcile operation increases the state ID value by 3 when performed in ArcMap. If performed programmatically and not within an edit operation, the state ID will increase by a value of 2. The diagram simply illustrates intuitively what is happening in the software.

Named Versions (t1, t4, and t7)

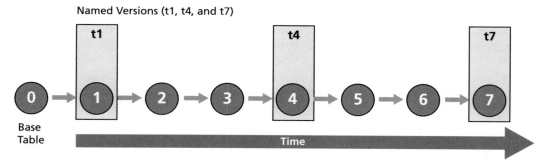

Figure 6-11

Default version lineage: How reference states are maintained in the database.

The "new version" at the top of figure 6-10 shows the life cycle of a long transaction. The transaction begins as changes from "state 1" of the default lineage. Maintenance updates reflected in that version are represented by new states in the edit session (1a, 1b, and 1c). During the edit session, the default version accepts new changes from other completed versions. The new-version active edit session is not aware of the posted changes to the default lineage (2, 3, and 4), since it includes only those changes from default state 1. Once the new version is complete, it must be reconciled with the default lineage. The reconcile process compares the changes in the new version (1a, 1b, and 1c) with changes in the default lineage (collected in default state 4) to make sure there are no edit conflicts. If the reconcile process identifies conflicts, these conflicts must be resolved before the new version can be posted to the default lineage. Once all conflicts are resolved, the new version is posted to the default lineage forming states 5, 6, and 7 (collection of all the default changes).

Figure 6-11 shows a typical workflow history of the default lineage. Named versions (t1, t4, and t7) represent edit transactions in progress that have not been posted back to the default lineage. The parent states of these versions (1, 4, and 7) are locked in the default lineage to support the long edit sessions that have not been posted. The default lineage includes several states (2, 3, 5, and 6) that are part of the default lineage created by posting completed changes.

Figure 6-12 demonstrates a geodatabase compress. Versions can be created to represent historical snapshots (regular points in time, events such as a parcel split, etc.), and any version can be materialized at any time. Very long default lineages (thousands of states) can affect database performance. The geodatabase drops the redundant states that are not named version parent states, thus decreasing the length of the default lineage and improving database performance.

A state lineage is a history of states that represent a series of database changes. The ArcSDE compress function reduces state lineage by consolidating unnamed state changes to the next higher named state.

Geodatabase versioning means making changes to a reference state in the database. How these states are maintained in the database is illustrated in figure 6-11. Once the geodatabase versioning concept is understood, it is useful to recognize how this is physically implemented within the database table structure (the structure of the tables and their relationship, or schema, in the database—sometimes referred to as the "database schema"). When a feature table within the geodatabase is versioned, several additional tables are created to track changes to the base feature table. An Adds table is created to track additional rows added to the base feature table, and a Deletes table is created to record deleted rows from the Base table. Each row in the Adds and Deletes tables represents changed states within the

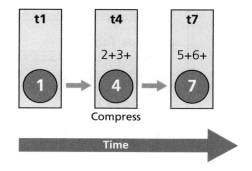

Figure 6-12

Geodatabase compress.

geodatabase. Additional tables support the ArcSDE tracking schema. As changes are posted to the default version, these changes are represented by the rows in the Adds and Deletes tables. Once there is a versioned geodatabase, the real-world view (default version) is represented by the Base table plus the current row state in the Adds and Deletes tables included in the default lineage (the Base table does not represent default). All outstanding versions must be reconciled and posted to default to compress all changes back to the Base table (zero state). Realistically, this is not likely to occur for a working maintenance database in an everyday environment because most operational databases always have open versions (work that has not been completed). So in the real world, thousands of projects are still in process—to get everyone in the company to complete their project in order to consolidate all the work into the Base table is not feasible.

These are actual geodatabase tables in figure 6-13: the Base table, Adds table, and Deletes table in a versioned geodatabase.

ArcSDE technology provides the tools to automate what has just been described. ArcSDE manages the versioning schema of the geodatabase and supports client application access to the appropriate views of the geodatabase. ArcSDE technology also supports export and import of data from and to the appropriate database tables and maintains the geodatabase schema that defines the relationships and dependencies between the various tables.

Geodatabase single-generation replication. The ArcGIS 8.3 release introduced a disconnected editing solution. This solution supported a registered geodatabase version extract to a personal geodatabase or separate database instance for disconnected editing purposes. The version adds and deletes values are collected by the disconnected editor and, on reconnecting to the parent server, can be uploaded to the central ArcSDE database as a version update.

Figure 6-14 provides an overview of ArcGIS disconnected editing with checkout to a personal geodatabase (PGD). The ArcGIS 8.3 release is restricted to a single checkout/check-in transaction for each client edit session.

Figure 6-15 provides an overview of ArcGIS disconnected editing with checkout to a separate ArcSDE geodatabase. Disconnected editing is restricted to a single checkout/check-in transaction for each child ArcSDE geodatabase. The child ArcSDE geodatabase can support multiple disconnected or local version edit sessions during the checkout period. All child versions must be reconciled before check-in with the parent ArcSDE geodatabase (any outstanding child versions will be lost during the child geodatabase check-in process).

Geodatabase peer-to-peer replication. The ArcGIS geodatabase checkout functions provided with disconnected editing can be used to support peer-to-peer database refresh. Figure 6-16 shows a peer-to-peer database checkout, where ArcSDE disconnected editing functionality can be used to periodically refresh specific feature tables of the geodatabase to enable a separate instance of the geodatabase environment. This functionality can be used to support a separate distribution

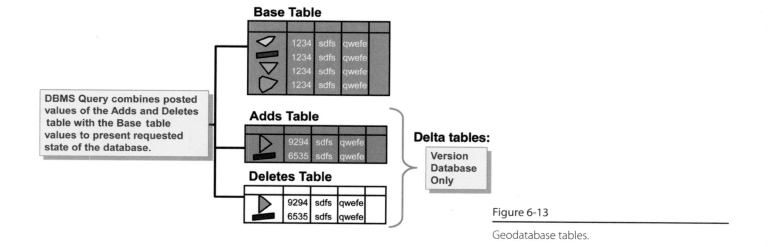

Figure 6-13

Geodatabase tables.

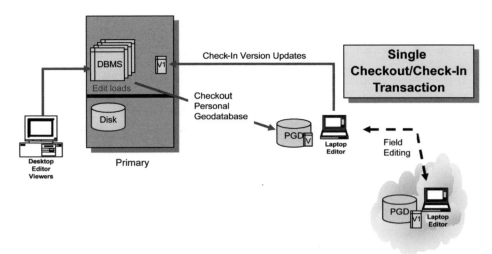

Figure 6-14

Geodatabase single-generation replication: personal geodatabase checkout. Diagram shows client connection to the geodatabase for checkout of a version to a local desktop personal geodatabase. Following the checkout, the user can disconnect from the network and work in a stand-alone configuration to view and pro-

vide updates to the personal geodatabase version. When returning to the network and connecting to the parent database, the remote updates can be uploaded for integration with the parent database.

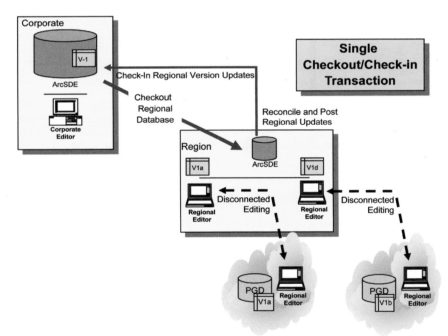

Figure 6-15

Geodatabase single-generation replication: database checkout. Diagram shows client connection to the geodatabase for checkout of a version to an ArcSDE geodatabase. Following the checkout, the database can operate as a disconnected child geodatabase separate from the parent goedatabase. Client desktop can connect to the child database and checkout a personal version for discon-

nected field operations, returning edits when connecting back to the child database. Once all sessions are retuned and reconciled to the child database, changes can be reconciled and posted to the child database version. Once all changes are posted, the child database can connect and the final consolidated remote updates can be uploaded for integration with the parent database.

view-only geodatabase that can provide a nonversioned copy of the parent default version.

The ArcGIS 9.2 software provides one-way incremental replication between the parent geodatabase and a child version replica. ArcGIS 9.3 supports one-way geodatabase replication to a file geodatabase. (Prior to ArcGIS 9.3, distributed geodatabase replication functions only worked between version geodatabase environments. A file geodatabase holds a terabyte of data per table, but can scale up to 256 TB per table. For some organizations, the maintenance and distribution database schema are different, and for those environments the geodatabase transition solution discussed in figure 6-6 would provide the optimum solution.)

Geodatabase multigeneration replication. With ArcGIS 9.2, geodatabase replication solutions were expanded to support loosely coupled ArcSDE distributed database environments. Figure 6-17 presents an overview of a loosely coupled ArcSDE distributed database concept.

ArcGIS multigeneration replication supports a single ArcSDE geodatabase distributed over multiple platform environments. The regional replica instances of the parent geodatabase provide an unlimited number of synchronized update transactions without losing local version edits or requiring a new checkout. Updates can be exchanged between parent and child geodatabase environments using simple XML datagrams that can

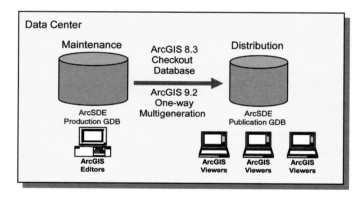

Figure 6-16

Geodatabase one-way multigeneration replication.

be transmitted over standard WAN communications. This new functionality enables a distributed geodatabase architecture in a connected central data center or over multiple remote sites connected by limited bandwidth communications (only changes are transmitted in GDB replication, then reconcile occurs each respective GDB *afer* the data transmission). It is important to carefully plan and manage update and backup operations to ensure distributed database replicas are maintained in a consistent state.

Administering distributed database solutions has traditionally required high-risk operations, bringing the possibility of data corruption and the use of stale data sources for what may be critical GIS operations. Organizations with operational distributed solutions credit their success to careful planning and detailed attention to administrative processes. Most successful GIS implementations deploy central consolidated geodatabase environments with effective remote user performance and support. ArcGIS distributed geodatabase technology may significantly reduce the risk of managing distributed environments. Whether centralized or distributed, the success of enterprise GIS solutions will depend heavily on the administrative team that keeps the system operational and provides an architecture solution that focuses on user performance needs.

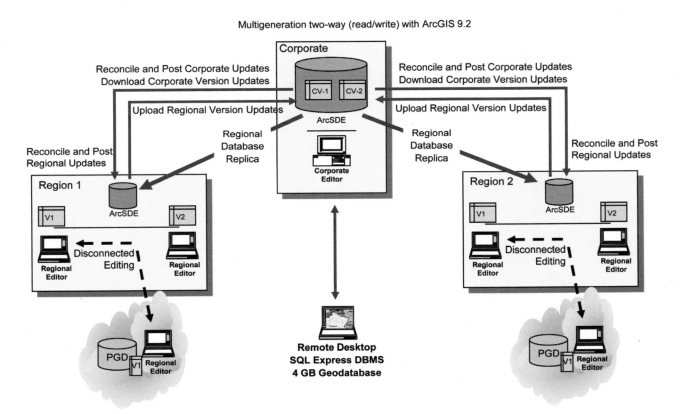

Figure 6-17

Geodatabase two-way multigeneration replication.

Part II

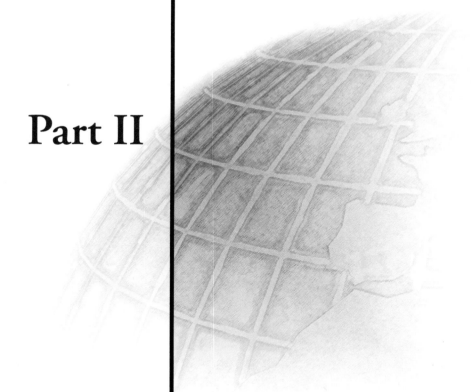

Understanding the fundamentals

7 | Performance fundamentals

A primary goal of the system architecture design process is to build a GIS that will allow for the peak operational performance you've planned. We have built a model (the Capacity Planning Tool) to help you identify the infrastructure to meet your performance needs. But before you can use it properly, you must understand the terminology of performance and the fundamental relationships underlying it. This chapter gives you that foundation, which is the underpinning of the model as well.

You can establish specific workflow and system performance targets, during initial planning, based on standard ESRI workflow models. The workflow models themselves are based on technology performance expectations derived from customer experience and the lab results of performance validation testing. These workflow models share the performance and scalability of the software—standard desktop and Web solutions that have been supporting ESRI customer operations reliably for years.

The terminology of performance is shared, too. Standard performance terms used throughout the corporate world (by business operations research to identify service staffing requirements) describe the performance and scalability of computer systems equally as well. This chapter introduces these performance terms used in relation to distributed computer platform and network environments. These performance relationships are the standard, and we can transmute them into numbers that serve to model system capacity needs. Later on, using these numbers in the Capacity Planning Tool, we can translate peak user workflow loads into specifications that will identify the platform(s) and network(s) capable of handling those loads.

The time it takes a system to perform a task (performance) and the number of tasks it can do at once (scalability) are two measures of the success of a system for GIS. Consistent and reliable performance of enterprise GIS operations depends on proper integration and interaction of a variety of software, hardware, and network communication components. A basic understanding of how computer systems perform work and how system components handle the shared workload is essential, not only for system architects but also for GIS managers and software developers. In turn, some knowledge of GIS is important for system architects and administrators. Understanding your strengths and your weaknesses is an important step toward success. This chapter provides the fundamentals that will help you understand system performance (how fast the system responds to a request) and scalability (how well performance can be maintained as workloads increase).

The chapters following this, on software and platform performance, look more closely at what it takes to actually design a scalable system. To make the best of your technology investment, you want to end up with a system that is "just right"—enough but not too much. You want to get the most out of what you spend. So you want the performance you need (a system that scales to support peak processing loads) but not more than you can use.

Distributed computer environments must be configured properly to support what you want to do with them. And even within the best configuration, there are many factors to either help or hinder meeting your system performance requirements. Many system resources are shared with other users, and there is much to affect user performance and productivity. In an enterprise GIS, with its distributed platform and network environments, the user experience is often the product of interactions among multiple components. Understanding them and how they work together—distributed processing technology—provides a fundamental framework for deploying a successful enterprise GIS. In short, you need to look at all of it, to understand how each piece fits together and really see the whole. It's astounding how frequently this fact, like the elephant in the room, is ignored—or misunderstood.

Figure 7-1

Understanding the technology.

Photo by Photolink/PhotoDisc/Getty Images.

Figure 7-1 reminds me of the story of six wise men who, although they were blind, went on a journey to "see" what an elephant is like. Having to rely on what he felt through touch alone, each blind man got his own impression, depending on his angle of approach. One, who happened to touch only the ear, was sure the elephant was like a fan. Another, touching the knee, said it was like a tree. Yet another blind man jumped to the conclusion that it must be a snake—because where he was coming from, he could only feel the trunk.

What a fitting tale to illustrate how a limited view of a something can lead to misunderstanding, a danger with enterprise GIS as well as elephants! Understanding your enterprise environment and how it performs and scales can be challenging. Technology is rapidly changing, along with the people you work with and the workflows themselves. We are constantly adapting to a rapidly changing world, as computer technology is becoming more and more a part of our everyday life. How this technology performs and scales has a big impact on our ability to do work, yet the more we learn about it, the more we realize how much more there is to understand.

If you think of the elephant as representing all of the technology in our work environment, you realize that what we assume versus what we truly understand is influenced by our experience. Working as a team, we can put together the bits and pieces that we understand with observations from others and get a clear picture of the whole. Documenting what we know, modeling how the different parts work together, and keeping an ear open to what others know better than we do helps us develop a realistic system design—and a system that will work, no matter how elephantine it may first appear.

ESRI has implemented distributed GIS solutions since the late 1980s. For many years, how to select the appropriate computer environment was not well understood, and customers relied on the experience of technical experts to identify their requirements and meet their needs. Each technical expert had a different perspective on what hardware infrastructure might be required to support a successful implementation and recommendations were not consistent. Many hardware decisions were made based on the size of the project budget rather than a clear understanding of user requirements and the associated hardware technology. The fundamentals as they applied to system design and GIS were not well understood at first. They had to be learned through experience.

Learning from experience

We began to document what was understood about distributed processing systems in the early 1990s by developing system performance models. The models were used to translate peak user workflow loads to the hardware specifications required to support those loads. System performance models have been used by ESRI consultants to support distributed computing hardware solutions for GIS implementations since 1992. They have also been helpful in identifying the cause of performance problems with existing computing operations. Just like systems themselves, the performance models serving as stand-ins for them have benefitted from some fine-tuning over time.

Using performance models initially to document our understanding of the technology, we soon saw their value in offering consistency of support with our system architecture design consulting services. Updated regularly to address changes in technology, these system performance models expanded over the years, supported by the *System Design Strategies* technical reference document on the ESRI Web site, which published the design process, an overview of the technology, and semi-annual updates of the platform performance models for more than 16 years. Provided as design guidelines early on, system performance models later developed into the platform sizing engineering charts discussed in chapter 9.

The initial performance models were developed for the technology at the time, desktop GIS applications with file and GIS database data sources. UNIX and Windows application compute servers were employed to provide remote terminal access to GIS applications located in centralized data centers. A simple concurrent-user model was used for capacity planning. A concurrent-user model identifies the number of concurrent-workflow users that can be supported on

a platform according to the platform's baseline performance. A simple capacity relationship was used along with published vendor SPEC benchmarks to translate peak concurrent-user loads to other platform environments. The performance baselines, like the models, were also updated to address changes in technology.

The late 1990s introduced Web mapping services, and transaction-based sizing models were developed to enable capacity planning and proper hardware selection. Transaction rates were identified in terms of map displays per hour. These transaction-based capacity planning models proved to be much more accurate and adaptive than the earlier concurrent-user models, although customers who had become accustomed to identifying sizing requirements in terms of peak concurrent-user loads rather than peak map requests per hour didn't always take advantage of them. The transaction-based models were more adaptive insofar as they were more accurate specifically in addressing changes in platform performance, allowing you to adapt to future changes more readily. The concurrent-user models were not as accurate, and it was important to update the platform sizing charts on a periodic basis for them to be useful. (The transaction-based sizing models have proven to be accurate over many years without updating the baseline performance platform. The SPEC benchmarks are an example of transaction-based sizing metrics.)

The new model

The release of ArcGIS Server 9.2 in 2006 introduced some new challenges for the traditional sizing models (for both the concurrent-user and the transaction-based models), and an effort to review lessons learned and take a close look at the road ahead was in order. The result is a new approach to capacity planning that incorporates the best of the traditional client/server and Web services sizing models and provides an adaptive sizing methodology for future enterprise GIS operations. We needed a capacity planning model that would accommodate a wide variety of workflow loads in a single design environment. So we leveraged all the lessons learned over the years in building a new capacity planning tool. This new capacity planning methodology is much easier to use and provides metrics to enable performance validation early during system implementation and delivery.

The new capacity planning model was developed and shared with the objective of helping software developers, business partners, technical marketing specialists, and ESRI distributors better understand the performance and scalability of ESRI technology. Those colleagues in turn have used it to offer customers the best possible GIS solutions for their enterprise GIS operations. In this book and in workshops worldwide, GIS users and organizations now have the option of using the model in the form of a tool on a CD, the Capacity Planning Tool (CPT).

What is capacity planning?

How do we address performance sizing? Figure 7-2 identifies some key factors that contribute to overall system performance. System architecture design is about managing implementation risk. The fact that there are many software and hardware factors involved contributes both to risk and to overall user productivity. As with the elephant, the bits and pieces of the system must work together to meet processing and communication needs. You can use performance models to represent the relationships between the various system performance factors.

There are many performance factors that contribute to overall user productivity. Improvements in critical system performance factors can improve user productivity and affect the total system performance capacity. Performance cannot be guaranteed by proper software or hardware selection alone. Proper software technology selection and a balanced hardware investment, based on projected peak user workflow loads, will satisfy system performance requirements and save money and time through properly targeted investments.

The performance sizing models are implemented in the Capacity Planning Tool, to be described in detail later. Performance sizing models can be used for capacity planning and establishing implementation performance milestones. These models provide a way to identify under what circumstances you may need more bandwidth from the network or more computing power from the hardware to keep your system running. You can tune your system for optimal performance later. Right now, you use these models to set proper performance targets that will get you under the limbo bar. You use them throughout the GIS implementation to demonstrate you are on target for a successful production rollout, with an infrastructure built to support your peak production processing loads. The performance models can translate initial test and deployment observations to expected system performance during peak operations. These models provide you with tools to ensure that the first time your GIS goes live, it will do what it's supposed to do; no more, no less.

The performance bar shown below represents the overall project budget. This must be spent wisely, purchasing the technology required for success, or the project will fail. While the project budget must be sufficient to purchase the right technology, it must be a reasonable sum as well; otherwise the technology will not be adopted at all. The conundrum goes full circle, beginning and

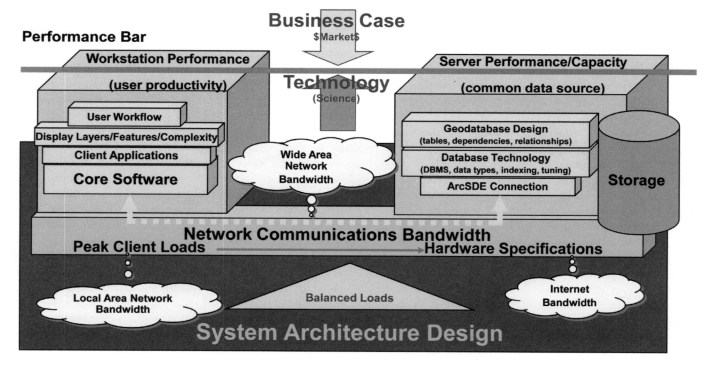

Figure 7-2

System performance factors.

ending with the right technology; you have to get it right to make it right, and vice versa.

How can these performance sizing models work when the technology is so complex? By providing measurable performance milestones the performance models can help you manage implementation success. The performance models also help you manage future growth and provide information to help you spend your budget intelligently. Why spend money on technology you don't need? You want only what you need just in time for when you need it, and the models are intended to get you within this "just right" range. Using the models to estimate growth, you can also gauge at what point your needs will justify purchases in the future. You can validate your selected performance models by observing system performance throughout deployment, providing timely information with which to manage success.

Think of the performance bar as the bar in a limbo dance; the lower it is, the higher the implementation risk. The tall dancers represent organizations with complex data models and heavy user workflows. The short dancers represent the simpler GIS implementations. The tall dancers have strong IT departments with excellent database administrators (DBA), while the short dancers may not be able to afford a good DBA. Therefore, the risk may be the same for both. For many years we have used the same performance models to support a wide range of users; both short and tall, with similar success.

What is system performance?

For users, system performance is how long they have to wait from the *click* to seeing the display, which is under one second these days for ordinary tasks. For the system architect, system performance is a collection of the times between the *click* and the display, when processing and interactions between components take place—how long *each* response takes weighs in as a factor in determining system performance and in planning for it. Knowing this, hardware vendors build computers with the appropriate component resources to optimize performance for a broad range of customers. Computer platforms are built on several component technologies. These component technologies include processors to execute program instructions, physical memory to provide processor access to the required program code, disk storage to provide online access to program code and data resources, and communication channels to bring these components together. Each component technology contributes to the overall computer performance.

In much the same way, distributed computing solutions (enterprise computing environments that include GIS and other systems) rely on several hardware platforms and network connections, each of which contributes to overall system performance. Hardware for a computing environment must be carefully selected to meet system performance needs *overall*.

The primary objective of the system architecture design process is to provide the highest level of user performance for the available system hardware investment. Each hardware component must be selected with sufficient performance capacity to support the required processing needs. The current state of the technology can limit system design alternatives. Understanding the processing loads at each hardware component level provides a foundation for selecting the appropriate technical solution.

In a simplified overview of the components in a stand-alone workstation and in a distributed client/server processing configuration, figure 7-3 provides an example of how overall performance hinges on component

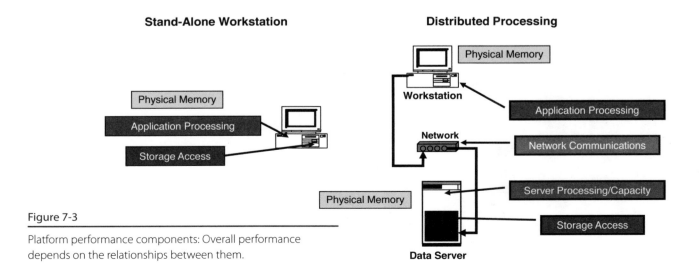

Figure 7-3

Platform performance components: Overall performance depends on the relationships between them.

relationships. Each component participates sequentially in the overall program execution. Memory is the place where program code and data wait for program execution. Processing is supported from platform memory—sufficient memory must be available to make way for the software and data required for program execution.

The total processing time of a particular application transaction will be a collection of the processing times from each of the system components. A computer vendor optimizes the component configuration within the hardware to allow for the fastest computer response to an application query. The customer IT/systems department is responsible for optimizing the organization's hardware and network component investments to provide the highest system-level response at the user desktop. Understanding peak user workflow needs and the software processing required to meet them lays the foundation for a proper compute environment, which leads to expected system performance, which can immediately boost user productivity.

System performance fundamentals

Many parts of a system contribute to overall performance. The change in hardware performance over the past 10 years has contributed significantly to user productivity. A map display that took 6 seconds to process in 2000 took less than 0.6 seconds in 2008. This change in processing performance is huge and exerts a major affect on system performance and scalability. Figure 7-4

provides an overview of the hardware technology contribution to performance change.

ESRI customers have been upgrading their hardware environment every three to five years; performance and user productivity are the primary drivers behind these hardware upgrades. Understanding how this hardware change affects system peformance and scalability is fundamental to proper system architecture design.

Performance terminology

To understand system performance and scalability, you must first look at the relationships between the work that needs to be done and the time it takes to do it in. These relationships (figure 7-5) are found in the terminology used to describe performance.

The primary terms used to describe system performance are the same as those used in capacity planning: transactions, throughput, service time, utilization, processor core, response time, and queue time. We use much the same terminology to describe and manage a variety of activities in everyday life, but it's the practical application of these terms—and how they are related to each other—that you need to master in order to describe and manage computer technology.

- *Transaction* is the term used to describe a unit of work (for GIS workflows, a transaction may be represented by a map display). Each user workflow can have a different measure of work. The transaction provides a unit for measuring throughput, service time, queue time, and

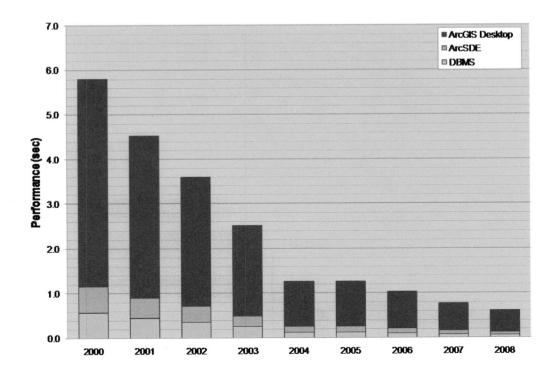

Figure 7-4

Time to produce a map.

Terms	Relationships

• Work Transaction (W_t)
• Throughput (T)
• Capacity (T_{peak})
• Utilization (U)
• Processor core (C_p)
• Service time (S_t)
• Queue time (Q_t)
• Response time (R_t)
	• Capacity = T/U
	• Service time (sec) = #Cp/Tpeak (DPM x 60 sec)
	• Queue time $Q_t = Q_r + Q_c$
	• $Q_r = (2^1$ x Utilization − 1) x Service time
	  Applied only when workflow exceeds 50% (1/2) capacity
	• $Q_c = 4^1$ x (Utilization − 1) x Service Time
	  Applied only when workflow exceeds capacity
	• Response time = $S_t + Q_t$

[1]Q factor will depend on the arrival time distribution. Factors are set based on consulting experience.

Figure 7-5

Performance terms and relationships.

response time—all measurements are relative to the workflow transaction.

- *Throughput* describes the rate of doing work, and is expressed in terms of a transaction rate (transactions per minute).
- *Service time* is the transaction processing time (usually identified in seconds).
- *Utilization* is a measure of platform capacity consumed by a specific throughput, and is represented as a percentage of platform capacity. Peak platform throughput is achieved at 100 percent utilization.
- *Processor core* is the computer component that executes the program code. (Computer processing capacity is signified by a vendor-published performance benchmark.) A single processor core can execute one transaction request at a time, and during peak loads the transaction workload is distributed to the available processor core for processing. If all processors are busy, the transaction must wait for the next available processor to service the request. A platform is identified by its hardware configuration (number of cores, number of sockets or chips, level 2 memory, front-side bus speed, etc.). We discuss choosing the right platform technology in chapter 9.
- *Response time* is the total system time required to process the transaction. Response time includes all the component service times plus any system wait times.
- *Queue time* is the term used to identify system wait times caused by random transaction arrival distributions and wait times due to overcapacity conditions.

The relationships between these concepts are simple —one might even go so far as to suggest there is a trace of common sense behind the logic of these relationships. Understanding how we apply and use the terms

in everyday life might help us appreciate how simple and logical all the puzzle pieces fit together to create an effective and efficient environment for enterprise GIS.

It is easy to focus on the wrong performance metrics. Understanding the real performance factors contributing to platform performance and system capacity is the key to capacity planning. Collecting the wrong performance factors leads to nothing but poor planning.

A system capacity performance profile
A system level performance profile is taking what vendors do at the platform level and applying the same capacity planning fundamentals toward designing systems for productive distributed enterprise operations. Workflow performance is the sum of all the service and queue times across the system (not just within a hardware platform). Many user workflows flourish concurrently in an enterprise GIS environment, each sharing system resources—server platforms and networks—to get the work done. Held in common, these resources (shared networks, enterprise servers, etc.) get busy, and when they reach over 50 percent capacity, temporary overcapacity conditions occur (due to the random arrival transaction request distribution) and user productivity is reduced.

Figure 7-6 shows the results of an ArcGIS Desktop test series performed on a system powered by two 4-core Windows server platforms (Windows Terminal Server and separate geodatabase platform). The chart shows the results of a series of 10 performance tests, identifying the time to process a single batch process and incrementing the number of batch processes to support each test measurement. The batch process consisted of a script of repeated random map display transactions, each about the same size. The 10th test series (on the right) provides the transaction performance results for 10 concurrent batch processes.

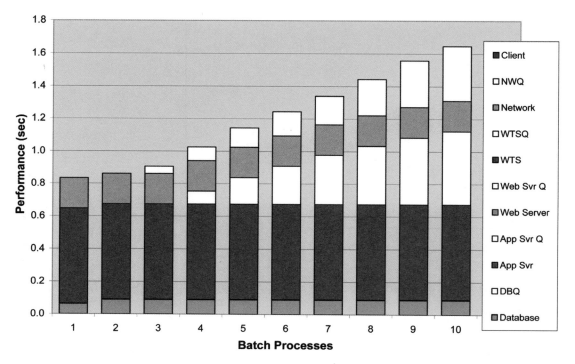

Figure 7-6

Workflow performance profile (4-core 2-chip test platform): Results of an ArcGIS Desktop series performed on a system powered

by a Windows Terminal Server (WTS) show each component's part in the time it takes to process a single batch process.

The user terminal client displays communicate with applications running on Windows Terminal Servers located in the data center. Remote communications are supported over a shared T-1 (1.5 Mbps) WAN communication. The chart highlights the Windows Terminal Server platform service times (blue) and queue times (white). Overall response time includes all service times and queue times (database, WTS server, and network).

Optimum system performance is experienced with one to two batch processes (0.82-second response time). System performance starts to degrade as the load increases beyond two batch processes. The system is operating at close to peak capacity when handling 10 concurrent batch processes (performance slows to about 1.62 seconds, about half the performance experienced during light system loads). The queue time approaches the value of the service time on the Windows Terminal Server platform and the network bandwidth, when supporting peak capacity loads. (Queue time represents process execution wait time due to random arrivals.) A total of four core processors are supported on the Windows Terminal Server platform. When inbound request rates exceed 50 percent utilization (average processing load exceeds two cores) there will be moments when more than four requests arrive at the same time. When these events occur, some of the requests will need to wait for processing. The wait time is called *queue time*.

There is a well-established guideline that predicts random arrival rates for large population environments (the formal area of study is called *queuing theory*). We can use such general guidelines in our models to estimate processing delays resulting from random arrival times.

Platform utilization

Computer processing is all about doing work: a computer executes procedures identified in a computer program. The core processing unit (CPU) includes a processor core (C_p) that performs the work. The amount of work a computer can perform depends on the design and speed of the processor core. The processor performs work at a constant speed and efficiency: processor speed is measured in megahertz (a measure of cycle frequency) and processor efficiency is calculated according to how many program instructions are handled by each processor core per cycle. Like some very organized and efficient folks you may know, computers only do work when there is work to be done. If there is none, the processor core is idle. Computer utilization is a measure of the percentage of time the processor core is not idle, but rather being used to do work.

Figure 7-7 shows the results of the previous test series in terms of Windows Terminal Server platform utilization.

Platform utilization is shown on the left axis. As we increase the workload by adding more clients, the percent of utilization increases. When our load is just a single client or batch process, and the process is executed sequentially (one instruction at a time), then only one instruction is available for execution at a time. Three of the processor cores will be waiting for work while one executes the work unit instruction. As we increase the number of concurrent client processes, we increase the amount of work available for execution. Once our load includes more client processes than processor cores, all processor cores will be busy doing work—and platform utilization reaches full capacity. The utilization rate levels off as we approach platform capacity.

Platform capacity is a measure of how much work gets done when all processor cores are fully employed (busy the whole time). The units of capacity are often called "transactions," units of work used to measure system capacity. For test purposes, it is convenient when transactions are about the same; then the processor core takes about the same amount of time to execute each work transaction. Of course, this is not the case in the real world (all transactions are not the same), so these measurements are somewhat arbitrary (although very important). We have a similar challenge in trying

to quantify how much work a person does in a day. In some cases work units may be defined to measure productivity, but most of the time we zero in on the hours employees spend at the workplace, not how much time they spent working while there. How much work is done depends on the work process. Many of the terms of measurement used to define computer performance are similarly based on averages, estimates, and just good sense born of experience.

Service time is the time required for the processor core to execute the unit of work. The processing time shown on the chart is 0.674 sec, and will be the same no matter how busy the system gets. (There's more about the relationship between service time and utilization in the platform throughput discussion that follows.)

Response time is the total time it takes for the system to complete a unit of work. Response time is shown by a line on the chart with the performance units on the right-hand axis. During light system loads the response time is 0.674 sec (this is simply the total processing time to do a unit of work and is equal to the service time). As the system load increases, the response time gets longer. Response time increases when all the processor cores are busy doing work. If the processor is busy when the unit of work arrives, execution must

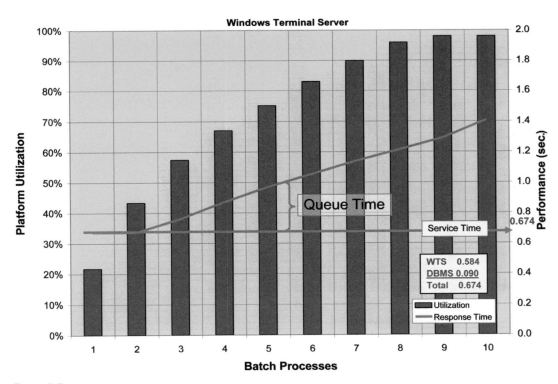

Figure 7-7

What is platform utilization? (4-core 2-chip test platform): The results of the same WTS test series, this time showing a measure of the percentage of time the processor core is being used.

wait until the next processor core is available to do the work. This wait time is called *queue time* and is related to the transaction arrival time distribution.

Platform throughput

Throughput is a measure of the rate at which work is being done. The processor cores do the work (execute the program code). You will first need to decide on how you will measure a unit of work, and then you can count the number of work units executed over a period of time. If you don't have a way to measure a unit of work, then you don't have a way to identify service time or response time. Performance is related to how long it takes to process a unit of work.

Figure 7-8 shows the platform throughput from the Windows Terminal Server test series. Platform utilization is shown on the left axis of the chart and throughput is shown on the right axis. As we increase the client load, platform utilization and throughput increases (notice a single client will use only one processor core). As we reach platform capacity, the throughput will approach a peak output rate. Adding more clients once the system reaches peak capacity does not increase the

rate work is being processed—it simply takes longer to get each unit of work done. The peak capacity for this system is a throughput of 411 displays per minute (DPM).

Computing platform service time

Figure 7-9 shows the platform utilization on the left axis and the throughput on the right. There is a very simple relationship between platform capacity, utilization, and throughput. Platform capacity is equal to throughput divided by utilization. If a platform supports a throughput of 89 displays per minute at 22 percent utilization, then it should be able to support a capacity of 405 DPM at 100 percent utilization (89/0.22).

There is also a simple relationship between service time and platform capacity. Platform capacity is reached when the processor core is busy all the time (100 percent utilization). Capacity is identified in terms of throughput (DPM) at 100 percent platform utilization. Service time is the time it takes for a processor core to execute a unit of work. If each platform core can do work at the same time, 4 cores can do 4 times the work of 1 core if there is enough work to do to keep them busy. When the platform is working at full capacity

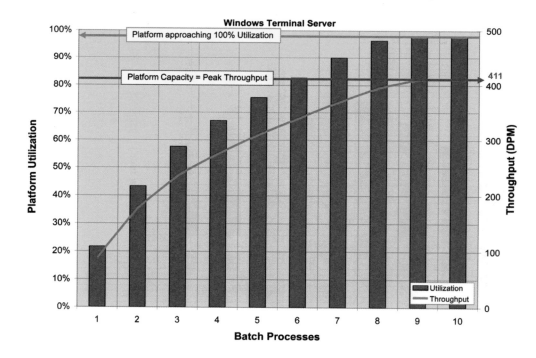

Figure 7-8

What is platform throughput? (4-core 2-chip test platform): The same WTS test series, charted here to measure the rate at which work is being done, shows the peak system capacity as a throughput of 411 DPM.

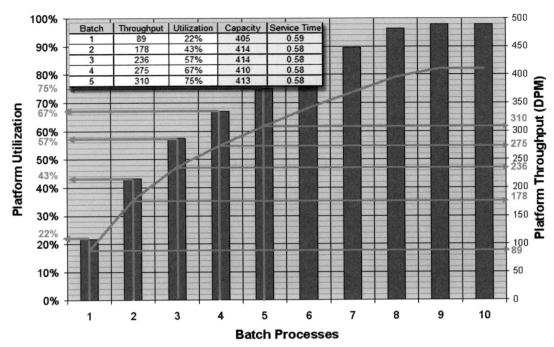

Batch	Throughput	Utilization	Capacity	Service Time
1	89	22%	405	0.59
2	178	43%	414	0.58
3	236	57%	414	0.58
4	275	67%	410	0.58
5	310	75%	413	0.58

Figure 7-9

Computing platform service times (4-core 2-chip test platform): The time it takes for a processor core to execute a unit of work equals the number of cores multiplied by the platform utilization then divided by the measured throughput.

(100 percent utilization), peak throughput equals the number of cores divided by the service time. You can also use this relationship to compute service time when the platform is not at full capacity (use the relationship between throughput and utilization to replace capacity). Service time equals the number of cores multiplied by the platform utilization then divided by the measured throughput. This is a valid relationship for all throughput levels, as shown in figure 7-9.

Response time
Response time is the total amount of time it takes to deliver a unit of work (client display refresh time). Response time includes unit processing time (service time) and any system wait times. Service time, platform capacity, peak system throughput—these three metrics are constant for a given platform configuration and defined unit of work. Response time includes all processing times (service times) plus all other time waiting to be processed (queue time) when all the core processors are busy supporting other people's work.

Figure 7-10 (page 112) shows workflow performance as the system load is increased from 1 to 10 batch processes. GIS users have an interest in response time because it is directly related to user productivity. If we measure

user workflow productivity in terms of client displays per minute, workflow productivity can be influenced by the display response time. For example, to maintain a workflow productivity of 10 displays per minute a user must submit a new display request every 6 seconds (60/10). If the display response time is 2 seconds, this leaves 4 seconds for the user to think about the display (think time) and submit the next display request. If the display response time is 4 seconds, and the user needs 4 seconds between display requests to do his or her work (think time), workflow productivity would be reduced to 7.5 displays per minute (60/8). On the other hand, if the response time were reduced to 1 second, then workflow productivity could increase to 12 displays per minute and the user would still have 4 seconds to think about each display.

Queue time
Queue time is the time a transaction is waiting to be processed as it travels through the system. There is a discipline within operations research that deals with queuing theory. Queuing theory is about standing in line to be serviced, how long the lines get, how long you have to wait, and how many service providers you need to support different kinds of user workloads. Smaller

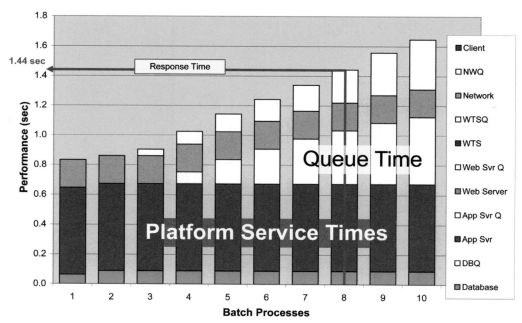

Figure 7-10

What is system performance? As the workflow load on the system is increased from 1 to 10 batch processes, response time includes all processing time (service time) and all wait time (queue time).

populations are much more complex than large populations. Large populations with random arrival times are the easiest to predict.

Two related conditions can determine how long we need to wait. One is based on random arrival times during normal workloads. The other condition is when the average arrival rate is greater than the peak transaction rate or system capacity. Figure 7-11 lays out these performance factors.

Random arrival average queue times. In the real world, we often wait to be serviced. The length of lines varies depending on utilization of the service (how busy it is) and how the service is provided (scheduled appointments, come when you like, show up at 8 AM). From my military service days, I can remember sick call when everyone had to show up at 7 AM. One doctor would meet with each patient for five minutes and provide appropriate medicine. Service time was five minutes, but response time (service time + wait time) could be several hours.

Fortunately for most computer workflow environments, we are dealing with a very high population of transactions (hundreds of little packets representing pieces of display transactions) distributed over a very short period of time (display cycle time of 6 to 10 seconds). Such a large population over such a very short period of time suggests that we can use the more simple queuing theory relationships to estimate wait time for this

environment, and that these estimates will tend to be consistent. For example, a guideline for estimating initial collisions in network environments is 25 to 35 percent capacity. This improves to 40 to 60 percent capacity in switched buffer cache environments (initial small delays can be ignored). As a general rule, I suggest

Random arrival average queue time (Q_r)

- Arrival rate varies based on random arrival times
- Service time is constant based on hardware performance
- Response time = average arrival rate delays + service time

$Q_r = (2 \times \text{Utilization} - 1) \times \text{Service Time}$
Applied when workflow exceeds 50% capacity

Capacity — Productivity tradeoff (Q_c)

- Throughput = workflow x response time
- Throughput peaks at 100% capacity
- Response time = throughput / workflow
- Work units are waiting on core service availability

$Q_c = 4 \times (\text{Utilization} - 1) \times \text{Service Time}$
Applied only when workflow exceeds capacity

Figure 7-11

Queue time performance factors.

using 50 percent capacity as the point where we might have to wait due to random arrival time. Based on our test experience, this wait time will increase to equal service time by the time you reach 100 percent capacity.

Arrival rate exceeds peak throughput capacity. This is not a real case in a steady state model environment. In a real environment, you can expect more random arrival conflicts during busy periods. If utilization were to continue above 100 percent, the queue would continue to get longer and longer. What happens in a real environment is response times get longer as the queue grows, and once the peak arrival period is past the queue starts getting shorter again. As a result, the amount of time operating at peak capacity is extended to service the additional arrivals. You can model this condition by increasing the wait time by the percent of overcapacity.

There are many ways to manage wait time. You can increase the system capacity so peak loads do not exceed 50 percent (this could increase hardware cost and software licensing). You can manage arrival distribution by scheduling work times, providing customer information on peak wait periods, and optimizing the number of service providers (processor cores and software instances). It has been my experience that most customers configure systems to support peak capacity workflows, while being sensitive to higher hardware and software licensing costs. It is important to configure systems with capacity to support projected peak workflows, in order to avoid unreasonable performance delays. But there are probably situations wherein customers prefer to have more hardware regardless of the cost, so as not to risk productivity slowdowns.

How we size the network

Network communications follow the same performance factors and relationships experienced within computer processing environments. The words used to describe capacity, throughput, and service time are different but the relationships are the same (see figure 7-12). You will recall this terminology from chapter 3.

- Network capacity is called *bandwidth*, and is expressed as megabits per second, kilobits per second, or gigabits per second.
- Network throughput is called *traffic*, and is expressed in the same units as bandwidth.
- Network service time is called *transport time*, and represents the time it takes to transport the amount of traffic associated with the unit of work over the available bandwidth (Mbpd/Mbps).
- Network queue time includes wait time due to random arrivals and overcapacity conditions, plus round-trip travel time, called *network latency*.

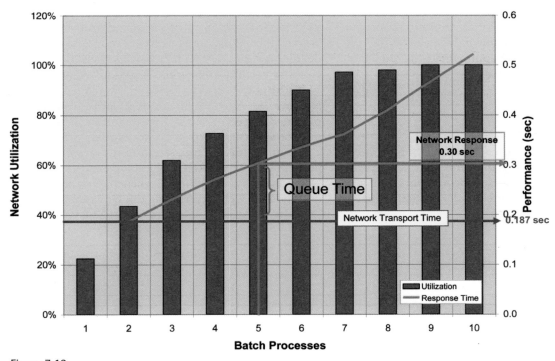

Figure 7-12

Sizing the network (ArcGIS Windows Terminal Server over 10 Mbps WAN bandwidth). The same performance factors and relationships experienced within computer processing environments apply to network communications as well.

Figure 7-12 shows a sample network load profile. Network transport time equals network response time during light loads (0.187 seconds). Queue time increases as the network traffic exceeds 50 percent capacity. Network response time with five batch processes is 0.30 second.

Network latency is related to communication chatter. It takes milliseconds to make a round-trip between the client and server over the network, and this is negligible when you use transactions with a small number of trips. Client/server map display transactions that have a geodatabase data source, however, make several hundred trips between the client and server, and this can add up to several seconds of additional travel time or latency when communicating over long distances.

Capacity planning models

Models that closely approximate real life are the most useful. In actual work situations, many hardware and network components affect system performance and scalability. We use mathematics and physics to translate those components into numbers that represent how they factor in. These factors are what the model uses to help you estimate and plan for an infrastructure to support the performance you need. Modeling system performance based on these fundamental elements provides a rich and adaptive capacity planning environment—which is what the Capacity Planning Tool is. These are the fundamentals or factors to consider when using a model to approximate the real life of system performance and scalability:

- Throughput capacity
- Service times
- Utilization
- Queue times

These same basic performance elements have been part of performance testing and tuning for many years. We were accounting for these basic performance elements in our legacy concurrent-user and transaction-based platform sizing models. These performance elements underlie the calculations in the Capacity Planning Tool described in chapter 10, which you will use for the process of system design in chapter 11.

The display-transaction models, used for sizing Web services since the mid-1990s, have proven much more adaptive and stable than the early peak-concurrent-user client/server sizing models. Several years of ArcGIS Desktop, ArcIMS, and ArcGIS Server performance-validation test results—from tests conducted using a common map display environment—provided a good foundation for establishing a platform-sizing-model strategy. Modeling system performance based on the fundamental elements of throughput, capacity, service times, utilization, and queue times provides a very rich and adaptive approach to system architecture design and capacity planning.

A common capacity planning model for ArcGIS Desktop and Server was published in the summer of 2006. This was the first time we brought the ArcGIS Desktop and ArcGIS Server models together into one common model (or one way to represent the models). Previous ArcGIS Desktop models were based on concurrent-user workflow loads (a unit of work was identified as a concurrent user). Previous ArcIMS and ArcGIS Server sizing models were based on peak platform displays per hour (unit of work was based on map display service times). With the current sizing models, the unit of work is defined based on average transaction service times. Figure 7-13 shows how the ArcGIS Desktop client/server and ArcIMS/ArcGIS Server Web services models were brought together by defining a new user-productivity relationship between concurrent-user workflow and map displays per minute.

Platform capacity is a function of map service time, and can be expressed in terms of peak displays per minute. The relationship between peak displays per minute and concurrent-user workflow is provided by a user productivity expressed in terms of user displays per minute (DPM). ArcGIS Desktop power users at 10 DPM match with the older client/server concurrent-user models. Web server sizing can be based on service popularity (displays per hour). Peak transactions per hour can be translated to displays per minute and converted to concurrent-users at 6 DPM.

Many software performance variables contribute to the processing time required to produce a map display: the software functions, the number of layers in the display, the number of features in a layer, the types of relationships and dependencies included in the geodatabase data model, and the quality of the display. Simple information products (map displays, tables, charts)

mean lighter processing, high user productivity, higher platform capacity, and less hardware. High-quality map products mean heavier processing, reduced user productivity, less platform capacity, more network traffic, and more hardware. Understanding how to include the right information product in the user workflow can make a big difference in user performance and system capacity. The next chapter will explore software performance sizing and examine the variables that take up processing time.

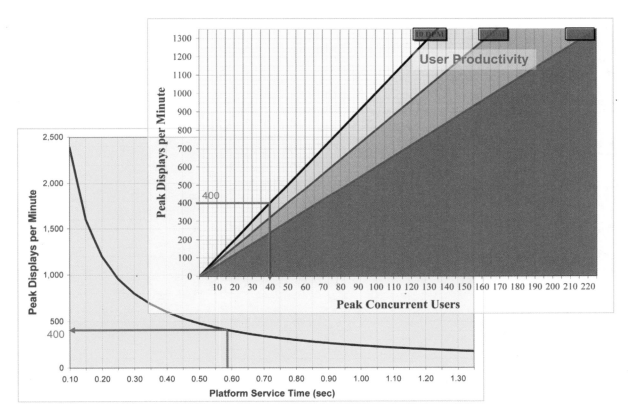

Figure 7-13

Platform performance (4-core server): The engineering charts above show how server capacity values are established based on platform service time. Each platform processor core can generate a display within the platform service time. Displays per minute

are a function of the service time and the number of platform processor cores (60 sec x 4 core / 0.6 sec = 400 DPM). Concurrent users are a function of user productivity (400 DPM / 10 DPM/user = 40 users).

8 | Software performance

OVERVIEW

This chapter reviews some of the software dos and don'ts in building a high-performance, scalable environment for a geographic information system (GIS). System performance and scalability can be limited by your software choice or the application design. Getting the software right from the start can save hundreds of hours of lost productivity during implementation. What are the basic factors determining software performance? The efficiency of the software underlying the GIS application and an efficiently managed data source on the database server are two primary factors in good performance.

There is always some kind of trade-off between efficiency and quality or functionality. Over time, user expectations have grown: we expect to see the display at the click of a finger. Yet historically, gains in software functionality have come at the price of a growing sequence of procedures that must be executed by the computer platform (processor core). This requires the hardware to work faster than before. And fortunately it does.

Regardless of how instantaneous software performance appears to us, literally millions of pieces of code are being executed—one after another, sequentially—a hundred times over or more for one second's worth of work. From the standpoint of sheer volume alone, taking into account ways to shorten this process, piece by piece, cumulatively makes sense. Software performance is about executing procedures that perform work in an efficient way. The procedures are defined in a program code that is executed by the computer. The program code includes instructions that define the sequence and steps of the procedure. The program instructions are executed one step at a time (sequentially) by the computer processor core. Longer procedures take longer to execute. Simple procedures with a smaller number of instructions (lines of code) take less time to execute.

ESRI software has developed over the last four decades to improve and expand the capabilities of the code. New procedures are added to the code with each software release, then modified again to enhance more efficient execution. Software matures with each release, improving the performance and functional capabilities of the code; older procedures are modified or replaced over time, all in the interest of providing better quality software.

Software just gets richer and more powerful. Assuming you can take the time to process the additional code instructions—for things like higher display resolution, 3D virtualization, and high-resolution satellite imagery—all of these great technology advances make the user experience more enjoyable and potentially more productive. So these software advances can be worth the additional code instructions and increased processing they require. Even so, the cost in performance must be accommodated or compensated for, something everyone, from developer and mapmaker to systems administrator and architect, would do well to consider. For example, a higher quality, dynamic map (made on demand) takes more processing, so if you can get by with a simple map that can be cached, you will be putting less of a load on performance.

How we store and manage the data makes a big difference. Preprocessing of map documents into an optimized, pyramid-file cache can reduce display processing requirements—the display is processed once and the results can be viewed many times with little additional processing. Since the cached images are standard files, they can also be cached on a local client machine. Bringing the data local again reduces processing requirements, eliminates the need to transport it over a network, and improves display performance.

Several features about the software contribute to system performance and scalability. These features include the ESRI core software, custom client applications deployed using the core software, the user display (number of layers, spatial features, and complexity), and the user workflow (procedures for scheduling heavy processing requirements).

The data source configuration is also a primary consideration. The connection protocol, the selected database technology, and the geodatabase design are important issues. To make an information product with real-time data integration (for example, the higher quality 3D map referred to earlier) requires more processing because the data models used to make it include complex table relationships and dependencies, to improve data integrity. The number of database tables, data model dependencies and relationships in the user schema, and proper care and tuning of the database environment—these are the factors we discover at the root of the majority of problems customers have with performance. Most GIS operational problems are caused by the same mistakes made over and over again—avoiding those mistakes from the beginning can significantly reduce implementation risk. Take heed of the best practices that follow; they account for the effect on your system design of software performance factors.

The purpose of this chapter is to help you recognize that how the software is deployed and used can affect overall system performance significantly. We depend on software developers and programmers to provide efficient software execution, but that's not the end of the story. System architects and users, too, play a big role in taking advantage of what's provided and what's possible. In designing any system infrastructure, for example, you need to understand what your user display needs are—and what they are not—in order to provide the exact display environment that promotes a productive workflow, no more, no less. Take into account existing infrastructure limitations and user needs, select the right technology for the job, and be aware of the effects on system performance of how it's used.

Programming and performance

Again, understanding a little history may help put system design options in perspective. The ESRI software programming environment has evolved over the years to promote more stable and adaptive development. Early software programs were developed as stand-alone executables with tightly compiled executions providing the software functions. Computer execution was slow compared to today's standards, so the simplicity and efficiency of the program was critical. Programs were organized into subroutines with common input and output to accommodate more efficient use of programming code. The final program was organized and compiled in the most efficient way to meet performance requirements.

Programming techniques changed in the 1990s to create a more open, component-based development environment. Software procedures came to be deployed as separate component objects with a common communication interface. Now these objects can be compiled as a group of code functions to support software licensing requirements. The operating system enables communication between the component objects within the software code. Maintaining, supporting, and developing code based on component objects has proved to be a much more efficient and adaptive programming environment, accelerating the advance of software technology and simplifying code maintenance. However, for system architects, this development introduced some new considerations: Software based on component objects includes an additional layer of communication instructions. Integrating object communications includes additional steps that must be performed by the computer to execute the software. In other words, component-based software performance is slower than the older, scripted code. Faster hardware compensates for this, but it is still something to take into consideration,

especially in light of other technology advancements that may increase processing requirements.

For example, Web services introduced additional hardware and communication components that also increased the number of instructions required to execute a program. For a given hardware technology, more instructions translate to slower performance. The software deployment strategies around service-oriented architectures (SOA) move enterprise applications one more step away from the vendor technology providing the services. This all comes with a need for more code and more computer processing, so applications in an SOA perform slower than compiled object code.

All in all, computers must work much faster than before to meet user performance expectations, and they do. But when the fees you pay for software licensing are based on the number of processor cores required to run your programs, it's critical to design and manage the system with the an eye toward optimal performance.

Map display performance

Performance is a fundamental consideration when establishing a map display format. "What's the most efficient map to make that will still meet user needs?" is the question to ask. The GIS planning team should promote keeping map service times within performance budgets. In order to keep within performance budgets and for better performance, follow the tips listed in this chapter whenever possible. You will have no trouble discerning when it is possible if you have followed the good planning advice in Roger Tomlinson's book, *Thinking About GIS*, and defined your GIS needs and described in detail the information products that will meet them. On so many levels, his book offers excellent advice to GIS managers about maximizing what a GIS can offer: Sometimes by generating less, you provide more. Make only the information products you need, and understand exactly how each will be used so that you can pinpoint what means of conveyance will do the trick. Is a simple map sufficient for your purposes or do you need one in high relief? Must it be made on the spot (dynamic) or can it be cached? In most cases, simple maps are easier to understand and use than complex, detailed maps. If so, limit the map to that—because simple map displays also travel better (perform and scale better) than ones of more complexity. By contrasting the "light" column with more complex workflows and applications, figure 8-1 shows how much faster simpler displays perform. (The way to "go light" is laid out in the bulleted points under the workflow performance summary on the next page.)

Map displays should start simple and use field visibility to show relevant data only. Use scale dependencies

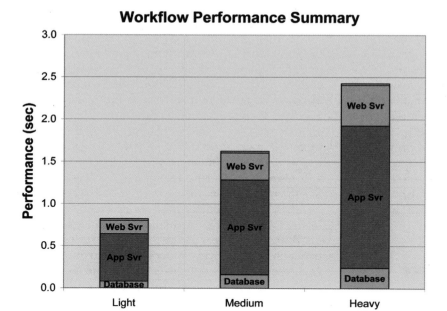

Only show relevant data:

- Start simple
- Use field visibility

Use scale dependencies:

- Use appropriate data for given scale
- Use the same number of features at all scales

Points:

- Single layer simple or character markers
- Use EMF instead of bitmaps
- Use integer fields for symbol values
- Avoid halos, complex shapes, and masking

Lines and polygons:

- Use ESRI optimized style
- Avoid cartographic lines and polygon outline

Text and labeling:

- Use annotation instead of labels
- Use indexed fields
- Use label and feature conflict weights sparingly
- Avoid special effects (fill patterns, halos, callouts, backgrounds)
- Avoid very large text size (60+ points)
- Avoid Maplex for dynamic labeling (avoid overuse)

Figure 8-1

Optimum display makes a difference: There are several best practices for making information products that can perform and scale. Most of the standard ESRI workflow service times are based on a light map display (we do provide a medium-level AJAX standard Dynamic workflow for ArcGIS Server). Service times for heavier user workflows must be adjusted to properly represent system processing loads, and in many cases the result may not be acceptable. Best practices must be followed when selecting workflow display environments to ensure acceptable workflow performance.

to display higher level data at higher map extents. As a general rule, the same number of features included in the display at all scales will provide a balanced user experience, both in performance and map display quality. Use the most efficient functions to display points, lines, and polygons. To achieve the best performance, efficiency in text and labeling is important also.

User productivity is related to the display response time. Lightweight displays perform quickly and provide easy-to-understand information products. Heavy displays, even when optimized, can require more than three times as much processing. Heavy displays also may be hard to read and understand, and require more network traffic, which may reduce response time during peak traffic loads. This is not always true; sometimes simple displays require lots of processing. You need to manage display performance by establishing targets during the planning process and by measuring your progress in meeting these set performance goals throughout the implementation.

Quality versus speed
High-quality maps require more processing, thereby reducing display performance. Here, high quality refers to map applications that are detailed and complex enough to demand more processing—and more time—than simpler maps. Simple maps, like the one on the right in figure 8-2, stick to low-resolution relief, solid colors, and simple annotation. High-quality maps, like the one on the left, qualify as complex because of the shaded relief, transparent layers, and Maplex labeling they would probably include. The high-quality

maps look nice, but there is a performance and productivity cost that should be considered when developing an application or publishing a Web service. Figure 8-2 shows quality versus speed by comparing the same map published as a heavy (left) and as a light (right) service. Also keep in mind that the quality improvement from using the antialiasing function can have a downside: it may increase the processing workload by as much as 50 percent.

Simple application strategies can provide big performance gains. A standard application workflow or Web service can use simple map displays for high-performance view and query, and provide high-quality mapping (if there is a need) on demand (when requested by the user). Trade-off between quality and speed is a primary consideration in selecting the right user display environment.

GIS dynamic map display process
Computers build maps much the same way geographers did before them. Geographic features are identified as points, polygons, and lines, and collected together as layers of geographic information. The software execution procedure for making a map begins with laying out the first layer; then adding layer after layer—following many, many steps (computer instructions)—until the map display is complete. These procedural steps are sequential instructions provided in program language for the computer to execute. Sequential programs are executed one instruction at a time (which can be very time consuming), and a single process is executed instruction by instruction by one processor core at a time.

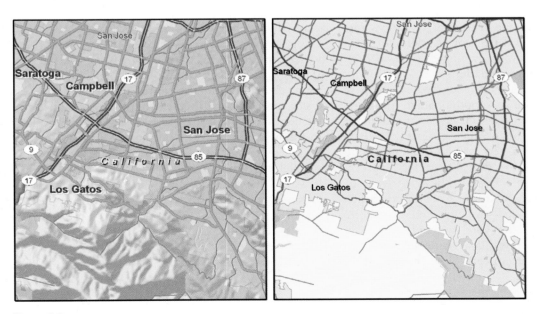

Figure 8-2

Quality (left) versus speed (right).
Data from ESRI Data & Maps, StreetMaps USA.

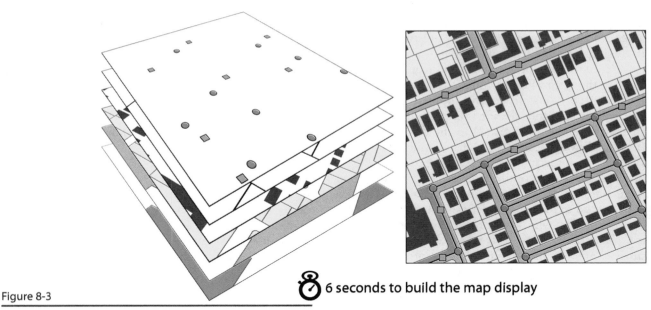

6 seconds to build the map display

Figure 8-3

Sequential processing: Traditionally, map layers are executed one instruction at a time—the fewer layers, the faster the service time.

Courtesy of the Office of Geographic and Environmental Information (MassGIS), Commonwealth of Massachusetts Executive Office of Environmental Affairs.

Figure 8-3 demonstrates the standard procedure for generating a map display, building a map one layer at a time starting with the bottom layer and stacking each geographic information layer on top until the map display is done. The total map processing time (map service time) includes the sum of all the processing time for each layer of the display.

Simple displays with a small number of layers perform faster than more complex displays with many layers. Map layers are executed one instruction at a time, and display performance will depend on the complexity of the map display and how fast a single processor core can execute the program instructions.

Hardware vendors are encouraging software developers to rewrite their software to support parallel program execution, so as to take better advantage of multicore processor technology. This is much easier to say than to do, since it requires a new way of thinking about and solving problems. Most of us like to follow procedures and deal with one step at a time. Programmers are like we are, and in many cases they are coding the procedures that we provided—following the same steps we would do if we did the work by hand. Doing all the steps in parallel (even if this were possible) is not a simple task. It requires teamwork—and some creative

process management. Figure 8-4 suggests the potential performance benefit if we could build the layers of our map in parallel.

This map display procedure divides the map display process into three sets of layers, and requests the three sets of layers be generated at the same time (parallel processing). Once all the geographic layers are collected from the data source, they are sorted into the proper sequence (stack) and mashed into the final display. This sounds quite simple to do, but we should recognize the thousands of lines of instructions and variety of functional interfaces fueling the current standard map display procedure in the existing software. Changing the procedures for making a map is a slow process—we all learn from our experience and we continue to improve our ways; there is simplicity in following a procedure one step at a time. Organizing the same process to complete a procedure by using parallel execution processes can introduce complexity—and in some cases be less efficient (slower performance and more problems)—than the sequential execution. There are some hidden risks: the complexity of putting the parallel pieces back together in the final map display may require even more processing than the traditional sequential way.

Fortunately, transaction-based Web applications provide opportunities to exploit parallel processing. Some Web service options build in parallel: layers can be requested from separate Web services and combined in the Web application map display. Requesting all services at the same time can improve display performance. The performance improvement is not guaranteed, though, and you must make sure you actually can put all the pieces back together in the final display.

GIS is not unique; many compute-intensive computer programs are executed sequentially. All of the SPEC compute-intensive benchmarks are sequentially executed, and the SPECint and SPECfp results for multiple CPU platforms demonstrate that these benchmarks do not take advantage of more than one processor core. The SPECrate throughput test procedures are executed with several concurrent benchmark instances to take full advantage of the capacity of multiple-core platforms. (See chapter 9 for more about SPEC benchmarks.)

In 1995 in California, the Orange County Transportation Authority (OCTA) had a need to improve execution performance of its transportation flow analysis program. The transportation flow analysis calculated freeway traffic volumes based on changes in housing and business development throughout Orange County. The results were used to plan and budget freeway upgrades and maintenance operations. The existing analysis took 48 hours to execute when using its existing single CPU UNIX server. OCTA wanted to purchase a platform solution that would complete the same analysis in less than 12 hours.

The 1995 CPU technology was more than 50 percent faster than its existing platform. OCTA was able to break the analysis into four separate batch processes they could run in parallel; then they manually integrated the four results together to complete the analysis. The platform solution was a four CPU Hewlett Packard machine. HP provided the machine for testing, and OCTA was able to demonstrate that it would support its processing timeline.

Figure 8-5 provides a summary of sequential versus parallel processing. Most of the ArcGIS software is sequential and user display performance will depend on single-core performance (faster processor cores reduce map display service times). The first workflow shows a single user generating a 20-layer display over a 1.5 Mbps network. The next four workflows show a single

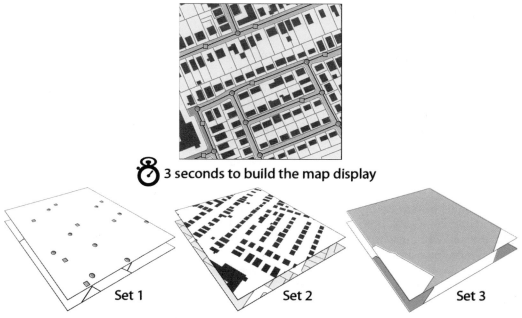

3 seconds to build the map display

Set 1   Set 2   Set 3

Figure 8-4

Parallel processing: Building the layers of a map in parallel brings the potential for performance gains but for complexity as well.

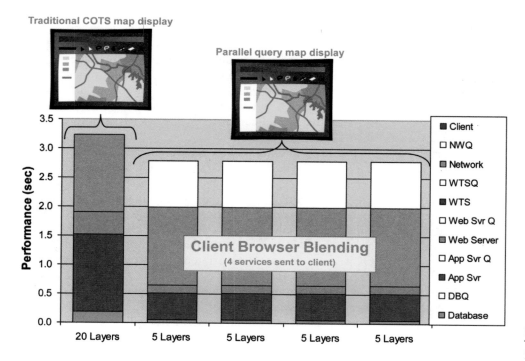

Figure 8-5

Sequential versus parallel processing.

user generating a 20-layer dynamic map display from 4 parallel map services (each sending a 5-layer image map service) over the same 1.5 Mbps network. Performance gains were mitigated by performance loss due to higher network traffic contention.

This display is provided for illustration purposes, generated using the Workflow Performance Summary provided with the Capacity Planning Tool that will be discussed in more detail in chapter 10. Platform loads were minimal (one user for each workflow), thus there were no platform queue times. The two simulations (which are traditional COTS map displays) are provided on separate networks, each set at 1.5 Mbps T-1 bandwidth. Client display blending was supported at the client display, which significantly increased the network traffic for the parallel display example (display traffic peaked at 1.2 Mbps increasing traffic throughput, which increased the network queue time). Parallel processing can sometimes improve display performance. Note, however, that this example does not consider any additional blending time required for the map display generated from the four parallel map services. In other words, the overall parallel response time may be slower than the sequential map display when we add the processing time to bring the pieces back together in the display.

Standard map display service times

Reducing the number of dynamic layers in a display can improve display performance. In many applications, the map reference layers can be provided by a separate

map service or by a local cache. ArcGIS Online offers a rich set of global, land-base data that can be used for reference layers in a desktop map display. Reducing the dynamic display layers by providing/using a pre-processed cached data source can dramatically improve map display performance.

Figure 8-6 shows a range of display service times for an ArcGIS Desktop dynamic map display. The ArcGIS Desktop service time is shown on the left axis and the ArcSDE service time is provided on the right. The DBMS service time is equal to the ArcSDE service time in the baseline models. The ArcGIS Desktop 9.2 dynamic map services times provided in the Capacity Planning Tool are based on a 20-layer map display. A service time estimate for an ArcSDE pyramid image layer is equivalent to 10 vector layers (Desktop = 0.3 sec., SDE = 0.036 sec., DBMS = 0.036 sec.).

Figure 8-7 shows a range of display service times for an ArcGIS Server dynamic map display. The ArcGIS Server 9.2 AJAX standard map services times provided in the Capacity Planning Tool are based on a 20-layer map display. A service time estimate for an ArcSDE pyramid image layer is equivalent to 10 vector layers (Web = 0.14 sec., Server = 0.42 sec., SDE = 0.036 sec., DBMS = 0.036 sec.).

The ArcGIS 9.3 release provides an expanding number of options for leveraging parallel Web service data sources and different layers of cache to improve client performance. Workflow map processing times will be different, depending on the type and number of layers in the dynamic map display. Understanding the

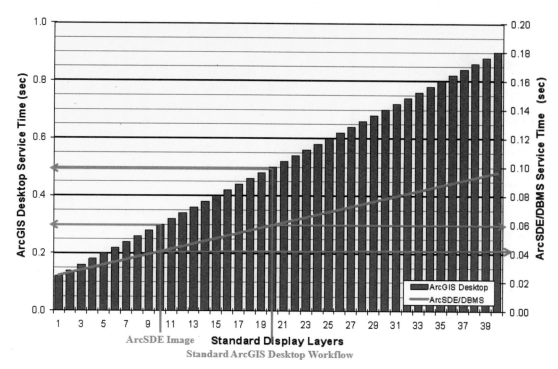

Figure 8-6

ArcGIS Desktop dynamic display service times.

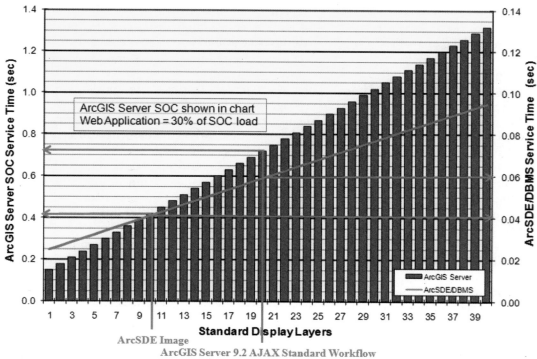

Figure 8-7

ArcGIS Server display service times.

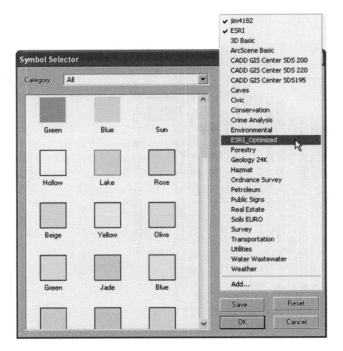

Figure 8-8

Optimized lines and polygons improve performance.

relationship between number of layers in a display and the display service time can inform your design decisions, enabling you to set achievable performance targets that will get the productivity you need out of a GIS.

Optimizing lines and polygons

Figure 8-8 shows how to select the ESRI optimized lines and polygons for best display performance. Outlines for all fills are simple lines instead of cartographic lines. Picture fills are EMF-based instead of BMP-based.

For best performance, avoid cartographic lines and polygon outlines, since these features use extended, compute-intensive functions that reduce display performance. ESRI-optimized lines and polygons can improve drawing performance by more than 50 percent.

Selecting the right image format

Figure 8-9 compares the size of Web Mapping Service (WMS) image files produced with a common 600 x 400 pixel resolution. The size of map images contributes a major portion of the network traffic per display.

Notice the importance of selecting the right image format for your map service. JPEG images are about the same size when supporting both vector and raster imagery. PNG24 imagery is lighter for the vector display but quite heavy, when supporting raster imagery. BMP images are very heavy and network traffic would

probably slow display response times. Selecting the right map service image format can reduce network transport requirements and improve user performance.

Figure 8-10 compares file size for the same vector map image published at two different display resolutions. The smaller pixel resolution 600 x 400, and the larger pixel resolution 1200 x 800.

The image-file size more than doubles when you double the display resolution. For optimum performance, use a small image resolution for standard view and query, and provide higher resolution only on demand (when someone needs to print the image or look more closely). Using smaller output image resolution significantly improves user performance.

Providing the right data source

The efficiency of the data source can make a big difference in performance and scalability. Efficiency in this case is measured in terms of the amount of processing required for query and rendering of an average map display, and in some cases the amount of network traffic required to support each display transaction. Several types of GIS data sources support today's GIS applications; they range from simple shapefiles and images to data managed in a database management system. (The single shapefiles and images are simple and convenient to use, but as they grow in number, they become difficult to manage and interface with other data sources.) ArcGIS Server technology does introduce new performance opportunities: ArcGIS Online, as mentioned, is one source of data that has been optimized by preprocessing; image file caches can be cached and used by clients to improve their own display performance. But at the same time, Web environments can be more difficult to manage and control.

Most enterprise GIS operations today rely on an integrated geodatabase environment that enforces data integrity and provides a variety of data management functions. Figure 8-11 (page 128) lists several key areas within a geodatabase that contribute to system performance and scalability. Building a high-performance geodatabase starts with planning and organizing the database design, and continues with maintenance throughout the life of the system.

A large percentage of system performance and tuning problems addressed over the last 10 to 15 years have focused on design and maintenance of the geodatabase. Many of the performance problems were simple and quite easy to recognize, once the database configuration was understood. The fruits of our labor and experience have produced the following best practices:

Optimizing workflow. Plan data loading and data maintenance during off-peak hours, and establish a

Raster images

JPEG = 90 KB
16 sec @ 56 Kbps
0.6 sec @ 1.5 Mbps

PNG24 = 237 KB
42 sec @ 56 Kbps
1.6 sec @ 1.5 Mbps

BMP = 704 KB
126 sec @ 56 Kbps
4.7 sec @ 1.5 Mbps

Vector images

JPEG = 82 KB
15 sec @ 56 Kbps
0.5 sec @ 1.5 Mbps

PNG24 = 28 KB
5 sec @ 56 Kbps
0.2 sec @ 1.5 Mbps

BMP = 704 KB
126 sec @ 56 Kbps
4.7 sec @ 1.5 Mbps

Figure 8-9

Map application image type.

Courtesy of the Office of Geographic and Environmental Information (MassGIS), Commonwealth of Massachusetts Executive Office of Environmental Affairs.

1200 x 800 resolution

600 x 400 resolution

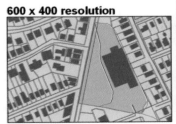

JPEG = 90 KB
16 sec @ 56 Kbps
0.6 sec @ 1.5 Mbps

JPEG = 567 KB
101 sec @ 56 Kbps
3.8 sec @ 1.5 Mbps

Figure 8-10

Map application image size.

Courtesy of the Office of Geographic and Environmental Information (Mass-GIS), Commonwealth of Massachusetts Executive Office of Environmental Affairs.

queue for reconcile/post/compress processing. Heavy batch processing consumes system resources, and performing these functions during peak periods can degrade performance.

Database configuration. Keep the configuration simple, and make the popular data easy to find. Finding the tables requires more processing than finding rows in tables. Reducing the number of data tables accessed during a display query can improve performance.

Identifying problems. Collect performance information and listen to users. Understand your system performance parameters and establish performance targets. Take time to analyze the available performance metrics and the information they provide.

Versioning methodology. Keep the state tree optimized, avoid long lineages, and follow a disciplined maintenance plan.

Editing operations. Edit cache (a function supporting a more efficient editing environment) moves critical data to the local machine, significantly reducing network traffic and loads on the server during edit operations. Taking advantage of opportunities to cache data on the client application machine can improve user productivity, reduce network traffic, and offload shared server resources.

Collect performance statistics. Establish a performance baseline and monitor platform statistics. Platform performance during light loads can be used to identify capacity

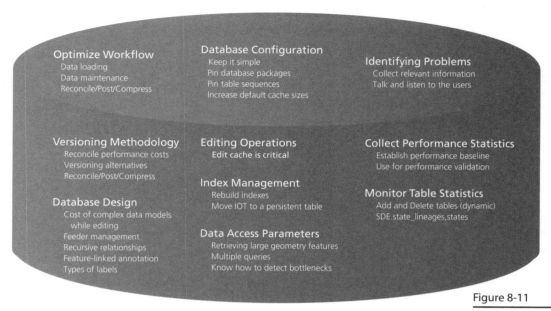

Figure 8-11

Geodatabase performance factors.

limitations and to forecast performance requirements for peak workflows. Performance statistics can identify potential problems in time to resolve them before they affect workflows. Learn how to monitor performance metrics.

Database design. Data models that are simple improve performance. Heavy quality assurance functions should be performed as periodic background processing rather than as part of a user display transaction. Relationships and dependencies triggered within a sequential display transaction can reduce user productivity. Understand the cost of complex data model functions, and find ways to run them as background processes or restricted workflows. For view and query users, make a copy: complex data-maintenance environments can be replicated to simple distribution servers to improve performance.

Index management. Keep the indexes current. Indexes reduce the processing load by narrowing the data search. A poor index is like an unorganized file cabinet—it takes forever to find anything. Locating data lost in a large database takes much longer than if it was properly indexed.

Monitor table statistics. Understand the size constraints and limits on tables, and monitor table statistics. Add and Delete tables are dynamic, and long state lineage tables affect performance. Performance problems are often understood once they are identified, so keep track of areas of the database that are rapidly changing in size.

Over and over again, I am amazed at the simple nature of the issues that affect performance and scalability. So many times I find that just understanding the technology and exercising common sense can reveal what's causing the performance problem almost instantly. Once the problem is identified, the user usually knows how to fix it. In fact, in most cases the user fixes the problem more quickly than the system tuning consultant who helped find the cause.

GIS operations are rich in data, growing in size every day as we find easier and more efficient ways to record and document where things are on this earth. Geography is about working with these spatial resources to better understand our world (organizing, relating, comparing, and displaying). The efficiency of the data source (effort required to find what you are looking for), the location of the data source (distance from the data source to where you want to work with the data), and the quality of the data (did you find what you are looking for) all have an important impact on performance and user productivity.

Building high-performance Web applications

Web services provide the most powerful technology for a scalable enterprise GIS. Why? A variety of deployment scenarios are available, ranging from centralized data center processing to distributed wireless communications. You can use Web applications to provide dynamic displays for browser clients, and you can use distributed configurations with background processing to support communications with mobile clients. Whatever configuration you choose becomes a variable affecting performance, as figure 8-12 shows by

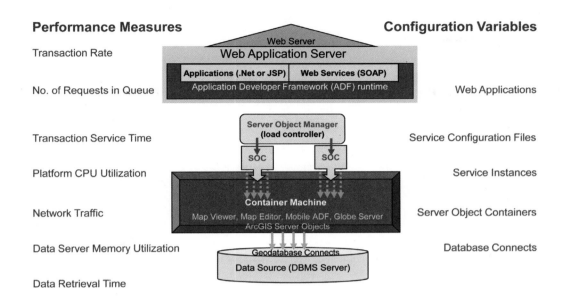

Figure 8-12

Web application performance factors.

comparing Web performance factors and their associated configuration variables.

Each component variable becomes a measure of performance, as shown in the performance tuning guidelines on the next page (figure 8-13). Seeing how each component measures up to its corresponding category of performance, you can modify map services (or the variables) to optimize Web site performance.

Web services popularity, adaptability, and scalability can all be measured in terms of the Web display transaction. Desktop and Web applications supported by published component Web services promote rapid technology change. Common HTTP protocols provide the enabling technology.

The same basic principles that determine how client/server applications perform influence Web performance too, but with the Web comes the unwelcome potential for more layers of hardware and communication exchange. Some of the more interesting Web solutions are those that provide local data and display executables on the local user display device, while carrying on communication and heavier analysis functions as background processes. Think about what makes this work; the basic principles are not complex: if you can separate the heavy processing and network transport time from the user interface, you can improve user performance and workflow productivity.

Several pitfalls within the Web architecture can contribute to causing performance bottlenecks. The performance fundamentals discussed in the previous chapter apply to all of the Web components, as does

the encouraging news from chapter 7: identifying the problem will often expose the solution. For instance, once you discover that a platform or network component is operating over 100 percent capacity, you know you need to upgrade or replace this component. To keep track of capacity utilization, you should monitor the following performance factors: service transaction rate, requests in the queue, service time, platform utilization, network traffic, memory utilization, and data retrieval time. Taking their measure can tell you a lot about a system.

The importance and success of a Web site is often measured in terms of peak transaction rates (popularity of the published service). Web site popularity might be expressed in terms of peak service transactions per day. A standard measurement for capacity planning is "peak service transactions per hour." Another term for transaction rate is "throughput," as defined in chapter 7. You can monitor throughput and platform utilization rates in order to establish service time and predict total system capacity, even during the initial system deployment.

The number of requests in the queue is a measure of queue time. Commonly, a Web site supports a transaction rate of more than 12,000 display requests an hour, which is about 200 displays a minute. Based on queuing theory, we can expect actual display request arrival rates to vary based on a statistical random arrival time distribution. Experience shows that as we approach 50 percent utilization, average wait times start to grow consistent with the peak number of requests in the

Performance Measures	Tuning Options
Transaction Rate	Transaction rate identifies the number of requests supported by the site configuration. Peak transaction rate is the maximum capacity of the site to support incoming requests. Peak capacity can be increased by reducing the time required to generate each map service (simpler information products) or by increasing the number of CPUs (more processing power). Sufficient service agent threads should be included to take full advantage of the available CPUs (usually 2–4 threads per CPU is sufficient). Once CPUs are fully utilized, additional threads will not improve site capacity.
Number of Requests in Queue	Each virtual server (SOM) acts as a processing queue for inbound requests. Requests are held in the queue until a service thread is available to process the request. If a request arrives and the virtual server queue is full, the browser will receive a server busy error. Queue depth can be increased to avoid rejecting browser requests.
Transaction Service Time	This is the CPU time required to process the published service once it has been assigned to a service agent for processing. Long service times can significantly reduce site capacity and should be avoided if possible. Simple map services (light data and minimum number of layers) can significantly improve site capacity.
CPU Utilization	Sufficient threads should be configured on the public service agents to support maximum CPU utilization. Peak site capacity is reached when CPU utilization reaches a peak level (close to 100 percent utilization). Increasing service threads beyond this point without increasing the number of CPUs will increase average client reponse time but is not recommended.
Network Traffic	Sufficient network bandwidth must be available to support information product transport to the client browser. Network bottlenecks can introduce serious client response delays. Bandwidth utilization can be improved by publishing simple map services, keeping image size from 100 KB to 200 KB, and ensuring sufficient bandwidth to support peak transaction rates.
Data Server Memory	Sufficient physical memory must be available to support all processing and adequate caching for optimum performance. Memory utilization should be checked once the system is configured to ensure more physical memory exists than what is being used to support the production configuration.
Data Retrieval Time	This is the CPU processing time on the ArcSDE server. Query time can be optimized by proper indexing and tuning of the ArcSDE database.

Figure 8-13

Performance tuning guidelines for Web mapping services.

queue. As we approach 100 percent throughput, the average wait time starts to exceed the service time. A large number of requests in the queue is an indication of degrading user response time.

Web services are carried on through loosely connected communications between the Web server and the browser clients. HTTP provides an optimistic data-streaming protocol designed to minimize communication chatter, which really helps when accessing the service over a high-latency connection.

Configuring the server instances
Understanding how to configure ArcGIS Server SOM (server object manager) service instances is a fundamental requirement of system architecture design insofar as it determines peak system capacity and optimum

service performance. You'll recall that an instance represents a single collection of software executables and map documents (MXD) published to service a Web request (each instance can process only one request at a time). The right maximum service instances must be defined properly to reach peak platform capacity. The right minimum service instances must be deployed for optimum system memory utilization. You also need to define how you're going to restrain heavy geoprocessing services by restricting the number of maximum service instances. High-available SOM configurations share the maximum-service-instance assignment across shared host platforms, so the maximum service instances for each host machine must be considered to avoid over-stressing platform capacity. It is important to clearly understand the terms used for assigning and controlling

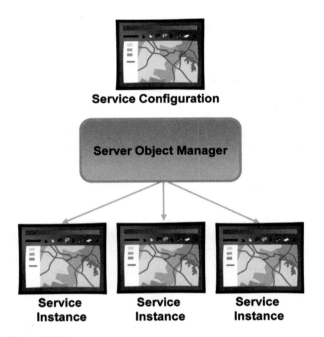

Service Configuration

Server Object Manager

Service Instance **Service Instance** **Service Instance**

Figure 8-14

ArcGIS Server instances.

service instances and how to configure ArcGIS Server for optimum performance and scalability.

For most standard map services, the peak number of SOC instances per platform host should not exceed 4 times the number of core processors. The standard ArcGIS Server platform is a 4-core server, so the maximum server object container (SOC) instances for that platform should not exceed 16 instances. Providing instances beyond this point can significantly slow response times during peak system loads, and all for a less-than-noticeable increase in service capacity.

Figure 8-14 demonstrates the configuration and deployment of Web services, defined in the SOM as separate service configurations. When publishing a service, the minimum and maximum number of service instances are defined for each service configuration file. This provides the SOM with limits for how each service is to be deployed on the host platforms.

During installation, a server object container agent is installed on each host machine. This provides the ArcObjects code needed to support SOC deployment. During startup, the SOM deploys the minimum number of service instances for each server configuration, distributing the instances evenly across the assigned host container machines. Service instances are deployed in SOC executables.

During operations, if concurrent requests for a specific service exceed the number of deployed service instances,

the SOM will increase the number of deployed service instances to handle peak request rates up to the maximum value allowed for in the service configuration. If the inbound concurrent requests for that service exceed the maximum deployed instances you've set in the service configuration, requests will wait in the service queue until an existing service instance is available for service assignment. If necessary, service instances for less popular service configurations will be reduced (closed) to make room for the popular service. Nonactive services can be reduced down to the minimum instances specified in their service configuration file.

Deployment algorithms within the SOM provide even distribution of service instances across the assigned host platforms. The deployment algorithms along with the service queue work to balance the ArcGIS Server processing load across the host container machines. The SOM should be used as the final load balance solution for the host container machine tier.

Let's see how this works in real life, by using an example that shows how Web services popularity, adaptability, and scalability are related to shifts in the SOM management of displayed SOC instances. The most common ArcGIS Server production configuration is composed of two map servers, each server configured to handle the Web application server, ArcGIS Server Web ADF, SOM, and SOC executables. We would configure each server the same to meet high-availability requirements. Let's say we have five public-facing map services and have decided to publish them in the following service configurations:

- Southern California Wildfires (min 0, max 16)
- Southern California Earthquakes (min 0, max 16)
- LA Traffic (min 1, max 16)
- Southland Weather (min 1, max 16)

We would have configured each Web application server host machine with a maximum capacity setting of 16 SOC instances.

During normal operations, only one of these services would be active (probably the LA Traffic), and each of the SOMs would increase its number of instances for this service to fulfill peak demand (up to 16 instances per SOM). Each set of SOC instances would be balanced across the two available Web application servers.

During the wildfire season, with the San Bernardino Mountains ablaze, popularity shifts to the Southern California Wildfires Web service. Demand is high, so the SOMs respond and reduce the LA Traffic service instances to the minimum (1 per SOM) and increase the Southern California Wildfires instances to the maximum (each SOM could deploy up to 15—for a total of 30—before reaching the SOC host maximum instances of 16 each). The system is now in an optimum

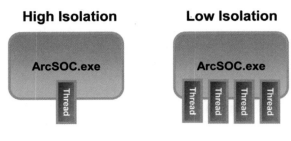

Figure 8-15

ArcGIS Server SOC isolation: Each SOC thread (service instance) is a pointer within the executable.

configuration to service the high demand of public wildfire service requests.

Selecting high isolation or low isolation

Figure 8-15 shows the number of instances in a deployed SOC. There are two types of SOC executable configurations, high isolation and low isolation (terms used by the ESRI ArcGIS Server software documentation). A high-isolation SOC is a single-threaded executable allowing for one service instance; in other words, a high-isolation configuration restricts a single SOC to one instance. A low-isolation SOC is a multithreaded executable supporting up to four service instances; a low-isolation configuration allows a single SOC executable to service as many as four concurrent instances (ArcGIS Server 9.3 SOC can support up to 8 threads). Each SOC thread (service instance) is actually a pointer within the executable tracking execution of the assigned service request (all requests share the same copy of the executables).

For example, a SOM deploying 12 service instances using high isolation would be launching 12 separate SOC executables each providing one instance thread. A SOM deploying the same 12 service instances using low isolation would be launching 3 separate SOC executables, each providing four 4 service instance threads. The low-isolation SOC configuration requires less host machine memory, but if one service instance (thread) fails, the SOC executable will fail along with its remaining 3 instances. When a high-isolation service instance fails, the SOC executable failure is isolated to loss of a single service instance.

ArcGIS Server physical memory requirements reduced significantly with the ArcGIS 9.2 release. The SOC executables make optimum use of shared memory and the SOM is able to both increase and decrease the number of service instances to optimize system performance. Standard vendor server memory recommendations of 2 GB per core (8 GB per 4-core server) should

be adequate for most ArcGIS Server deployments. ESRI currently recommends use of high isolation for all ArcGIS Server service instance configurations.

Selecting a pooled or nonpooled service model

Figure 8-16 sets forth the pooled service model. This is the best configuration for most service configurations. The current-state information (extent, layer visibility, etc.) is maintained by the Web application or with the client browser. The deployed service instances are shared with inbound users, released after each service transaction for another client assignment. The pooled service model scales much better than the nonpooled model because of this shared object pool.

Figure 8-17 shows the nonpooled service model, which should be used only when an application's function requires it. A nonpooled SOC is assigned to a single-user session and holds its reference to the client application for the duration of the user session. The current state information (extent, layer visibility, etc.) is maintained by the SOC.

The way the nonpooled service executes is similar to an ArcGIS Desktop session on a Windows Terminal Server. Both use the same ArcObject executable functions; both are restricted to one user able to take advantage of the service instance. ArcGIS Desktop on Windows Terminal Server would perform best for most implementations. ArcGIS Server implementation software license per application server platform (40-50 concurrent user sessions) would be less expensive.

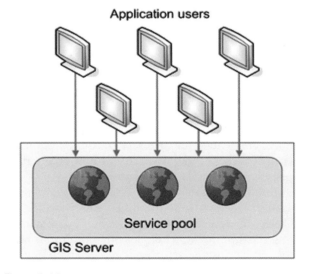

Figure 8-16

ArcGIS Server pooled service model: The best for most service configurations.

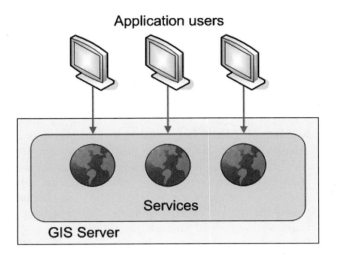

Figure 8-17

ArcGIS Server nonpooled service model: Use only when an application's function requires it.

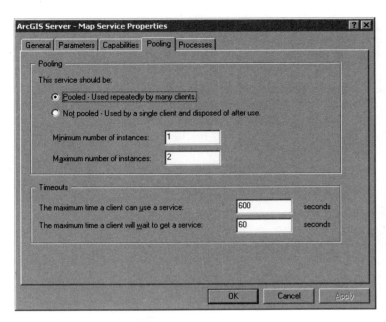

Figure 8-18

ArcGIS Server: Configuring SOM service instances.

Configuring SOM service instances

Figure 8-18 shows the user interface for configuring the SOM service instances. The service instance parameters are identified on the pooling tab within ArcGIS Server Map Service Properties.

Use the pooled service whenever possible. For optimum user performance and system capacity, a single pooled service can handle many concurrent clients, while access to a single nonpooled service is limited to a single user session.

The minimum number of instances would normally be "1" to avoid startup delays if requested by a single user. If the service was seldom used, this setting could be "0."

The maximum number of instances establishes the maximum system capacity the SOM can expose to this service configuration. It's worth repeating: configuring service instances beyond the maximum platform capacity will not increase system throughput, but it may very well reduce user display performance.

Heavy batch process. For handling a heavy batch service, the maximum instances should be a small number to protect site resources. A single, heavy batch service can consume a server core for the length of the service process time. Four concurrent batch requests on a 4-core server can consume all available host processing resources.

Popular map service. For a popular service, the maximum instances number should be large enough to take advantage of the full site capacity. Full site capacity could support 3-4 service instances per core. Two 4-core container machines could handle up to 32 concurrent service instances—the maximum instance setting for a popular map service.

Configuring host instance capacity

Implementing the right software configuration can improve user display performance and increase system throughput capacity. The right number of service instances to assign (to enable maximum throughput) will vary slightly, based on the type of service (reference heavy batch process and popular map service examples above). If all the service execution is supported on the host container platform, a single instance per core may be an optimum capacity setting (heavy batch process). For most map services, however, the execution is distributed across the Web Server, container machine, and database server, which shares the processing load across multiple platforms. Additional delays can occur due to random instruction arrival times (queuing theory, in chapter 7), which can account for up to 50 percent of the total processing time when approaching full system capacity. The test results in figure 8-19 (next page) show the optimum capacity setting at 3-4 service instances per core, using the same results you will see with the Capacity Planning Tool (in chapters 10 and 11). In both the tool and testing, the performance fundamentals presented in chapter 7 have been applied to find the right instance capacity. ESRI capacity recommendations are based on the results of performance tests and our basic understanding of performance fundamentals, backed up by customer operational experience.

The number of deployed service instances are increased in the chart below to show their impact on overall system throughput and capacity. A single instance (thread) can only take advantage of one of the two CPUs, and reaches less than 25 percent capacity during peak loads. As more threads (instances) are deployed, the peak throughput increases until an optimum configuration is reached for this service. In this case, the optimum trade-off between throughput and user response time was four users per CPU.

Figure 8-20 shows the user interface for establishing the host capacity settings. The host capacity setting restricts the number of maximum SOC instances each SOM can deploy on the assigned host container machines. The recommended capacity setting—or the maximum number of instances the parent SOM should deploy on each of the host container machines—is

3-4 service instances per core. A standard commodity Windows server has a total of 4 processor cores, so the optimum capacity setting would be 16 concurrent instances.

A standard, high-available, 3-tier configuration for ArcGIS Server could have 2 Web servers (each with a SOM), 3 host container machines, and a database server. If each of the host container machines was a standard 4-core Windows server, the optimum capacity would be 16 service instances on each host machine. In a high-available configuration, the SOMs are not cluster-aware, so separate configurations would be allocated for the shared environment, with a separate capacity setting needed for each parent SOM.

It is very easy to inadvertently allow too many SOC instances. With the same ArcGIS Server configuration identified above, full capacity of the host machine tier (3 servers, each with 4 cores) would require a total capacity of 48 service instances (16 on each host machine). If all the host capacity settings were 16 (not necessarily

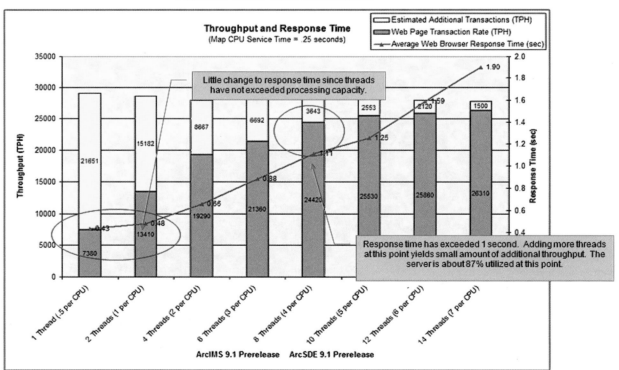

Figure 8-19

Selecting the right capacity: Test results show the optimum capacity setting at 3-4 instances per core.

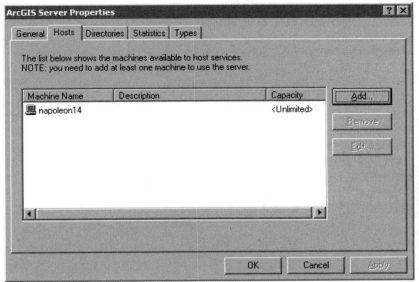

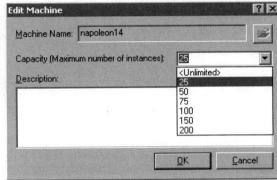

Figure 8-20

ArcGIS Server interface for establishing host capacity settings.

the right answer with both Web servers SOM operational), the fully operational configuration would allow a total of 96 concurrent service instances (48 per SOM, each SOM would deploy 16 on each host machine for a total of 32).

There are two potential performance problems when deploying too many service instances per host machine. When server loads exceed 100 percent utilization, an optimum capacity configuration would minimize response time by completing work in process (first in, first out) before assigning more requests (concurrent arrivals would not be assigned for processing until the current work is completed). When you deploy too many service instances (more than the number of core processors can service at the same time), the available CPUs must share their time to process all the assigned service requests simultaneously—and they do this in a very fair way. All assigned service requests are completed at about same time—all must wait until all processing is complete for them to be done. Too high a capacity setting would result in increased user response times during heavy system utilization (over 100 percent) due to a high number of deployed service instances executed by the limited number of processor cores.

Peak capacity should be set no higher than 4 instances per core (capacity of 16 for a 4-core host machine).

Selecting the right physical memory

Memory is like water, it doesn't appear to be the most important thing to have when you have enough. But if you don't have what you need, it can very quickly become the most important thing. Without enough physical memory, your application will begin to slow down, and possibly fail eventually. If you don't have enough swap space, your application may not start. Understanding what amount of memory you need to meet your requirements will keep you out of trouble.

There are several layers of memory used by the computer to support program execution. Software program executables are loaded into virtual memory during startup. Virtual memory includes physical memory and memory swap space.

Physical memory is made of solid state memory chips (no mechanical moving parts). Accessing data from physical memory is orders of magnitude faster than accessing data from disk storage. Memory speed on solid state memory chips is all about distance (speed of light). Advancing memory chip technology is about making smaller semiconductors (the material that supports solid state memory). Smaller semiconductors perform faster than larger semiconductors because the bits of data you are looking for are closer together.

Memory swap space is a dedicated volume of local storage space on disk allocated to memory processing. Memory swap space is used to store a copy of current program executables; physical memory uses it to reload program components during heavy memory utilization.

The software component instructions used to execute the program must reside in physical memory. If physical memory resources are adequate, program executables will be loaded into physical memory during startup and remain available for processing. As memory utilization gets busy, program components that are not being executed may get paged out of physical memory to make room for other active program executables. Inactive program executables do not need to reside in physical memory. They can reside in the memory swap space on disk. If an executable that was paged out is needed by the program, it will be fetched from the local storage memory swap space. Processor core resources are used to support paging, so increased paging reduces workflow performance. Paging is an indication of not enough memory and can increase rapidly as the processing load increases. At some point, physical memory will no longer have enough capacity to retain the active program executables and the program will crash (one of the program components being executed will be paged out). Many customers over the years have reported random process failures. Random process failures of normally good programs are often caused by an inadequate amount of physical memory.

Memory swap space is normally the size of the physical memory, and some operating systems recommend twice the size of physical memory. Memory swap space can be configured during system install.

Memory cache is part of computer design, and bits of memory cache are included in different components across the system to improve performance. There is memory cache on the processor chip (L2 cache), memory cache in the video card (VRAM), and memory cache on the network interface card. Hardware vendors continue to strive for ways to design computer platforms to be ever faster and more efficient.

Figure 8-21 makes the case for looking at system throughput as a function of memory utilization. The initial throughput increase occurs as service instances are added to take advantage of the available platform cores. Three to four instances per core will provide optimum throughput for standard map services (maximum of 4-6 instances per core). Maximum throughput will increase gradually as more service instances are deployed on the system; users will experience rapidly deteriorating response time with each slight gain in throughput. If available physical memory is exhausted, performance

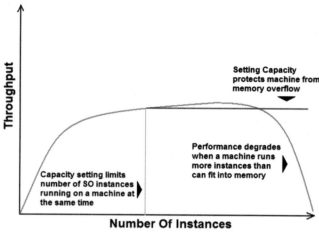

Figure 8-21

Platform memory requirements.

and throughput will fall off rapidly. A capacity setting should be established to support optimum throughput (good throughput with reasonable response times). Memory resources should be adequate to avoid access swapping or paging and support all active program executables in physical memory.

Avoiding disk bottlenecks

Disk storage is one of the few remaining computer components that still include mechanical moving parts. Each disk stores data in blocks on a magnetic platter. A drive arm moves across the disk to the assigned block location to read and write data. The disk spins at a fixed RPM (revolutions per minute). Average seek time is the average time to reposition the drive arm for each input/output transaction. Disk access is a mechanical process, one of the last mechanical devices left on the computer. As such, it is the slowest operation on a computer. For example, accessing several MB of data from physical memory may take milliseconds, while accessing the same data from a disk may take seconds. The actual storage performance is difficult to quantify, since location of the data you want may be located in memory cache. Storage systems are designed to take best advantage of file caching.

Figure 8-22 is a standard disk storage array configuration. Disk arrays are used to continue supporting data access in the event of a single disk failure, which improves disk access performance: copies of the data protect against disk failure, and striping data across several disks improves access performance.

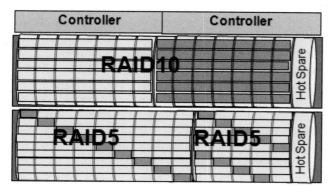

Figure 8-22

Avoiding disk bottlenecks with a standard disk storage array.

RAID 10 disk volumes provide two copies of all data, and data is striped across all disks for each volume of the array. This configuration is the best for high access to data. RAID 5 provides data protection using a parity disk. RAID 5 disk arrays can perform as fast as RAID 10 during normal loads, although with small array volumes they are more vulnerable to disk contention.

Disk arrays are designed to optimize the use of caching. Most disk storage solutions support controller cache. When writing to disk, the write is made to controller cache and the disk write is supported by background processing. In many new arrays, data is again cached at the disk level. Data is also cached on the file and DBMS servers. File servers will use local cache data before accessing the disk. Database management systems include a memory cache for high access files, such as indexes and common SQL calls. For many GIS operations, disk bottlenecks are minimized by the many layers of system cache and disk contention is not a primary concern.

Disk access performance problems can be identified by monitoring the server input/output (I/O) queue. Disk contention during peak operations will increase wait time due to disk bottlenecks (multiple requests for data on the same disk at the same time). Disk contention can be resolved by spreading data across larger RAID disk volumes or increasing use of the cached data sources. Faster RPM disks will reduce I/O time if necessary, and locating hot files on the outside of the disk platters can reduce seek time and improve data access.

ArcGIS technology is coming up with more and more solutions that leverage data caching. Data caching can improve performance by reducing processing requirements and accelerating response times. As hardware and network access performance improves, data

will be moving through these caches at an increasing rate. There is a potential for higher disk access demands as we see GIS implementations using new caching technologies. Some examples of technology advances that could put more demand on disk performance include the high data exchange in Web temporary output files; prerendering map and globe tiles; the larger files needed by geoprocessing services, and file data sources such as SHP, PGDB, FGDB (see Glossary), and imagery. Faster data access or data write performance may demand faster access to data on disk in the future. With traditional systems, the storage configuration was seldom identified as a performance bottleneck. Higher disk access demands, coupled with data distributed across a smaller number of disk platters, can increase the probability of disk contention, which may be a more significant problem in future system deployments.

ArcGIS Server cache: The performance edge

ArcGIS Server offers a variety of new opportunities for faster user display performance and increased system capacity. Taking advantage of data caching is the key to realizing these performance gains. Data caching won't work for everyone; for many workflows, the best solution is dynamic access to a central geodatabase environment for the complete map display. Nonetheless, data caching does work for much of the general population. Many very popular mapping services on the Web today owe their high performance and scalability to precached data sources.

Precached data sources process the map products in advance; "precooked," sort of like fast food, it's the quickest way to get a meal, if that's the meal that meets your needs. Just as fast food is not as fresh as food prepared on demand, there is a trade-off with preprocessed maps, too; they can be out of date, and may not reflect current conditions (weather, traffic). These are not limitations, of course, if the data is relatively static and hasn't changed since, say, a month earlier when it was cached.

Services can be partly cached to improve performance. Many GIS map displays include dynamic business layers (roads showing snow depth, electrical network showing latest posted work order, weather data, traffic data, etc.). Most GIS workflows (water, electric, gas, land permits, parks, etc.) are responsible for maintaining at least a minimum number of dynamic business layers. Yet much of the display uses relatively static reference layers (land-use/land-cover, road network, basemap data), which lend themselves to caching. (In many cases, static reference layers used to support standard operations are provided from a supplier and updated on a periodic basis, which means they can be optimum candidates for a file cache.)

Figure 8-23 shows a performance comparison between a light dynamic Web service (simple map display), medium dynamic Web service (medium complexity display), and a fully cached service. For dynamic map displays, the complexity of the display directly affects display performance. When the display is precached, display performance for light, medium, and highly complex display environments are all the same. Also, the cached data can be captured by a system cache between the data source and final application, and may be sourced from a local system cache much quicker than from the server. This makes the cached data source very attractive for implementations with high capacity and high performance needs.

Selecting the right technology: A case study

Figure 8-24 shows a hypothetical example of three configuration alternatives for deploying a simple national level application used for citizen residence declaration. The application provides a variety of address matching functions, landmarks, basemap data layers, and imagery

to help locate the citizen residence. Once the residence is located on the display, the citizen enters a simple point identification to identify his residence location on the map and completes a census (attributes) form. Citizen declaration for the entire country must be completed in three-months, so the architecture must be able to handle 2,000 concurrent users during any given hour (estimated peak hour concurrent user load).

Each of the three technology alternatives relies on a central geodatabase with application and database administration from a single central data center. Citizens access the system from 100 declaration center offices (DCO) located throughout the country. Systems must be connected so citizen declarations can be viewed from all locations, so the DCO locations are connected to the central data center over a WAN.

The three architectures demonstrate the adaptability of ArcGIS Server and the value of cached map services. The first solution supports the Web application from the central data center using a geodatabase data source. The second provides the same Web application from the central data center, using only the declaration point layer in the geodatabase; the static reference layers are precached. The third solution uses a Mobile ADF application to handle the citizen declaration, with the declaration layer supported by the geodatabase and the

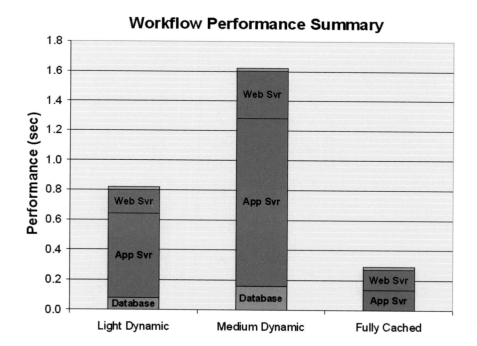

Figure 8-23

Caching brings performance advantages for Web services.

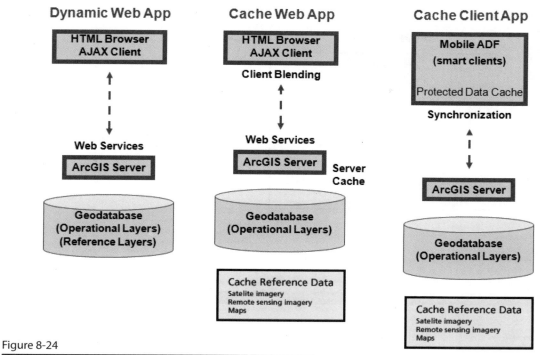

Figure 8-24

ArcGIS Server cache performance: Three configuration alternatives.

reference layers in a mobile cache. The declaration layer is synchronized with the central geodatabase with each citizen declaration.

Just looking at figure 8-24 does not tell you which of the three architectural solutions is the best choice. You must do a system architecture design analysis to identify which architecture best supports the business needs.

Dynamic Web application

Figure 8-25 (next page, top) provides the results of such an analysis for the dynamic Web application architecture. The dynamic Web application solution here requires 35 Web servers (8 Web application servers and 27 container machines) and three data servers. The three database servers could be supported as three regional child database instances with one additional corporate server as the parent. Other options include Oracle RAC or a larger single DBMS platform. In any case, this would be a very expensive hardware and software solution.

Peak network traffic requirements are also very high. The central data center would need to handle more

than 350 Mbps of traffic during peak periods, requiring a minimum of 700 Mbps bandwidth for good performance. The DCO site traffic, depending on the number of concurrent users, would range from less than 10 user sessions (1.75 Mbps) at the smaller sites to 100 user sessions (17.5 Mbps) at the larger sites. This is more traffic than most standard WAN bandwidths could bear. The hardware platform and network communication cost to support this solution? Very high.

Cache Web application

Figure 8-26 (next page, bottom) shows the system design analysis results for the central Web application alternative that uses precached reference layers. As we recall, cached layers are prerendered and maintained on the server storage network—layers that the system has at the ready to serve quickly.

With these "precooked" reference layers ready to serve at a moment's notice, the hardware server requirements diminish to 12 Web servers (5 Web application servers and 7 container machines) and a very light load on the data server (less than 25 percent utilization). For

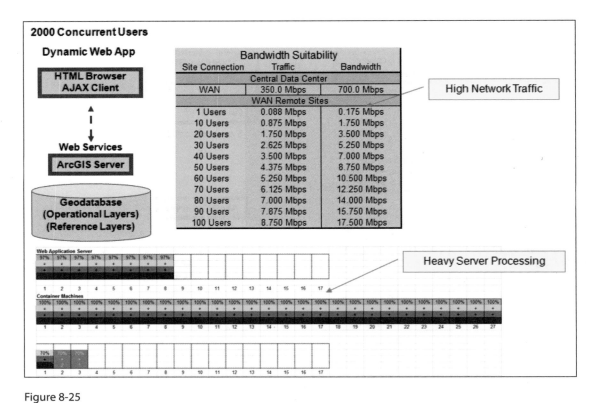

Figure 8-25

ArcGIS Server Web application with geodatabase: Results for
design analysis of dynamic architecture.

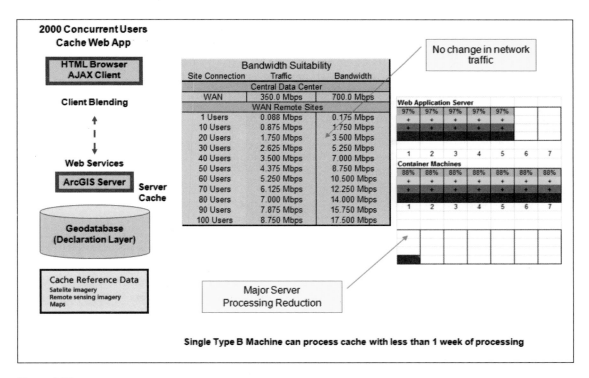

Figure 8-26

ArcGIS Server Web application with cache reference data: Results
for design analysis of centralized alternative.

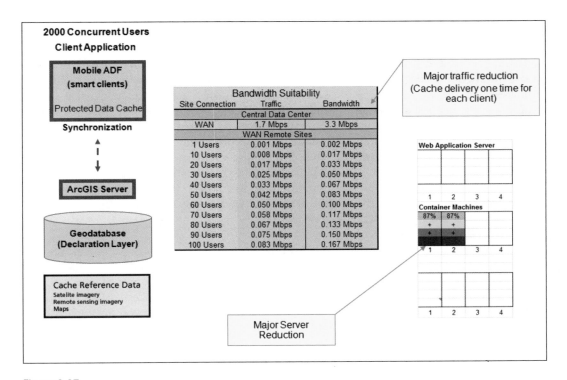

Figure 8-27

ArcGIS Server Mobile ADF cache: Results for design analysis of mobile solution.

the Web application with cache reference layer solution, this is about 33 percent of the hardware required for the previous dynamic Web application solution. Reference layers can be precached before the citizen declaration period because they will not change during the declaration project (no concerns about stale data). The Web applications of both configurations are still supported on the Web servers, with the declaration layer located in the geodatabase (declaration layer is a single point feature class used to identify citizen residence locations).

Network traffic is still high, since all data displays continue to be provided from the central Web application. The central data center must handle 350 Mbps of traffic during peak periods, requiring a minimum of 700 Mbps bandwidth for good performance. The DCO site traffic, depending on the number of concurrent users, will range from less than 10 user sessions (1.75 Mbps) at the smaller sites to 100 user sessions (17.5 Mbps) at the larger sites.

Because reference data is preprocessed and supported in an image cache, this second alternative significantly reduces the Web application map display processing time, thereby reducing the hardware platform and software licensing costs significantly. But the network traffic requirements remain high.

Cache client application

Figure 8-27 shows the system design analysis results for the ArcGIS Server Mobile ADF solution, the third and last alternative we began with in figure 8-25, and the one using the mobile cache data source. Noteworthy here is that the mobile application is a very light client; it can be deployed from a Web URL while the reference data can be located in a local client cache, and only the declaration layer must be generated as a service from the central Web site. In fact, the declaration layer data exchange with the server is done as a background process (in other words, client display performance is based only on local cached data, and that's very fast).

With the mobile solution, the reference data layers are precached and the declaration layer changes are synchronized across the network. A test of the caching timeline verifies that the complete reference layer cache can be generated within one week of processing. Mobile clients need to cache the reference data layers one time only (such as in the office before going into the field, or preinstalled at each site prior to the declaration period), in order to be able to access them wherever they may be from then on. For all client queries, these cached reference data files can be downloaded over the network, provided on a DVD or flash drive, or installed locally before the declaration period.

The Mobile ADF client application can be installed from a published URL, and the client notified to download and install any updates over the network. The Mobile ADF is a very light client; download is small and install is very quick. The declaration layer and schema can also be installed over the network, establishing background synchronization with the central geodatabase during the install.

During operations, the local mobile client application accesses the local reference data cache to identify the citizen residence location. Published Internet services could provide additional geocoding and points of interest locations to help in residence location. Once located, the citizen identifies the point location of the residence and submits it to the server. The declaration display is synchronized when declarations are submitted, providing clients a current view of all declarations. Coming from a local data cache, displays are very fast.

The Mobile ADF application alternative reduces hardware server requirements to two Web servers. A parent geodatabase server is required for the declaration layer and background synchronization processing. This amounts to merely 5 percent of the hardware required for the Web application with all its data in the geodatabase. For the mobile architecture, the reference layers are precached before the citizen declaration period and will not change during the project. All of the mobile client applications and local cache can be in place before the declaration period, ready for operations.

Network traffic is also very low. Peak traffic at the central data site is less than 2 Mbps. All synchronization communications are supported as background processes, and any traffic contention would not affect Mobile ADF user performance. If the networks were to go down, the client application could continue collecting citizen declarations. When network connection is reestablished, updates are automatically synchronized with the central geodatabase. All in all, the Mobile ADF architectural solution significantly reduces hardware cost and software licensing, and reduces network traffic loads and dependence on the network.

The Mobile ADF is only one example of the advantages of ArcGIS Server caching. ArcGIS Online (with ArcGIS Desktop and ArcGIS Engine) and ArcGlobe Services (ArcGIS Explorer) are other implementations of the same type of caching technology. ArcGIS Server caching does make a difference in performance and cost. Understanding when and where to use data cache is an important part of the planning process. Software functionality is important in general—it must support your operational needs. The available solutions are expanding, but with new technology the pros and cons

of your alternatives may still be under evaluation. The real answers come from customers that demonstrate how the technology can make a difference.

Caching is becoming more popular as the technology evolves, and promises to provide fast performance and quality display while retaining the integrity of the data source. As with fast food, we need to learn how to keep our prerendered data fresh and up to date when we do use caching. To decide when to cache, ask the following questions: when do we need to use dynamic map displays to support our business needs, and where can we take advantage of preprocessed map services?

Building the data cache

The promise of rapid user display performance and highly scalable Web services comes at the price of preparation. The precaching process is like producing a map atlas before it is published: you must generate in advance a complete set of source images for the reference display. The performance advantage is in generating each image *one time only*; after that, you share the image to build requested displays. Sharing the image is very fast (nothing has to be generated), and if accessed by a local GIS application, the image will be cached on the local machine. It can also be shared to network accelerators that maintain a shared-image cache for an entire organization. The potential benefits of cached reference layers—with current platform and network technology—are quite spectacular.

You make a cached map pyramid style, and the concept behind a cached map pyramid (figure 8-28) is simple: It starts with a high-resolution data source that can be used to build a cache map pyramid. The top layer provides a map extent generated at the lowest resolution (1:1,000,000). Each additional pyramid layer is made by representing every pixel by 4 pixels to increase

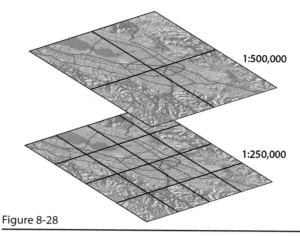

1:500,000

1:250,000

Figure 8-28

Cached map pyramid.

Data from ESRI Data & Maps, StreetMaps USA.

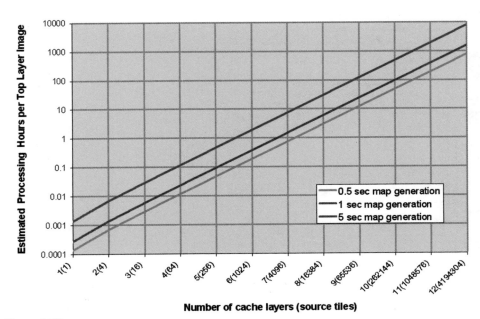

Figure 8-29

Map caching timelines.

the display resolution (1:500,000). The third layer resolution combines every 4 pixels again to provide the next layer resolution (1:250,000). The precaching process generates images for each of the map tiles and puts all the layers together into a compressed file cache. This can be lots of map images for ArcGIS Server to put together, and it can take some time.

The time it takes to create a map cache depends on the quality and number of maps that must be generated. The number of pyramid layers included in the map cache is a major factor in determining the generation time.

Figure 8-29 provides a rough estimate of the map cache time, based on the total number of cached tiles multiplied by an estimate of the time to produce an average tile. The chart includes three average map production timelines, each representing a different map complexity based on the average map service time. For caching 9 image extents with 9 layers at level 9, estimated caching time for a map with 5 seconds generation time would be about 900 hours.

Clearly, you don't want to pay that price of preparation more than once for the same map. So if you need approval for the look of a map you plan to cache, be sure to get it before you start caching. After the cartography is approved, the first step before actually building a map cache is to build a sample of the cache pyramid for a small area of extent. It's a short time to spend to gain a sample output that can be used as a test data source. This way, you can validate the output symbology and labeling, and then test the display performance with a prototype application. Also, you can use the sample output generation time to estimate the whole time it will take to complete the data cache. Once you get the sample right, you are ready to build the production map cache.

It is best to execute map cache jobs in sections so you can manage production time. It is also best practice to monitor large caching jobs so you are there if anything goes wrong. ArcGIS Server will automate the caching process, and manage it by using several batch process instances working in parallel to reduce the overall caching time. ArcGIS Server also integrates the final cache output into a single composite cached environment.

Figure 8-30 simulates part of what you'll see when using ArcGIS Server for caching. The cache generation runs as a batch process, and for the most part consumes one processor core. Concurrent cache generation processes can be run at the same time and ArcGIS Server will integrate the results of multiple cache generating batch services into a single cache pyramid structure.

When you have access to a multicore server, it is best to configure the cache service configuration to N+1 instances per host machine, where N is the number of host machine processor cores. The extra instance will ensure full use of the available processor core(s) when individual instances are waiting on data.

ArcGIS 9.3 significantly expands the options available for maintaining map and globe cache data sources. A data cache is simply preprocessing map images that are served on demand (high performance display and mini-

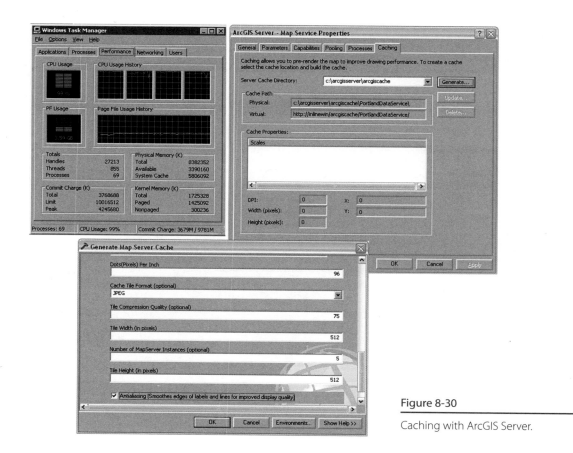

Figure 8-30

Caching with ArcGIS Server.

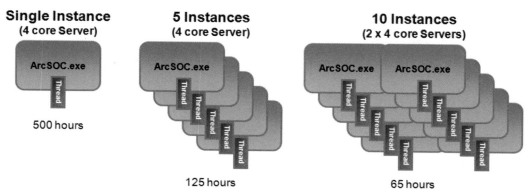

Figure 8-31

ArcGIS Server parallel caching options: Taking full advantage of
hardware resources saves time.

mum server processing—high scalability). Partial caching is possible, implemented by defining areas of high interest that will benefit from caching. Partial cache can be supported based on area of interest and level of resolution (more cached layers in high-interest areas). On-demand caching is also supported. Image is cached with first user access (cache image saved based on requested maps). The first map is dynamic; subsequent maps from the same location would be served from the cache.

Figure 8-31 shows examples of three different ArcGIS Server cache service configurations, to emphasize the point that you should take full advantage of available hardware resources when caching.

What cached service instance configuration should you use? That depends on the number of processor cores available on the host platform. If you configure one (1) instance on a 4-core server, your cache process takes advantage of only a single core. Cache time would be about 500 hours. If you configure N+1 or five cache service instances on the same 4-core platform, you can complete the same cache process in about 125 hours (4 times faster). If you have two 4-core servers, you can run 10 cache service instances and complete the cache process in about 65 hours.

Use ArcGIS Server to manage multiple cache service instances to reduce overall caching timelines. ArcGIS Server provides the tools to automate the caching process, with ArcGIS 9.3 supporting cache on demand—opening better ways to improve the currency and relevance of cached data sources.

This chapter covers several areas where software technology can make a big difference in system performance and scalability. The next chapter addresses hardware performance, and the importance of selecting and deploying the right hardware technology. Once we understand performance fundamentals, software performance, and hardware performance, we will be ready to put this all together with the Capacity Planning Tool. That tool provides a framework for us to actually engage in the system design process itself: to take what we understand about the technology, establish performance targets for each deployment milestone, and verify that these performance milestones are met at each stage of the implementation process.

9

Platform performance

Platform technology advances have accounted for almost all software performance gains since the early 1990s. All changes in standard performance models published for the ArcGIS software since 2000 can be attributed to improved hardware performance. Since we use these models for sizing purposes, understanding how to adjust them to account for these hardware performance differences is critical when addressing capacity planning, performance, and scalability needs.

As the graph on the next page shows, hardware got a lot faster in a very short span of time; this raised user productivity and performance expectations—and the expectation for more advancements. And sure enough, they came. The technology change from single-core to multiple-core platform environments (in 2006) expanded capacity by more than 100 percent, while reducing platform and software licensing costs. Commodity Intel platforms in 2007 were providing more than seven times the capacity of commodity platforms in 2004, doing so at a lower overall cost. Now it's a matter of course that enterprise application and data servers (among other things) must be upgraded to support increasing user productivity and the higher expectations that come with it.

This chapter takes a closer look at the kinds of resources offered by the current hardware technology; then through examples, applies these compute resources toward meeting software processing needs. This overview of current platform technology uses vendor-published benchmarks (SPEC compute-intensive throughput benchmarks) to compare available platform performance and capacity options. You will find platform service times, queue times, and response times identified for standard ESRI workflows along with a full set of platform sizing models. Models based on vendor-published benchmarks have served as tools in figuring out the platform requirements of ESRI customers since the mid-1990s. Today the platform sizing models work even better, for the simple reason that we've been able to make improvements in them because of all we've learned since the beginning. These models—and more important, the fundamentals underlying them—are incorporated in the Capacity Planning Tool described in chapter 10, practiced in chapter 11, and provided on the CD included with this book, for use in our own system design process.

In chapter 7, we reviewed the fundamental relationships among system components and processes, whose truths set the terms for capacity planning. In chapter 8, we looked at the issue of platform sizing in terms of the resources needed to process the commands of a selected software solution. Now we bring it all together for a complete picture of the principles underlying the Capacity Planning Tool. You can learn how to modify the CPT for your own use in the next chapter. But first, as promised in chapter 7, let's thoroughly review the vendor technology available for you to choose from, along with some simple engineering charts that lay out current hardware performance and scalability.

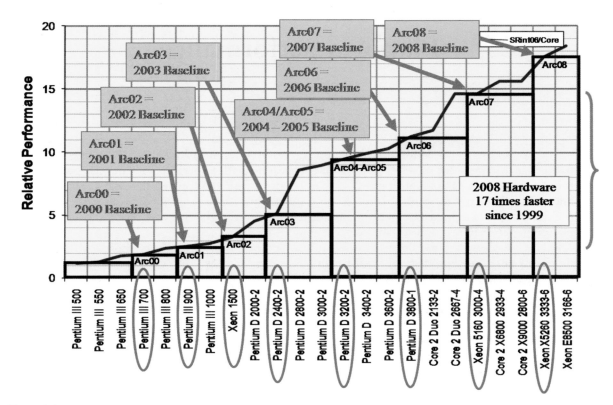

Figure 9-1

Change in relative platform performance over time, using PC Intel as the example.

The world we live in today is experiencing the benefits and challenges of rapidly changing technology. Technology advancements can directly improve our individual productivity. But we need to meet the challenges in order to take advantage of the benefits. To provide GIS users with a proper system design—one that will enable this productivity—you must seriously take into account user performance needs. It also helps to recognize that meeting these computing needs over the years has led to increased expectations. GIS users have always benefited from faster platform performance, and each year power users are looking for the best computer technology available to meet their processing needs. These user productivity requirements, which reflect user performance expectations, are summed up in the workstation platforms users select to support their computing needs.

System performance baselines

Over time, performance expectations change dramatically, as you can see in the list of workstation platforms GIS users selected during a 10-year period. These are the hardware desktop platforms selected by the majority of GIS users since the ARC/INFO 7.1.1 release in February 1997:

- ARC/INFO 7.1.1 (Feb. 1997)
  Pentium Pro 200 MHz, 64 MB Memory
- ARC/INFO 7.2.1 (April 1998)
  Pentium II 300 MHz, 128 MB Memory
- ArcInfo 8 (July 1999)
  Pentium III 500 MHz, 128 MB Memory
- ArcInfo 8.0.2 (May 2000)
  Pentium III 733 MHz, 256 MB Memory
- ArcInfo 8.1 (July 2001)
  Pentium III 900 MHz, 256 MB Memory
- ArcInfo 8.2 (July 2002)
  Intel Xeon MP 1500 MHz, 512 MB Memory
- ArcInfo 8.3 (July 2003)
  Intel Xeon 2400 MHz, 512 MB Memory
- ArcInfo 9.0 (May 2004)
  Intel Xeon 3200 MHz, 512 MB Memory
- ArcInfo 9.1 (June 2005)
  Intel Xeon 3200 MHz, 1 GB Memory
- ArcInfo 9.2 (August 2006)
  Intel Xeon 3800 MHz, 1 GB Memory
- ArcInfo 9.2 (June 2007)
  Intel Xeon 2 core (1 socket) 3000 (4MB L2) MHz, 2 GB Memory
- ArcInfo 9.3 (August 2008)
  Intel Xeon 2 core (1 chip) 3333 (6MB L2) MHz, 2GB Memory

In 1990, users were impressed with platforms that would generate a map display in less than a minute (more

than a week's effort if completed using earlier Mylar digitizing overlay techniques). Today users complain if they have to wait a couple of seconds for the computer to generate the same map display. The increase in user expectations—and productivity—has come despite the fact that, because of improved functionality, today's software programs are bigger and slower than they used to be. Faster platform performance has compensated for that, and combined with lower hardware costs, continues to improve the user experience. Changes in GIS user need for faster platform performance was not a software-driven phenomenon. This was a change in performance expectations brought about by faster and less-expensive workstation and server technology.

Figure 9-1 shows the radical change in relative platform performance since 1997, using Intel workstation performance as an example. Technology change introduced by hardware platform manufacturers represents the primary contribution to performance and capacity enhancements over the past 10 years. Note that relative performance at the left side of the figure are relative numbers; like the SPEC benchmarks, they are relative values—relative to the SPEC performance baseline reference platform. (More about SPEC benchmarks as guidelines for measuring performance comes later in this chapter.)

The boxes on the chart (figure 9-1) represent the performance baselines selected to support the ESRI sizing models throughout the years. These performance baselines have been reviewed and updated each year to keep pace with the rapidly changing hardware technology.

User productivity

Our understanding of hardware performance is backed up with real user experience. We've put what we've learned from users over the years into platform sizing models that have helped other users make the right hardware decisions. Modeling what users require to meet their needs is a task that inextricably links the two concepts, user productivity and platform sizing. In the context of system design, user productivity is a measure of the rate of doing work. By definition, a user is a person whose work involves using a computer. Therefore, the rate at which the technology involved can process work becomes a key part of the measure of user productivity. For example, in 1992 it could take up to 60 seconds for the UNIX workstation to generate a map display, and the GIS user would take up to a minute to review the map display (user think time) before submitting the next map display request. Both rates factor into user productivity, and therefore into predicting the platform performance necessary to enable that productivity. A few definitions are in order:

A *user workflow* is a computer software process flow that supports people in doing their work. The user workflow includes computer processing time and any process queue delays (this is the display response time discussed in chapter 7) and the user think time (when the computer waits for user input while the user looks at the display). The display cycle time is the sum of the response time and the user think time. User workflow productivity is the rate at which the person is doing work, represented in displays per minute (DPM/client).

A *batch process* is a workflow that executes without any user think times. Batch processes are often used to simulate user workflows, creating the same sequence of user displays without including the user think times. Batch processes are also used to do heavy processing jobs that do not require intermittent user input (geodatabase reconcile-and-post processing, automated replication services, automated map production, etc.).

Client/server sizing models used during the 1990s were developed based on our understanding, at the time, of the nature of the GIS processing load. A batch process can be used to evaluate server performance and capacity. Understanding the relationship between the GIS batch process loads and an equivalent concurrent

user workflow load (users/batch process) provided a foundation for the concurrent user platform sizing models. In showing how the equivalency between the number of users and one batch changed over the years, figure 9-2 traces a history of the concurrent-user client/server sizing model in terms of how we've defined a real user. (Note the arrow at the bottom: it's a reminder that the processing performance of the CPU has been increasing with each new platform release.)

We first used the concurrent-user sizing model in 1992, as part of our system design consulting services. We've learned so much of what we know through our efforts to meet users' needs, and most of what we understand about our technology is backed up by user experience. I remember one experience in particular, no doubt because the story of it has been repeated hundreds of times by GIS customers over the years, in the process of using the models to inform their hardware purchasing decisions. I participated in designing a system for Pennsylvania Power and Light (PP&L) in 1995, and recommended four UNIX application compute servers as adequate to handle peak concurrent user loads of up to 100 concurrent users. The design document showed that these servers would reach peak capacity at

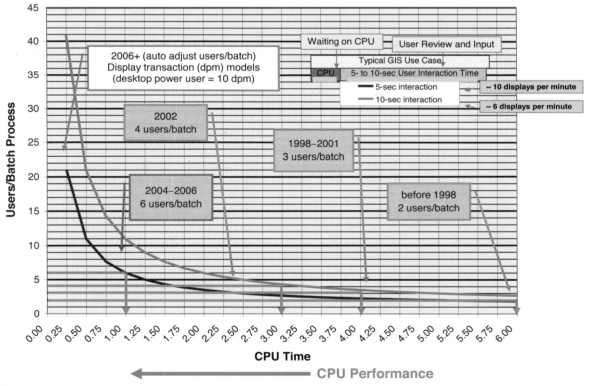

Figure 9-2

What is a real user? Early desktop sizing models were based on concurrent user work units expressed in terms of users per batch process. Current capacity planning models are based on user display work units expressed in terms of workflow service times. Workflow service time loads (DPM) are equal to peak concurrent users multiplied by user productivity (user DPM).

8 concurrent users, and user productivity would diminish to 33 percent of peak rate, as the number of concurrent clients per platform increased to 25 users. (The HP application server platforms at that time were three times faster than the performance available from UNIX workstations—the current performance baseline.)

A couple years later (in late 1997), I received a call from PP&L saying that their GIS users were complaining about a slowdown in display performance. I asked the database administrator (DBA) to share with me how many users were on the servers when they would start to slow down. He sent me a graph showing the server would start slowing down at 8 concurrent users, and at about 12 concurrent users he would start receiving performance complaints. This was documented in our earlier design recommendations based on our platform sizing models, and this discussion led to the customer purchasing new server hardware in 1998.

HP claimed their new server would easily support 6 to 10 concurrent users, suggesting that an 8 CPU platform could support up to 50 concurrent users before reaching full capacity. I conceded that the new machine might be able to support more than 2 concurrent users per CPU, although I was not comfortable with the HP claim. HP provided a new 8-way server for PP&L to test, and I was invited to attend a system-wide live performance test of the new platform. PP&L suspended operations for the test, and used the remote field crews to run live workflows to test the new server capacity. I sat with the DBA to watch the server CPU performance. The number of concurrent users was increased as we watched the CPU utilization climb to 10 percent, 20 percent, then higher. The CPU reached full capacity at about 27 concurrent users, which would be about 3.5 concurrent users per CPU.

Basically, the sizing models in our initial 1993 consulting effort predicted the results experienced in 1995. This also helped us realize that faster hardware can affect the accuracy of concurrent user (users/batch process) sizing models. So we've been making adjustments in them ever since, according to what we learn from users' real experience. The most useful model is the one that most closely represents reality. Most of what we understand about our technology is backed up by the reality of user experience.

As the platform CPU processing performance increased, we had to adjust the concurrent user/batch process sizing models. We adjusted our concurrent user platform sizing model from 2 to 3 users/batch processing in 1998 (the HP server was faster than most of their competitors). The model was increased to 4 users/batch process in 2002 and 6 users/batch process in 2004. The concurrent user models were tightly coupled to the

platform baseline CPU performance—updating the models with new performance baselines each year was very important to support accurate sizing recommendations.

We learned quite a bit about GIS workstation performance in the 1990s, and helped hundreds of customers with successful system design consulting services. Our measure of success was that customers could support their full implementation (often completed two to three years later) on the recommended hardware platform configuration. We had several return customers that used our services to support upgrades to new hardware.

We learned that a real user is different than a batch process. The difference is represented by user think time. You can simulate a real user workflow with a batch process by inserting pauses in the workflow to represent user think time. A lot of our current testing includes these pauses to provide a more accurate platform load profile. The number of concurrent users per batch process is the number of map service times (processing time for one map display) that can fit in the user display cycle time (total time between map requests). The user think time is the display cycle time minus the map service time. Figure 9-2 shows two lines, one for a 5-second user think time (10 displays per minute) and the other a 10-second user think time (6 displays per minute).

Web services were first introduced in 1996, causing the consultants among us to rethink our capacity planning models. It was very difficult for customers to identify the number of concurrent users that might access their public Web service site. Web performance monitoring tools would track the number of concurrent display requests per hour, and we were able to measure the processing time (map service time) for a published map service. Each platform CPU (core) could process a map service in parallel, so peak platform capacity could be calculated based on the measured map service time.

Figure 9-3 shows platform capacity based on map service times. The chart to the left shows capacity in peak displays per hour, and the chart to the right shows peak capacity in peak displays per minute. The peak capacity per core can be computed by dividing the hour or minute (time unit) by the map service time.

Measuring platform performance

Customers support GIS environments with a variety of hardware platforms, each with very different performance capabilities. I was helping City and County of Honolulu select the proper server to handle a new enterprise permit application back in 1995. They planned to

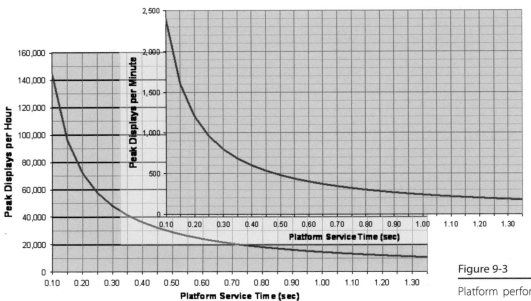

Figure 9-3

Platform performance capacity (4-core server platform).

use an expensive IBM server purchased in 1992 to support more than 200 concurrent users processing permits. I pointed out that my laptop was more powerful than their old IBM server—no way would they expect my laptop to handle 200 concurrent permit application users. The difference in hardware performance introduces some unique capacity planning challenges.

Many software vendors have had a difficult time supporting customer performance and scalability, and this continues to be a problem with vendors today. Platform selection based on previous customer experience—without adjusting for change in platform performance—is simply wrong. Technology implementation without understanding hardware capacity needs is a high-risk endeavor. Understanding how to identify and compensate for hardware performance differences is critical when addressing capacity planning, performance, and scalability needs.

During my first year at ESRI (1991), we were implementing a change in ARC/INFO desktop licensing, replacing a multiuser license based on platform capacity with a new FLEXlm license manager based on peak concurrent users. I conducted a series of ARC/INFO tests on Sun workstations to identify a relationship between server capacity and concurrent users. Figure 9-4 presents a summary of my findings. This simple model was used to migrate ESRI customers to the new license environment.

The relationship simply states that if one can determine the amount of work that can be handled by server A (clients supported by server A) and identify the relative performance between server A and server B, then one can identify the work that can be handled by server B. This relationship is true for single-core servers (servers with a single computer processing unit) and for multicore servers with the same number of cores. This relationship is also true when comparing the relative capacity (throughput) of server A and server B.

The next challenge was to identify a fair measure of relative platform performance. We were looking for a single number to represent how much work a computer can do relative to another computer. (This turned out to be the SPEC number explained on page 153.) Selection of an appropriate performance benchmark and agreement on how the testing will be accomplished and published are all very sensitive hardware vendor marketing issues. Fortunately, I was able to find a set of common performance benchmarks published by each of our hardware vendors that supported what I was looking for—platform benchmark specs.

The Standard Performance Evaluation Corporation (SPEC) is a consortium of hardware vendors established in the late 1980s for the purpose of setting guidelines for conducting and sharing relative platform performance measures. Their mission is "to develop technically credible and objective benchmarks so that both computer designers and purchasers can make decisions

Theory of Relative Performance

The relative performance of two servers is directly proportional to their compute capacity

$$\frac{\text{Performance of Server A}}{\text{Performance of Server B}} = \frac{\text{Clients of Server A}}{\text{Clients of Server B}}$$

Figure 9-4

How do we handle change?

on the basis of realistic workloads." ESRI's system design historically has used the following benchmark results: SPEC92 for 1992–1996; SPEC95 for 1996–2000; SPEC2000 for 2000–2006; SPEC2006 for the year 2006 and beyond (the SPEC performance benchmarks are published at www.specbench.org).

The SPEC benchmarks were updated in 1996 and 2000 to accommodate technology changes and improve metrics. The latest SPEC2006 release was published in 2006, providing the platform relative performance measures for the 2007 (Arc07) performance baseline. (There is normally a 6–12 month overlap in benchmark testing and published results once SPEC introduces the new benchmarks.) See box at right and figure 9-5 to see how we calculate the benchmark ratio.

This is how we measure performance. The SPEC compute-intensive benchmarks have been used by ESRI as a reference for relative platform capacity planning metrics since 1992 (the benchmarks conducted and published by the hardware vendors tell us how much work each specific platform can do relative to the SPEC baseline platform). The system architecture design platform sizing models used in conjunction with these relative performance metrics have informed ESRI customer capacity planning since that time. Hardware vendors conduct the tests, but it's up to us software vendors to make good use of the information.

2000–SPEC 2006 conversion
Common test results for more than 45 vendor platforms were published on the SPEC2000 and the SPEC2006 benchmark sites. I used these test results to calculate a transfer function between the two sets of benchmarks for capacity planning purposes. Older platform benchmarks would be published on the SPEC2000 baseline while newer platforms would only be available on the SPEC2006 baseline. Below, the ratio of the published benchmark results (SPECrate_2000/SPECrate_2006) are plotted on a graph to calculate a mean value for the transfer function. A published SPECrate_int2000 benchmark divided by 2.1 will estimate an equivalent SPECrate_int2000 benchmark value. A published SPECrate_int2006 benchmark multiplied by 2.1 will estimate an equivalent SPECrate_int2006 benchmark value. It is interesting to note that the older platform technology performs best on the SPEC2006 benchmarks (the relative SRint2000/SRint2006 throughput for the new Intel Xeon 51xx series platforms is over 2.2, while the relative throughput for the older Intel Pentium D platforms is under 2.0).

SPEC provides a separate set of integer and floating point benchmarks. CPUs can be designed to optimize support for integer or floating point calculations, and performance can be very different between these environments. ESRI software test results, since the ArcGIS technology releases closely follow the integer benchmarks, suggesting the ESRI ArcObjects software

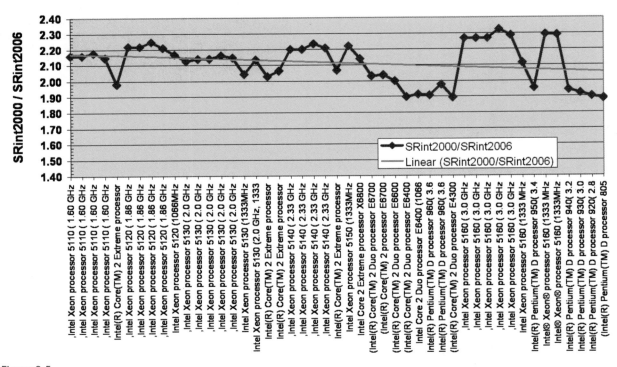

Figure 9-5

SPEC2000 to SPEC2006 benchmark ratio: The red line identifies an average transfer function for estimating SRint2006 equivalent benchmarks from older SRint2000 published values (SRint2006 = SRint2000 / 2.1).

predominantly uses integer calculations. The integer benchmarks should be used to estimate relative platform performance for ArcGIS software technology.

SPEC also provides two methods for conducting and publishing benchmark results. The SPECint2006 benchmarks measure execution time for a single benchmark instance and use this measure for calculating relative platform performance. The SPECint_rate2006 benchmarks include several concurrent benchmark instances (maximum platform throughput) and measure executable instance cycles over a 24-hour period. The SPECint_rate2006 benchmarks are used for relative platform capacity planning in the ESRI system architecture design sizing models.

There are two results published on the SPEC site for each benchmark, the conservative (baseline) and the aggressive (result) values. The conservative baseline values are generally published first by the vendors, and the aggressive values are published later following additional tuning efforts. Either published benchmark can be used to estimate relative server performance, although the conservative benchmarks would provide the most conservative relative performance estimate (removes tuning sensitivities).

Figure 9-6 provides an overview of the published SPEC2006 benchmark suites. The conservative SPECint_rate2006 benchmark baseline values are used in the ESRI system architecture design documentation as a vendor-published reference for platform performance and capacity planning.

SPEC2006 Comprises Two Suites of Benchmarks

CINT2006: Compute-Intensive Integer
- Twelve CPU-intensive integer benchmarks (C and C++ Languages
- Base (SPECint_base2006, SPECint_rate_base2006)
- Peak (SOECint2006, SPECint_rate2006)

CFP2006: Compute-Intensive Floating-Point
- Seventeen CPU_intensive floating-point Benchmarks (C++. FORTRAN, and a mixture of C and FORTRAN)
- Base (SPECfp_base2006, SPECfp_rate_base2006)
- Peak (SPECfp2006, SPECfp_rate2006)

Sun UltraSPARC-II 296 MHz Reference Platform

Figure 9-6

SPEC2006 platform relative performance benchmarks.

Impact of platform performance

With vendor technology changing rapidly over the past 10 years, improved hardware performance has enabled deployment of a broad range of powerful software and continues to improve user productivity. Most productivity increases experienced over this decade have been promoted by faster computer technology. But technology today is fast enough for most user workflows, so faster compute processing alone (faster hardware) is becoming less relevant as a factor in increasing user productivity. Most user displays are generated in less than a second. Access to Web services over great distances is almost as fast. Most of a user's workflow is think time—the time a user spends thinking about the display before requesting more information. The quality of information provided by the technology can make a user's think time more productive.

Most user productivity gains in the future are likely to come from more loosely coupled operations, disconnected processing, mobile operations, and more rapid access to and assimilation of distributed information sources. System processing capacity is very important, but system availability and scalability are most important.

Hardware processing encountered some technical barriers during 2004 and 2005, which slowed the performance gains between platform releases. There was little user productivity to gain by upgrading to the next platform release (which was not much faster), so as a result, computer sales were not growing at the pace experienced in previous years. Always striving toward more capacity at a lower price, hardware vendors' focus on mobile technologies, wireless operations, and more seamless access to information has also borne fruit. Competition for market share has been brutal, and computer manufacturers have tightened their belts and their payrolls to stay on top. As a result, 2006 brought some surprises: the growing popularity of the Advanced Micro Devices (AMD) technology and a focus on more capacity for less cost. Among the big surprises were new performance gains; Intel raised the bar with a full suite of new dual-core processors (double the capacity of the single-core sockets), offering significantly higher performance without the comparable rise in cost. Hardware vendor packaging (Blade Server technology) and a growing interest in virtual servers (abstracting the processing environment from the hardware) may prove to further reduce the cost of ownership and provide more processing capacity in less space.

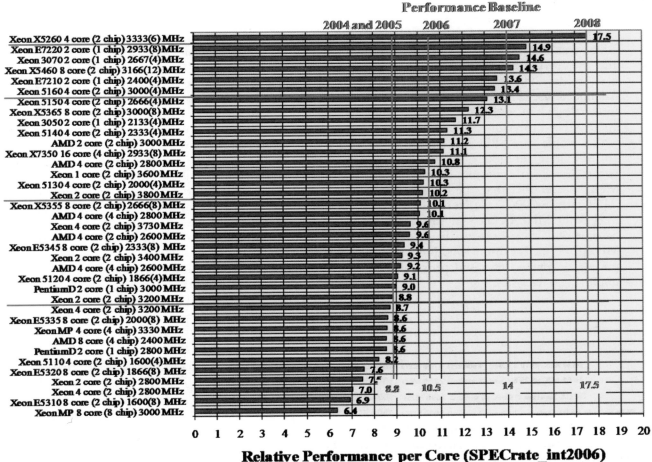

Figure 9-7

Platform performance makes a difference: Windows.

Figure 9-7 provides an overview of vendor-published, per-core benchmarks for hardware platforms supporting the Windows operating system. SPECrate_int2006 benchmarks are published for a range of server configurations. Basically, the numbers in the figure are relative performance values: Intel Xeon 4 core (2 socket) 3000-4 MHz platform with value of 13.4 is 1.52 times faster (13.4/8.8) than the Intel Xeon 2 core (2 socket) 3200 MHz platform with value of 8.8. (Keep in mind that the benchmark results are all relative numbers from the baseline platform. They do not have units and do not represent throughput values, they are only relative to each other.) A single GIS display is a sequential execu-

tion and display performance is a function of processor speed, which can be represented in terms of platform throughput per core. Platform throughput per core is calculated by dividing the published throughput benchmark value by the total core included in the platform configuration. The Intel Xeon 3200 MHz platform with 2 MB L2 cache (single-core SPECrate_int2000 = 18 / SPECrate_int2006 = 8.8) was released in 2004 and remained one of the highest-performing workstation platforms available through 2005. The SPECint_rate2000 benchmark result of 18 was used as the Arc04 and Arc05 performance baseline. The year 2005 was the first since 1992 that there was no noticeable platform performance change (most GIS operations were supported by slower platform technology).

Some noticeable performance gains came early in 2006 with the release of the Intel Xeon 3800 MHz and the AMD 2800 MHz single-core socket processors. In May 2006, we selected an Arc06 performance baseline of 22 (SPECrate_2006 = 10.5). Since May, Intel released its new Intel Xeon 4-core (2 socket) 3000 MHz processor, a dual-core socket processor with a single core SPECrate_int2000 benchmark of 30 (SPECrate_int2006 = 13.4) and operating much cooler (less electric consumption) than the earlier 3.8 MHz release. The Arc07 performance baseline of 14 (SPECrate_int2006 = 14) was selected based on the Intel 3000 MHz technology. The 2008 performance baseline of 17.5 was selected based on the Intel 3333 MHz platform. The future looks promising, as hardware pricing is coming down and performance per core is improving once again.

Figure 9-8 shows vendor-published, single-core benchmarks for hardware platforms supporting UNIX operating systems. The UNIX market has focused on large "scale-up" technology (expensive high-capacity server environments). These server platforms are designed for large database environments and critical enterprise business operations. UNIX platforms are traditionally more expensive than the Intel and AMD "commodity" servers, and the operating systems typically provide a more secure and stable compute platform.

IBM (PowerPC technology) is an impressive performance leader in the UNIX environment. Sun is also a powerful competitor with a significant hardware market share and many loyal customers, particularly in the GIS marketplace. (Note that, among UNIX platforms, ArcGIS Server is supported on Sun Solaris—SPARC platforms—only.) Many GIS customers continue to handle their critical enterprise geodatabase operations on UNIX platform environments.

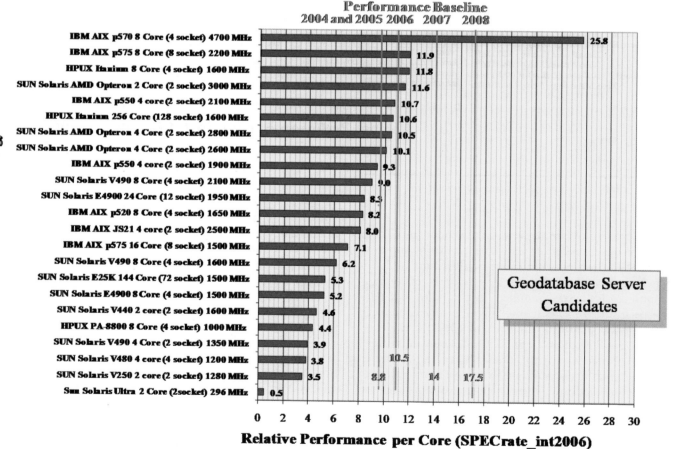

Note: ArcGIS Server supported on Sun Solaris (SPARC platforms) Only

Figure 9-8

Platform performance makes a difference: UNIX.

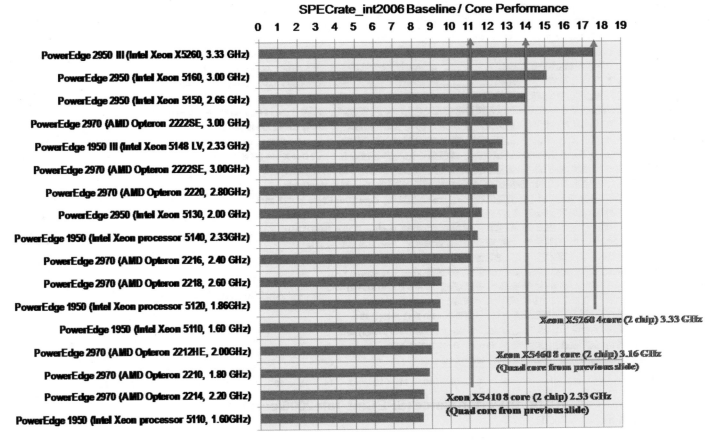

Figure 9-9

Identifying the right platform.

The efforts of hardware vendors to reduce cost and provide more purchase options make it especially important for customers to understand their performance needs and capacity requirements. In the past, new hardware included the latest processor technology, and customers would expect new purchases to increase user productivity and improve operations. With the technology we see today, many of the new duo core designs incorporate older CPU technology in a new chip configuration.

Figure 9-9 lists SPECrate_int2000 benchmarks for Intel platforms for sale by Dell in 2007. These same Intel platforms are available from HP, IBM, Sun, and several other hardware vendor sources. These platforms are normally sold without sharing the performance benchmark information. Typically, customers select the model number, processors, memory, and disk requirements without having the performance and capacity information available for their consideration.

There is more variation in the performance of today's platform environments than ever before. Platform performance affects platform capacity, software licensing,

and user productivity, yet it is an unknown factor until put to the test. Platform performance is not something that is obvious when buying today's hardware; you need to do your homework to identify what platforms will meet your capacity needs. Consider the following as an example of what even a little bit of research might tell you. This is an example of purchasing a Dell server from their Web site in September 2006. Selecting an application server would typically include these decisions:

The ideal configuration would include the right processor, memory, and hard drive. Configuring a Dell PowerEdge server with two Intel Xeon 5160 2 core (2 chip) 3000 MHz processors, 8GB 667 MHz memory, 64 bit standard windows operating system, and dual RAID 1 146 GB disk drives costs just under $9,000. This was the standard recommended application server platform (Windows Terminal Server, ArcGIS Server, ArcIMS, etc.) from mid-2006 through 2007).

Processors: The 5160 3000 MHz processors provide the best performance/core. Selecting the Intel Xeon 5050 3000 MHz processor reduces the overall cost by about $1,600. What is not shown when making the

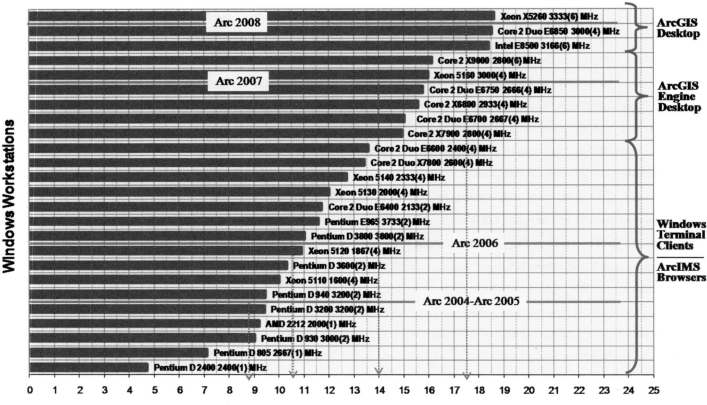

Figure 9-10

Workstation platform recommendations.

processor selection is the relative performance. Performance of the Intel Xeon 5150 processor is 47 percent of the Intel Xeon 5160 processors, so you would need to purchase two of the slower server configurations to match the premium solution— the second server would cost more than an extra $8,000 to support the same number of concurrent users. You can also save about $1,100 if you purchase only one 5160 processor, but again you have half the overall server capacity. (Another thing not shown, but is something to keep in mind: higher MHz means higher energy bills.)

Memory: Recommended memory is 2 GB per core for Web servers or 8 GB for a 4-core server configuration. Reducing to 4 GB memory saves $800, while increasing to 16 GB memory increases the price by almost $5,000. Knowing what memory you need can make a big difference in the bottom line.

Disk configuration: Reducing disk storage to RAID 1 73 GB disks would save about $150, and increasing disk storage to RAID 1 300 GB drives would cost an additional $1,300. The 146 GB drives are a good buy.

Knowing what you need—and what you don't need—is important.

ArcGIS Desktop platform selection

Selecting the right desktop workstation directly enhances user productivity, particularly if the desktop workstation is used to run GIS software. You can gauge the user productivity you can expect by comparing the relative performance of workstations, those currently for sale in the marketplace or possibly still available within your company inventory. ArcGIS Desktop power users look for the highest performance workstations. Standard office machines can be handled with desktop workstations that have less processing power.

Figure 9-10, an overview of supported ArcGIS workstation platform technology, charts the Intel platform performance changes experienced over the past eight years. The new Intel Xeon 3333 MHz dual-core processor is more than 30 times faster and has almost

60 times the capacity of the Pentium Pro 200 platform that supported ARC/INFO workstation users in 1997. The advance of GIS technology is enriched by the remarkable contributions provided by ESRI's hardware partners.

Recent release and support for Windows 64-bit operating systems (OS) have introduced some performance enhancement opportunities for ArcGIS Desktop workstation environments. The increasing size of the OS and increasing number of concurrent executables supporting GIS functions translate into more memory requirements. The Windows 64-bit OS supports more memory and provides improved memory access, which is an advantage for ArcGIS Desktop users. Recommended physical memory for an ArcGIS Desktop workstation with an ArcSDE data source is 1 GB, and up to 2 to 4 GB may be required to support large image and/or file-based data sources.

Most GIS users are quite comfortable with the performance provided by current Windows desktop technology. Big performance improvements for power users and heavier GIS user workflows come with the new Intel dual-core technology (5260 processors with 6 MB L2 cache). Dual-core technology is becoming the standard for new desktop workstations and laptops. Although a single process will see little performance gain in a dual-core processor environment, significant user productivity gains will come by enabling concurrent processing of multiple executables. For instance, many processes can be running in the background when you are working on your computer (e-mail, online backups, security scans, print or plot spooling, etc.). GIS power users often have geoprocessing running in one window while working on another mapping project in another—having an extra core processor to handle these background processes can significantly improve productivity.

Server platform sizing models

The platform sizing tools we have developed and maintained over the years to help ESRI customers with system architecture design planning and proper selection of vendor hardware come in various formats. Developed from the same models introduced in chapter 7, these tools can be used in an engineering style format, for help in platform selection. The engineering presentation format is very useful in showing the relative performance and capacity differences between available hardware choices. This is how we represented our sizing models before the Capacity Planning Tool, and it is still a useful tool for many.

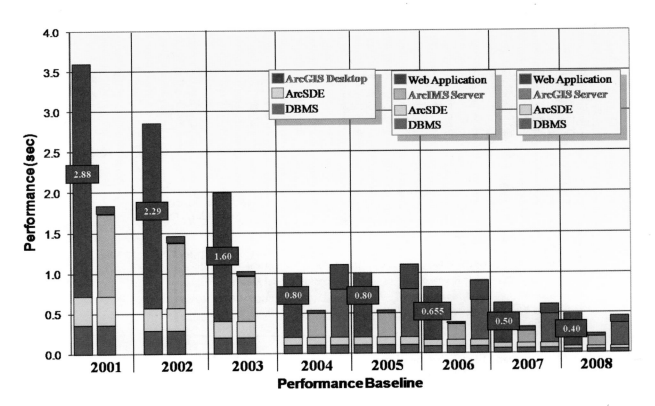

Figure 9-11

Software baseline performance summary.

Once you choose your architecture strategy, you can use a chart for sizing purposes based on the configuration decision you just made.

Figure 9-11 shows the change in ArcGIS Desktop, ArcIMS, and ArcGIS Server performance when using the different technologies to generate the same map display. The results of these display times were used over the years to establish a performance baseline for determining hardware specifications.

Software performance has changed with each release, although primary performance differences showing in the chart above were due to hardware performance gains. Software performance gains were made as the code matured, and performance losses came along with additional functionality and quality improvements, as we've discussed. Performance was better in some quarters than others, usually improving over time to stay within performance expectations.

It's worth repeating: All of the adjustments to the baseline-performance model for each software technology (ArcGIS Desktop, ArcIMS, ArcGIS Server) since 2001 represent changes in hardware performance. Differences in software performance over time were marginal, and in most cases could be ignored. Hardware performance increased over time, so by 2008 the hardware was over seven times faster than it was in 2001, which accounts for most of the performance change.

The platform sizing charts (engineering charts) are good tools with which to identify hardware capacity. In them, you will see standard workflows used to represent the performance capabilities of available vendor hardware. (These are the software performance specifications for standard GIS user workflows supported by the current ESRI software technology and they will be described in terms of the Capacity Planning Tool in the next chapter.) The engineering charts that follow are based on the standard ESRI workflow service times, and you can use them to identify performance capacities of the vendor platforms.

Windows Terminal Server platform sizing

Windows Terminal Server supports centralized deployment of ArcGIS Desktop applications for use by remote terminal clients. The same software configuration options introduced in chapter 4 appear here in figure 9-12 for easy reference. You can use the platform sizing chart (figure 9-13) for selecting the right hardware platform technology based on your peak workflow

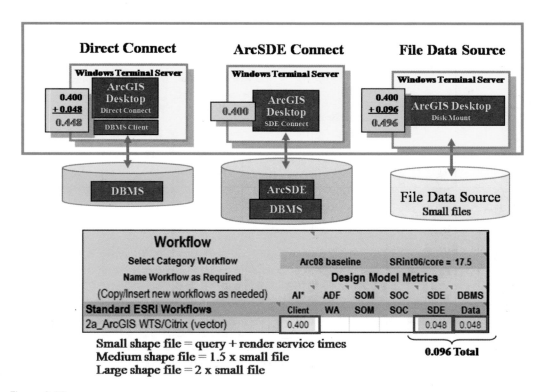

Figure 9-12

Windows Terminal Server architecture options: Note that all service time units are in seconds.

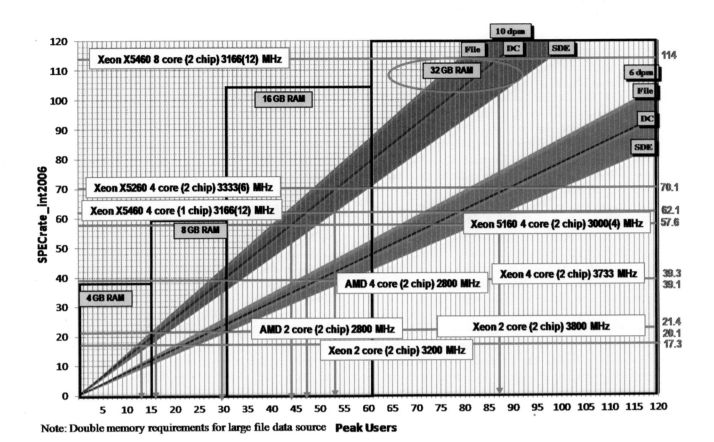

Figure 9-13

Windows Terminal Server platform sizing chart.

needs. The platform solution will depend on how the system is configured. For each configuration, the top box in figure 9-12 provides platform services times for standard ESRI workflows; the bottom box, software service times. Figure 9-12 also summarizes the baseline software service times for the standard ESRI workflow sizing model. Location of the software on the different platform configurations can be used to identify the specific baseline platform processing service times (platform service time = sum of the software service times supported by the assigned platform).

For the rest of this chapter, you can use the standard platform sizing chart introduced in figure 9-13 as a tool to identify the number of peak concurrent users that can be supported with a selected vendor platform configuration. Hardware platforms are represented on these sizing charts as a horizontal line. The location of the platform on the chart is determined by the vendor-published SPECrate_int2006 benchmark results for the represented platform configuration.

The standard ESRI workflow models are used to generate these charts. These are the models we referenced for proper hardware selection earlier in chapter 7.

The CPT integrates the models used to generate these charts into a comprehensive system design; this integration was done manually in our system architecture design process before we had the CPT. Insofar as they provide a different view, these charts are still useful for comparing the various platform candidates. For managers making a single platform decision, these charts may provide all they need.

The peak displays per minute represent user productivity, and are used with the Arc08 component service times to identify the specifications required to support each specific GIS workflow. The platform performance specifications are represented by the vendor-published SPECrate_int2006 benchmark on the vertical axis of the sizing chart.

The two diagonal fans on the sizing chart stand for two different user productivity rates (10 displays per minute in blue and 6 displays per minute in green). A standard ArcGIS Desktop power user (consistent with user workflows in the Arc05 sizing models) would be the equivalent of 10 displays per minute, while a lighter Web user could be represented by 6 displays per minute. The concept of user productivity is a new parameter introduced with the Arc06 sizing models, and as user workflows require more complex processing and rich map displays, user productivity may decrease accordingly. (For example, a display generated in 1 second may be OK for a workflow with 10 displays per minute, while a richer display generated in 2 seconds could reduce user productivity to 5 displays per minute.) All the more important then to investigate what a user really needs to see—their information display requirements—so that you create applications that provide simple map displays when simple will do. Such limited display requirements foster optimum use of available system resources.

Each fan on the chart in figure 9-13 includes three lines, each of them representing one of the three Windows Terminal Server configuration options (file data source, geodatabase Direct Connect, and ArcSDE Connect). These three configurations appear in figure 9-12. We recommend using an ArcSDE direct connect architecture when accessing a DBMS data source.

When using the platform sizing charts provided in this section, follow the selected platform line from the left margin until it intersects the diagonal line representing your workflow configuration. Then, drop a vertical line to the bottom of the chart to identify

the peak number of users supported for that workflow, according to that platform configuration (engineers like these kind of charts). The recommended solutions are included on the chart for easy reference.

Windows Terminal Server with geodatabase direct connect

If you know how many peak concurrent users your organization needs to support, you can use the platform sizing chart (figure 9-13) to identify server platforms that are up to the task. In the Windows Terminal Server with geodatabase direct connect architecture, the standard Xeon X5260 4-core (2-chip) 3333(6) MHz server platform (2008 performance baseline) can support up to 53 concurrent ArcGIS Desktop users. This is a significant gain over the 44 concurrent users supported by the Xeon 5160 4-core (2-chip) 3000 MHz platforms that represented the 2007 technology baseline. The new Xeon X5460 8-core (2-chip) 3166(12) MHz platforms can support up to 87 users, which will require upgrade to 32 GB memory. The additional memory cost may suggest that a better purchase decision might be two of the faster Xeon X5260 4-core (2-chip) 3333(6) MHz platforms; they will support 106 concurrent users.

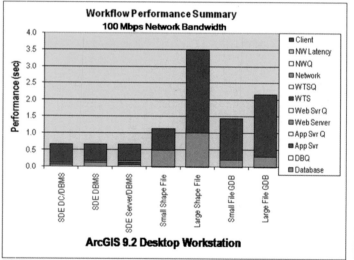

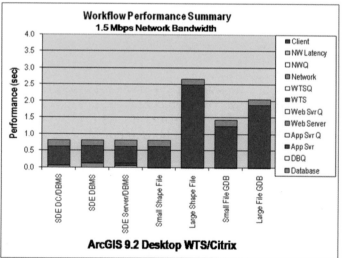

Figure 9-14

Performance profiles of ArcGIS Desktop standard ESRI workflows.

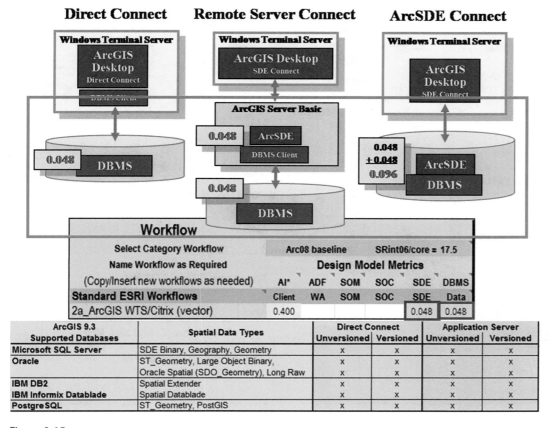

Figure 9-15

GIS geodatabase architecture alternatives.

ArcGIS 9.3 Supported Databases	Spatial Data Types	Direct Connect		Application Server	
		Unversioned	Versioned	Unversioned	Versioned
Microsoft SQL Server	SDE Binary, Geography, Geometry	x	x	x	x
Oracle	ST_Geometry, Large Object Binary,	x	x	x	x
	Oracle Spatial (SDO_Geometry), Long Raw	x	x	x	x
IBM DB2	Spatial Extender	x	x	x	x
IBM Informix Datablade	Spatial Datablade	x	x	x	x
PostgreSQL	ST_Geometry, PostGIS	x	x	x	x

Performance profiles of the standard ESRI ArcGIS Desktop workflows included in the Capacity Planning Tool are charted in figure 9-14. This chart shows the variety of workflows supported by the capacity planning models, and the difference in service times generated by each configuration. The same models are used to support the ArcGIS 9.3 release. More on how to use these models (including definitions of shapefile sizes) in chapter 10.

GIS data server platform sizing

Figure 9-15 identifies recommended software configuration options for the geodatabase server platforms. The geodatabase transaction models apply to both ArcGIS Desktop and Web mapping service transactions. Normally a geodatabase is deployed on a single database server node, and larger capacity servers are required to support scale-up user requirements. ArcGIS Server 9.2 introduced distributed geodatabase replication services that can be used to replicate the contents of a geodatabase to another geodatabase easily. ESRI also provides functional support for Oracle and IBM DB2 distrib-

uted database environments. (ESRI has not validated the performance and scalability of vendor multimode DBMS clusters.) The ArcGIS 9.3 release adds support for PostgreSQL and one-way replication to the file geodatabase.

Several database data types support the geodatabase. The geodatabase can be supported by each of the 11 different database environments listed in the table above. For each software release, performance validation testing is completed for the whole set of supported data types. The standard ESRI workflow service times provide target performance metrics appropriate for platform sizing when using the full range of supported data types. A current list of supported geodatabase environments is published on the ESRI Web site at http://support.esri.com.

You can employ the standard ESRI workflow service times to support implementation with all data types, with the exception of Oracle Spatial (SDO_Geometry). For use in the model, the standard ESRI workflow ArcSDE and DBMS service times should be doubled to more realistically represent workflow loads handled by Oracle Spatial (SDO Geometry) data types.

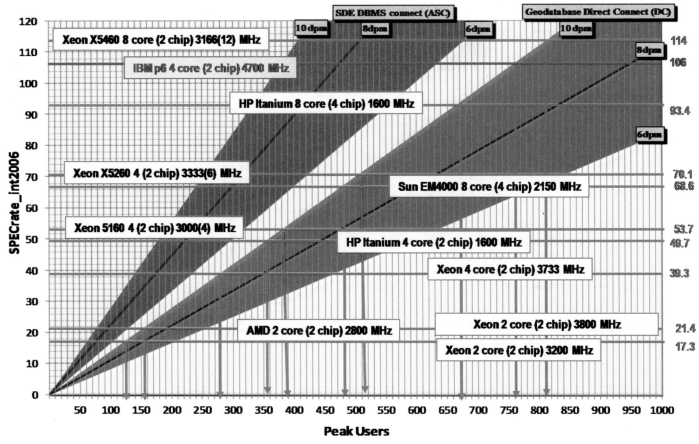

Figure 9-16

Geodatabase server platform sizing (up to 8 core capacity).

Geodatabase server platform sizing

Figure 9-16 provides a chart for geodatabase server platform sizing for standard systems. The chart shows two fans, one used for platform selection when supporting ArcSDE on the DBMS platform, and the other when configured using a geodatabase direct-connect architecture. The geodatabase direct-connect fan can also be used for remote ArcSDE server platform selection (ArcSDE and DBMS loads are the same for these configurations). Each fan on the chart includes three lines (10 dpm, 8 dpm, and 6 dpm) representing user productivity. We recommend use of the 10 dpm productivity line for conservative platform selection. The chart includes a range of current vendor platforms, from 2-core through 8-core capacity, supporting less than 1,000 concurrent desktop users. Additional vendor platform configurations can be included on the chart, represented by a horizontal line defined by the vendor published SPECrate_int2006 benchmark baseline.

The standard Intel Xeon 2 core (2 chip) 3800 MHz platform—about the same performance as the AMD 2 core (2 chip) 2800 MHz server platform—can support up to 150 concurrent power users. The Intel 4-core (2-chip) 3000 MHz platform can handle up to 380 concurrent users. The Sun EM 4000 8-core (4-chip) 2150 MHz platform can support up to 460 concurrent power users. The Xeon X5260 4-core (2-chip) 3333(6) MHz server platform can support up to 520 concurrent users. The Xeon X5460 8-core (4-chip) 3166 MHz platform can support up to 810 concurrent users, while the IBM p6 4-core (2-chip) 4700 MHz platform can handle up to 760 concurrent power users. (Note that the remote ArcSDE server is not required when using an ArcGIS Desktop or Web service direct connect architecture.)

ESRI recommends use of the ArcSDE direct-connect architecture (presented in the previous charts) to reduce database platform loads, improve system capacity, and reduce software license costs. Performance with the ArcSDE direct-connect architecture is comparable to performance with the ArcSDE DBMS (ArcSDE service installed on the DBMS) architecture.

These platform sizing charts demonstrate the value of supporting the geodatabase with an ArcSDE direct-connect architecture. The direct-connect architecture reduces database hardware and software costs by up to 50 percent, with minimum increase in client hardware or software pricing, and in many situations (when client workstation is faster than the central data server) performance and user productivity is improved.

File data server

Figure 9-17 provides an overview of the file server platform sizing model. (This same model was presented in section 7, and file data servers were introduced in chapter 3. Note that ArcGIS Desktop service time varies with the size of the GIS data files.) Configured using standard operating system NFS or CIFS communication protocols, file servers provide a computer connection to remote shared files over the network. File servers are not compute-intensive in themselves, but application access to files across the network can introduce lots of network traffic. (The assumption in the model is that 1 display can happen every 6 seconds, and that the average network traffic per map display is 50 Mb.) This network traffic must pass through the file server network interface card (NIC)—the server NIC bandwidth will reach capacity before other server components. File access is also a very chatty protocol, with lots of sequential communication between the query process executed by the client application and the server network connection agent. For this reason, file access is not recommended over high-latency network connections.

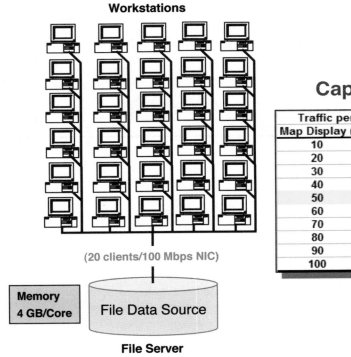

Workstations

(20 clients/100 Mbps NIC)

Memory
4 GB/Core

File Data Source

File Server

GIS File Servers

Capacity Planning Matrix

Traffic per Map Display (Mb)	File Server NIC Bandwidth (Mbps)			
	10	100	200	1000
10	6	60	120	600
20	3	30	60	300
30	2	20	40	200
40	2	15	30	150
50	1	12	24	120
60	1	10	20	100
70	1	9	17	86
80	1	8	15	75
90	1	7	13	67
100	1	6	12	60

(Peak Concurrent Clients)

Figure 9-17

GIS file server platform sizing.

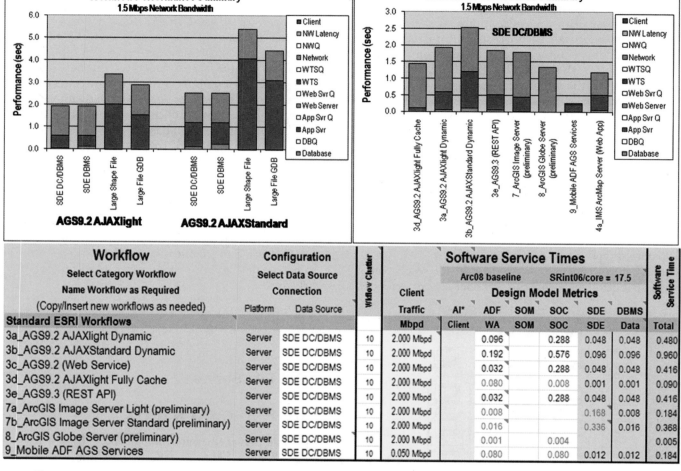

The following table accompanies the figure:

Workflow		Configuration		Wfllow Chatter	Software Service Times							Software Service Time
Select Category Workflow		**Select Data Source**			Arc08 baseline		SRint06/core = 17.5					
Name Workflow as Required		Connection			Client	**Design Model Metrics**						
(Copy/Insert new workflows as needed)		Platform	Data Source		Traffic	AI*	ADF	SOM	SOC	SDE	DBMS	
Standard ESRI Workflows					Mbpd	Client	WA	SOM	SOC	SDE	Data	Total
3a_AGS9.2 AJAXlight Dynamic		Server	SDE DC/DBMS	10	2.000 Mbpd		0.096		0.288	0.048	0.048	0.480
3b_AGS9.2 AJAXStandard Dynamic		Server	SDE DC/DBMS	10	2.000 Mbpd		0.192		0.576	0.096	0.096	0.960
3c_AGS9.2 (Web Service)		Server	SDE DC/DBMS	10	2.000 Mbpd		0.032		0.288	0.048	0.048	0.416
3d_AGS9.2 AJAXlight Fully Cache		Server	SDE DC/DBMS	10	2.000 Mbpd		0.080		0.008	0.001	0.001	0.090
3e_AGS9.3 (REST API)		Server	SDE DC/DBMS	10	2.000 Mbpd		0.032		0.288	0.048	0.048	0.416
7a_ArcGIS Image Server Light (preliminary)		Server	SDE DC/DBMS	10	2.000 Mbpd		0.008		0.168	0.008		0.184
7b_ArcGIS Image Server Standard (preliminary)		Server	SDE DC/DBMS	10	2.000 Mbpd		0.016		0.336	0.016		0.368
8_ArcGIS Globe Server (preliminary)		Server	SDE DC/DBMS	10	2.000 Mbpd		0.001		0.004			0.005
9_Mobile ADF AGS Services		Server	SDE DC/DBMS	10	0.050 Mbpd		0.080		0.080	0.012	0.012	0.184

Figure 9-18

Performance profiles of ArcGIS Server standard ESRI workflows.

Web mapping server platform sizing

Web mapping services are provided by ArcIMS and ArcGIS Server software technology. The ArcIMS image service and ArcIMS ArcMap Image Service are provided by the ArcIMS software, and the ArcGIS Server map services are provided by the ArcGIS Server software. All three Web mapping technologies can be deployed in a mixed software environment (they can be installed on the same server platform together). All three mapping services can be configured to access a file data source or a separate ArcSDE geodatabase. Geodatabase access can be through direct connect or an ArcSDE server connection.

Performance profiles for the standard ESRI ArcGIS Server workflows included in the Capacity Planning Tool are provided in figure 9-18. The chart shows the variety of workflows you can use in the capacity planning models, and the difference in service times generated by each configuration.

Web two-tier architecture

For smaller two-tier Web mapping deployments, see the recommended software configuration options and standard ESRI workflow service times in figure 9-19. This two-tier architecture supports the Web server and spatial servers on the same platform tier, and is recommended for implementations that need only one- or two-server platforms. For example, a single Xeon X5260 4-core 3333(6) MHz platform can support peak trans-

action rates of up to 34,000 map displays per hour. A high-available (2-clustered platform) solution can support peak transaction rates of up to 68,000 map displays per hour. In our experience, for most customers this is enough capacity to satisfy peak user requirements.

Figure 9-20 is a two-tier capacity planning chart for two levels of ArcGIS Server deployments. This sizing chart identifies peak display transaction rates that can be supported on the Web server platform. For Web

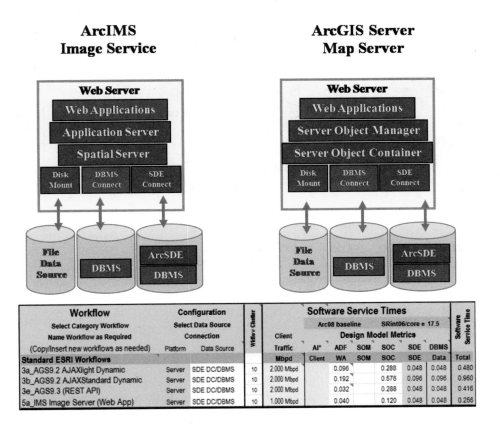

Figure 9-19

ArcGIS Server two-tier architecture.

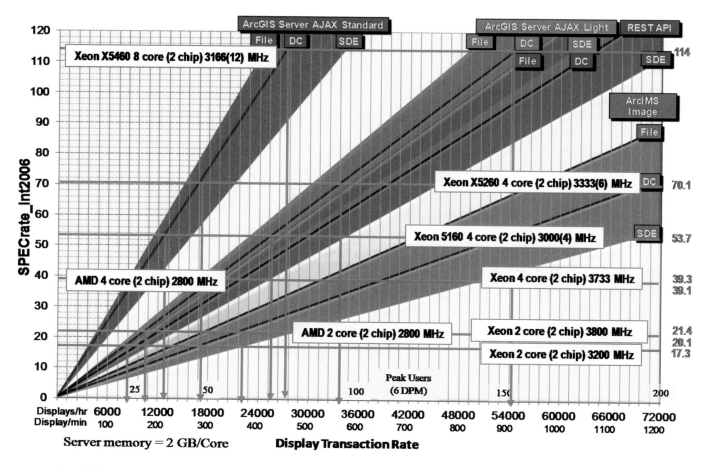

Figure 9-20

ArcGIS Server platform sizing: Web two-tier architecture.

services, peak users are those likely to call for six displays per minute. ArcGIS Server AJAX standard and ArcGIS Server AJAX light workflows are shown on the same chart along with the ArcGIS Server REST API. The ArcGIS Server AJAX light workflow is based on simple map services, similar to those provided with legacy ArcIMS implementations. The ArcGIS Server AJAX standard workflow represents more complex data products enabled by the ArcGIS Server technology. (For the chart, platform memory recommendations are 2GB/core.)

Three data configurations are shown on each fan (File, DC, ArcSDE); the direct-connect architecture is recommended for most implementations. Peak transaction

rates (displays per hour and displays per minute) are provided on the x-axis. Peak concurrent users (based on six displays per minute per user) are also shown, for reference purposes. The older scripted ArcIMS Image Service software can provide roughly twice the transaction capacity of the ArcGIS Server AJAX light software (multiply peak displays per hour on x-axis by two).

The Xeon X5260 4-core (2-chip) 3333(6) MHz server platform can support up to 34,000 ArcGIS Server map displays per hour. This is 12,000 more peak map transactions per hour than one of the most powerful application server platforms sold just two years earlier, the Intel Xeon 3.8 GHz servers supporting ArcIMS image service.

Web three-tier architecture

Map Server (container machine) sizing: Figure 9-21 shows the recommended software configuration options and standard ESRI workflow service times for larger, three-tier Web mapping deployments. This configuration option locates the Web server and spatial servers on separate platform tiers, and is recommended for implementations with a large number of concurrent Web mapping clients. Server platforms can be added to support even larger peak capacity requirements.

Figure 9-22 is a three-tier capacity planning chart for ArcGIS Server AJAX light and standard deployments. This sizing chart identifies peak display rates for the map server/container machine platforms. Again, a peak user is one who is likely to call for as many as six displays per minute, and platform memory recommendations are 2GB/core.

An Xeon X5260 4-core (2-chip) 3333(6) MHz server platform can support up to 43,000 ArcGIS Server map displays per hour. Entry-level ArcGIS Server AJAX light

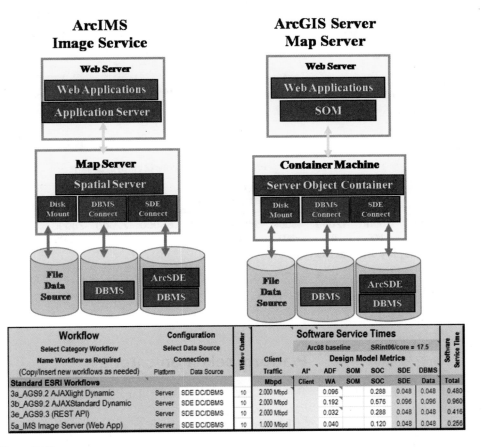

Workflow		Configuration			Workflow Chatter	Software Service Times							Software Service Time
Select Category Workflow		Select Data Source				Arc08 baseline		SRint06/core = 17.5					
Name Workflow as Required		Connection			Client	Design Model Metrics							
(Copy/Insert new workflows as needed)		Platform	Data Source		Traffic	AI*	ADF	SOM	SOC	SDE	DBMS		
Standard ESRI Workflows					Mbpd	Client	WA	SOM	SOC	SDE	Data		Total
3a_AGS9.2 AJAXlight Dynamic		Server	SDE DC/DBMS	10	2.000 Mbpd		0.096		0.288	0.048	0.048		0.480
3b_AGS9.2 AJAXStandard Dynamic		Server	SDE DC/DBMS	10	2.000 Mbpd		0.192		0.576	0.096	0.096		0.960
3e_AGS9.3 (REST API)		Server	SDE DC/DBMS	10	2.000 Mbpd		0.032		0.288	0.048	0.048		0.416
5a_IMS Image Server (Web App)		Server	SDE DC/DBMS	10	1.000 Mbpd		0.040		0.120	0.048	0.048		0.256

Figure 9-21

ArcGIS Server three-tier architecture: Larger three-tier implementations.

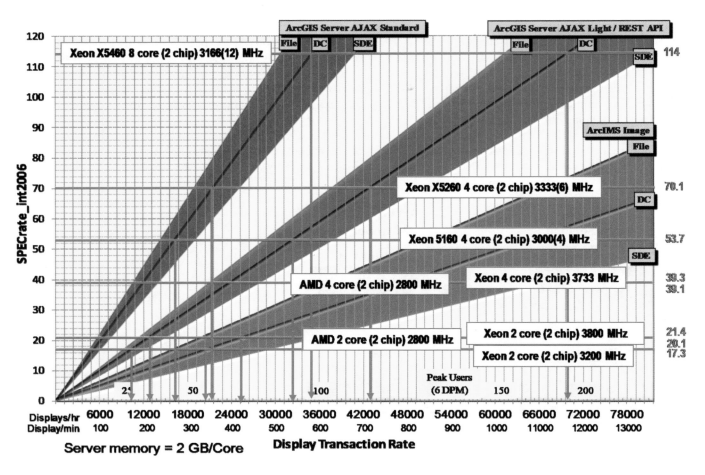

Figure 9-22

ArcGIS Server platform sizing: Web three-tier architecture.

with the Xeon X5260 4-core (2-chip) 3333(6) MHz platform supports 17,000 more peak map transactions per hour than ArcIMS image service supported by Intel Xeon 3.8 GHz servers just two years ago.

Web server platform sizing: Figure 9-22 can be used to identify platform capacity for the Web Server tier. Standard .NET and JAVA Web application servers using the AJAX ArcGIS Server ADF components typically require about a third of the processing capacity as the container machines identified above. You can use the container machine SDE connection line to identify platform capacity, then multiply by a factor of three to identify capacity for the Web server. Web server sizing for the REST API is about 10 percent of the container machine processing capacity.

Platform selection criteria

These are the factors to consider in selecting the proper hardware:

Platform performance: Platform must be configured properly to support user performance requirements. Identifying proper platform configurations based on user performance needs and the ESRI design models establishes a solid foundation for proper hardware platform selection.

Purchase price: The cost of hardware varies depending on the vendor and platform configuration. Pricing should be based on the evaluation of hardware platforms with equal performance capacity.

System supportability: Customers must evaluate system supportability based on vendor claims and previous experience with vendor technology.

Vendor relationships: Relationships with the hardware vendor may be an important consideration when confronted with complex system deployments.

Total life cycle costs: The total cost of the system may depend on many factors, including existing administration of similar hardware environments, hardware reliability, and maintainability. Assess these factors based on previous experience with the vendor technology and your evaluation of vendor total cost-of-ownership claims.

The primary purpose of the capacity planning process is to understand what hardware you need to promote productive GIS operations. *Platform performance is the key to ensuring a successful GIS deployment.* Hardware vendors cannot guess what your platform performance needs are—you must calculate what's necessary to satisfy your peak processing needs and be able to tell the hardware vendor exactly the platform configuration you are looking for.

The hardware vendor can tell you the purchase price, how reliable and supportable the platform is, and some estimates on total life cycle cost. It may be necessary to review several design options—all the ones that support your performance needs—before making a final purchase decision.

Establishing specific hardware performance targets during the process of hardware source selection significantly improves your chances of getting it right. Proper system architecture design and the right hardware choice build the foundation for successful system deployment. The next chapter describes the Capacity Planning Tool in detail, explaining how it functions and how to use it.

Part III

Putting it all together

10 | Capacity planning

OVERVIEW

In January 2006, using a Microsoft Office Excel spreadsheet, I began building a simple model of the fundamental relationships that determine the performance and scalability of a system for GIS. The spreadsheet was quite capable of dealing with simple data integration, user display, and the required calculations. The challenge was developing an easy-to-use interface that customers could adapt; a versatile user interface that would allow system architects to configure their environment, enter their user requirements, select their favorite vendor platforms, and see the results.

This chapter describes the Capacity Planning Tool (CPT) developed from this effort: how it works and how you can use it in planning your enterprise GIS. The system architecture design methodology in this book has already led to thousands of successful GIS deployments by ESRI customers over the last two decades—this same methodology is implemented in the two basic CPT modules. With the CPT, you can

- define your own workflow performance targets (or use standard ESRI workflow models) to represent your specific GIS operational environment;
- identify your peak concurrent user needs using the Requirements Analysis Module templates in the CPT;
- define your available network bandwidth;
- select your favorite hardware platform technology using the Platform Selection Module.

The Capacity Planning Tool will complete the system design analysis by translating user workflow needs to specific network and platform capacity requirements. In other words, all of the design analysis is completed by the tool as you identify what you want out of the system.

By allowing you to try out various platform solutions beforehand, the Capacity Planning Tool can help you make the right technology decisions before spending your money. The CPT is not just a faster way to calculate—it is a tool that incorporates experience gained from more than 15 years of system design consulting and ESRI customer support. The CPT is an adaptive configuration tool that can be used with standard ESRI workflows, or, if you prefer, you can define your own performance targets.

In an effort to make this tool useful for a broad range of users, we provide an open interface that you can modify to represent your own situation. This means that all of the formulas used to create the tool are not locked in the spreadsheet, which will help if you want to introduce your own ideas into the tool. It is possible to lock the spreadsheets, but this would limit adaptability of the user workflows and site configurations. The one cautionary note, however, is this: be careful not to change the formulas that make the model work. The primary user entry cells—the cells where you enter requirements representing your user needs—are identified in white. Most of these cells are drop-down menus that provide a selection set for the cell. You can use separate workbook tabs or save multiple workbook documents in a project folder to document your capacity planning analysis. The support for separate CPT2006 lookup tabs or workbook documents (hardware and workflows) makes the CPT more adaptable and improves application control and stability.

Though the CPT may look complex, you'll find using it quite simple once you get the hang of it. Learning to use the tool is much like learning to ride a bike; it seems difficult at first, but very soon you are on your way.

For a review of the fundamentals on which the CPT is modeled, you can refer to chapter 7; it reflects the thinking behind how such a tool can be made to model the real world. When combining throughput and service time performance metrics with platform utilization, we have a very simple relationship between what we see in the computer room (platform utilization) and the peak workflow the computer systems support in the workplace. This simple relationship makes for a more obvious correlation between what happens in the real world and what is being modeled. Performance models based on throughput and display service times represent what we see in the real world, providing a more accurate and adaptive model than one simply based on concurrent user peak loads.

Many customers are concerned about user productivity and display response times. Throughput, system utilization, and service times answer capacity planning and budget questions, but there is more to consider when talking about user response times. Understanding queuing theory fundamentals, and applying random arrival times across the components of the system, provides the rest of the story. Map display response time is the sum of all service times plus queue times (wait times) across the system. This important relationship between display service times and response times was discussed in chapter 7 as part of the performance fundamentals.

This and most of what we have learned about performance and scalability can be explained by standard operations research theory. For many years, I have shared these simple relationships in the "System Design Strategies" technical reference document, updating what we came to understand about the technology and making this information available on the Web. The challenge has always been to find a comprehensive yet simple way to document and share what we know, making virtual the required-system-loads analysis and platform sizing calculations; a focused way of offering customers what they need to understand to make the right technology decisions. The result is the Capacity Planning Tool that follows.

This chapter introduces you to the Microsoft Office Excel functions you need to use in working with the Capacity Planning Tool (CPT). Chapter 11 will then show you how to use the CPT to complete your own system design process. You may, like many other people throughout the world, already use Excel documents in your work—you will find the functions you need to configure the CPT similar to (or even the same as) what you would use in many other areas of work. These functions are defined on page 189 and include the following:

- Select row
- Copy row
- Insert row
- Delete rows
- Update summation formula range
- Move or copy sheet
- Save document

All of the rest of the analysis is included in the CPT. Your job is to identify the workflows that apply to your specific environment, identify the peak user loads for these workflows, and make your platform selection. The CPT will calculate the resulting performance and capacity information, providing you with a final system design based on your selections.

Capacity Planning Tool user interface

Let's get started by understanding what we are looking at when viewing the CPT. The user interface is designed as an information management product, providing

managers with a comprehensive overview of their enterprise workflow requirements, the selected platform solution, and a computed workflow performance summary. When you're finished with the system design process, this set of displays is what you can show upper management. To produce it, you have entered your user requirements and selected the vendor platforms; the CPT has completed the rest of the analysis and displayed the solution.

The final view for each deployment phase is shown on page 182. One would normally break down this view by module, and show each module on a separate slide when presenting an executive summary. In describing how to use the CPT, we will do something similar, breaking down the CPT into its modules and highlighting specific parts, in order to allow the captions to fleshout the visual story. The two modules include Requirements Analysis (10-2) and Platform Selection (10-4). You will see these modules in the City of Rome presentation in chapter 11. The City of Rome is a fictional case study we will use for our system design process in chapter 11, so let's use it here to better understand how to view the management information that will make the CPT useful.

GIS users are located in city hall, the operations facility, and five remote regional offices. There are also Web services published for public user access. The user locations are connected to the central data center through the LAN backbone, the WAN, and Internet communications, respectively, as shown in figure 10-1. The computer platforms are located in the central IT department computer facility located at city hall.

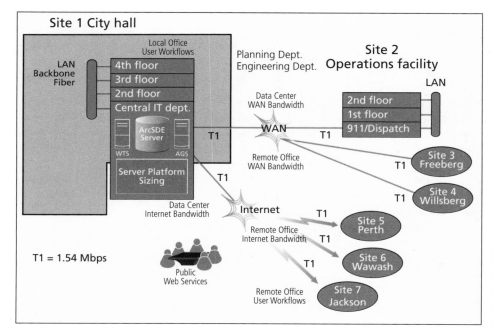

Figure 10-1

City of Rome user locations. The computer platforms are located in the central IT department computer facility at city hall.

10-2 City of Rome user requirements analysis

An example of the management information view of the user (or workflow) Requirements Analysis Module of the CPT is provided in this chart. It shows the projected peak user workflows (approved by the department managers) planned for the year 2 deployment.

	A	B		D	E	F	G	H	I
3	Labels	**Requirements Analysis**					WEB TPH =>	WEB Users	Network
4		**Types of Workflows**			User Requirements		13,000	36	Bandwidth
5						Peak Workflow		Network	Mbps
6		Workflow		Data Source	Users	DPM/Client	DPM	Mbps	DPH
7	LAN –	Local Clients		LAN Clients = 64		3 sec	LAN =52.5 Mbps		1,000.0
8	1.0.1	R_Server DBMS Batch Process	SDE DC/DBMS	1	60.94	61	0.001	3,656
9	1.0.2	R_ArcGIS Desktop Editor	SDE DC/DBMS	19	10	190	15.833	11,400
10	1.0.3	R_ArcGIS Desktop Viewer	SDE DC/DBMS	44	10	440	36.667	26,400
11	WAN –	WAN Clients		WAN Clients = 102			WAN = 4.8 Mbps		6.000
12	Site 1 –	Site 1 - Operations		Site Clients = 32			Traffic =1.5 Mbps		1.500
13	2.1.4	R_Remote WTS ArcGIS Desktop Viewer	SDE DC/DBMS	32	10	320	1.493	19,200
14	Site 2 –	Site 2 - Freeberg		Site Clients = 30			Traffic =1.4 Mbps		3.000
15	2.2.5	R_Remote WTS ArcGIS Desktop Viewer	SDE DC/DBMS	30	10	300	1.400	18,000
16	Site 3 –	Site 3 - Willsberg		Site Clients = 40			Traffic =1.9 Mbps		3.000
17	2.3.6	R_Remote WTS ArcGIS Desktop Viewer	SDE DC/DBMS	40	10	400	1.867	24,000
18	Internet	Internet Clients		Internet Clients = 98			Internet =10.1 Mbps		6.000
19	Public	Public		Site Clients = 36			Traffic =7.2 Mbps		256.000
20	3.4.7	R_AGS9.3 Map Server	SDE DC/DBMS	36	6	216	7.200	12,960
21	Site 4 –	Site 4 - Perth		Site Clients = 2			Traffic =0.1 Mbps		1.500
22	3.5.8	R_Remote WTS ArcGIS Desktop Viewer	SDE DC/DBMS	2	10	20	0.093	1,200
23	Site 5 –	Site 5 - Wawash		Site Clients = 40			Traffic =1.9 Mbps		1.500
24	3.6.9	R_Remote WTS ArcGIS Desktop Viewer	SDE DC/DBMS	40	10	400	1.867	24,000
25	Site 6 –	Site 6 - Jackson		Site Clients = 20			Traffic =0.9 Mbps		1.500
26	3.7.10	R_Remote WTS ArcGIS Desktop Viewer	SDE DC/DBMS	20	10	200	0.933	12,000
27									1.500
28		Total Workload			264		1,886	67.354	

The three city communication network environments (LAN–**Row 7**, WAN–**row 11**, and Internet–**row 18**) are shown in gray. The remote user sites (Operations–**row 12**, Freeberg–**row 14**, Willsberg–**row 16**, Perth–**row 21**, Wawash–**row 23**, and Jackson–**row 25**) are represented by green rows, each located within its respective city network environment. On the far right in **column I**, you see the network communication bandwidth for each user site or location (gray rows for the data center bandwidth connections and green rows for the remote site connections). An additional green public location is identified in **row 19** for public Internet users; note that their network bandwidth is set well above 50 percent of the total Internet traffic, to account for the many different user connections represented. You could add a separate public user site and its appropriate network connection if you have a need to evaluate public user response times.

The number of peak users (**column E**) is identified for each of the user workflows named under a site or location in **column B**. Also listed for each workflow is the desired user productivity (**column F**), measured in displays per minute (DPM) per client. (Workflow display service times used by the CPT analysis are de-

fined in advance on a separate workbook tab; the CPT Workflow tab is discussed beginning on page 210.)

"Client summation" is the term I'm using for a formula that accounts for the cumulative effect of site clients and network users. Located in **column D** is the summation formula—**cell D7** formula is SUM(E7:E11)—for each user site location and network; the client summation totals are centered over **column D and E** for display purposes. The summation formula adds the peak users identified for all the client workflows included in each network or site location. These summation ranges must be adjusted to include all site workflows when inserting new site locations. (How to adjust these user summation ranges is explained on page 189.)

After the client workflows are selected and the peak users and displays per minute/client (measures of user productivity) are entered, the CPT formulas calculate the total DPM per workflow (**column G**) and the network traffic (megabits per second) per workflow (**column H**).

Each user site location includes a traffic summation, which is a summation formula located in **column G** for each network and site location (traffic summation total—in Mbps—is centered on column G and H for display purposes). The summation formula adds up the traffic generated by all the client workflows within each network or site location. Again, when inserting new site locations, you must adjust these summation ranges to include all site workflows (how to do it is discussed on page 189).

Once all of the traffic summation ranges are set, the CPT productivity and traffic cells will change color to highlight any performance problems or infrastructure constraints identified during the CPT analysis.

10-3 Workflow Performance Summary—sample

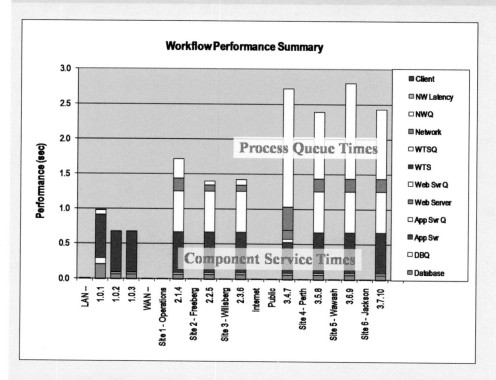

The CPT2006 tab includes a Workflow Performance Summary located to the right of the requirements analysis module (see top right of figure 10-5). The Workflow Performance Summary provides a graphic display of component service times, queue times, and response times for each client workflow—these values are calculated based on the selected workflow, identified peak users, and user productivity values. (Note: CPT2006, above, refers to the latest SPECrate_int2006 benchmark values used to represent server performance, introduced by SPEC in 2006.)

The Workflow Performance Summary is a tool used to support platform and performance tuning decisions during the design process. The graphic shows the platform service times and queue times for each workflow based on your selected platform configuration. Service times are shown in different colors for each platform tier (network is shown in gray), with any associated queue time shown in white directly above each hardware component. The horizontal axis identifies each workflow by network (LAN, WAN, Internet) and by remote site (Site 1, Site 2, etc.) along with a workflow label. The three numbers represented in the workflow label identify the network (1: LAN, 2: WAN, 3: Internet), the site location (1, 2, 3, etc.) and the workflow number (1, 2, 3, etc.). Thus, 2.2.5 represents workflow 5 located on the WAN at remote site 2.

10-4   Platform Selection Module

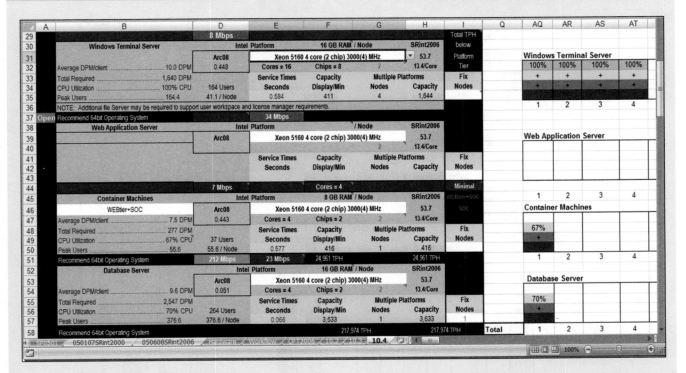

The CPT2006 Platform Selection Module represents the central data center server platform configuration as a series of server tiers, similar to the way hardware platforms might be represented in a data center server configuration drawing. You see it in the figure above: a CPT overview of performance based on the selected platform configurations. Platform selections are made from drop-down lists identified in the white blocks (**column E:G**) for each platform tier. These platform options represent existing or planned server configurations. The CPT calculates the required number of servers and the server performance metrics (CPU utilization, throughput, service time, system capacity, and number of server nodes).

The resulting Platform Configuration Summary at the right (Windows Terminal Server cells **AQ32:AT36**, Web application server cells **AQ40:AT43**, container machines **AQ47:AT50**, and database server cells **AQ54:AT57**) displays the number of server nodes and the CPU utilization when supporting the peak users identified in the Requirements Analysis Module. (Peak users for each platform tier are also identified in **column B**). Platform performance specifications used by the CPT are supported on a separate workflow tab discussed on page 210.

10-5 Capacity Planning Tool user interface—sample

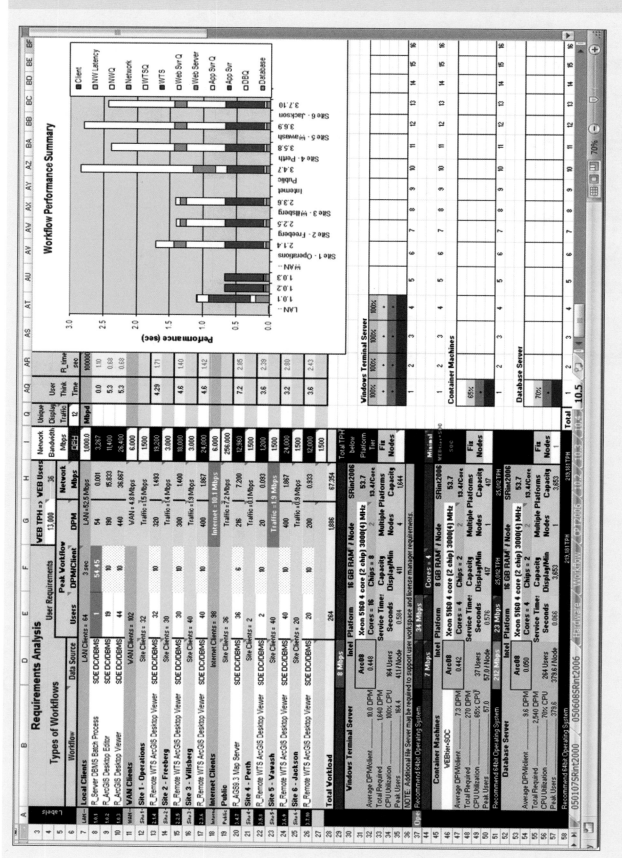

10-5   Capacity Planning Tool user interface—sample

This is the CPT information management view you will use when building your system solution. The Capacity Planning Tool management view includes several modules configured on a Microsoft Excel worksheet. The primary modules include the Requirements Analysis Module (**rows 5 through 28**) on the top half of the spreadsheet and the Platform Selection Module (**rows 29 through 58**) on the bottom half. An information product is provided with each of these modules (**columns AQ through BF**). A Workflow Performance Summary is provided with the Requirements Module and a Platform Configuration Summary is provided with the Platform Selection Module.

The Requirements Analysis Module (**rows 5 through 28**) includes user workflows and data source, peak users and productivity, network traffic, response time, and batch displays per minute. Workflow performance is displayed in the Workflow Performance Summary (**columns AQ through BF**). This module is located on the top half of the CPT2006 management information view.

The Platform Configuration Module (**rows 29 through 58**) provides drop-down menus for selecting the vendor platform (white cells on platform tiers in **columns E:G**) and a switch (**white cell** in **B46** of the container machine) to select two- or three-tier Web configurations. The recommended platform configurations (based on capacity requirements) are displayed on the lower right in the Platform Configuration Summary (**columns AQ** through **BF**).

Each of these modules is discussed in more detail as we continue through the chapter.

At first glance, what I'm calling the Capacity Planning Tool is simply a spreadsheet, automated by the computer. But in using it, you bring it to life as a problem solver, as a planner, and as a learning experience. In the CPT, performance sizing models, developed over 17 years of consulting, connect the requirements analysis with the performance and platform sizing results. (You may find various uses for the CPT other than platform sizing and network sizing. Its potential in being useful to you will be determined by how closely you can make it model the reality of your own situation.) As a manager, consultant, and teacher of system architecture design for GIS, my primary objective is to provide managers with an information product that shares our understanding of performance and scalability of ESRI software—too many customers make the same mistakes over and over again because they don't understand the technology. We have integrated the CPT into our consulting service offerings, and provide the same technology to our business partners and distributors to help them be more proactive in supporting our customers. As a teacher, I have integrated the CPT as an integral part of our "System Architecture Design Strategies" training course, so that students may learn the fundamentals that can make them successful and use the CPT as a framework to expand their understanding of the technology.

The CPT models (the background functions that make it work) are laid out in the performance fundamentals discussed in chapter 7. Understanding the performance factors and their associated relationships (represented in the CPT models, figure 10-6, next page) provides insight on how to identify and solve real life operational performance issues. The payoff begins when customers using the CPT are able to understand and resolve their own performance issues.

Development of this tool is an evolutionary process, providing a framework in which to represent what we understand about our technology. The CPT offers a place to record what we know, model what we expect to see as we deploy the technology in a real environment, and continue to learn more about what we don't know. The CPT is an adaptive model based on fundamental principals that contribute to performance and scalability of enterprise GIS operations.

The following figure (10-6), as well as its accompanying description, exposes some of the models on which the CPT is based. It will give you an idea of how the CPT can be used to define many different GIS environments, and can be modified to represent your business needs. After that, we will take some time to understand each of the CPT modules, and learn how to configure the CPT to represent your workflow needs.

10-6   Capacity Planning Tool model representation

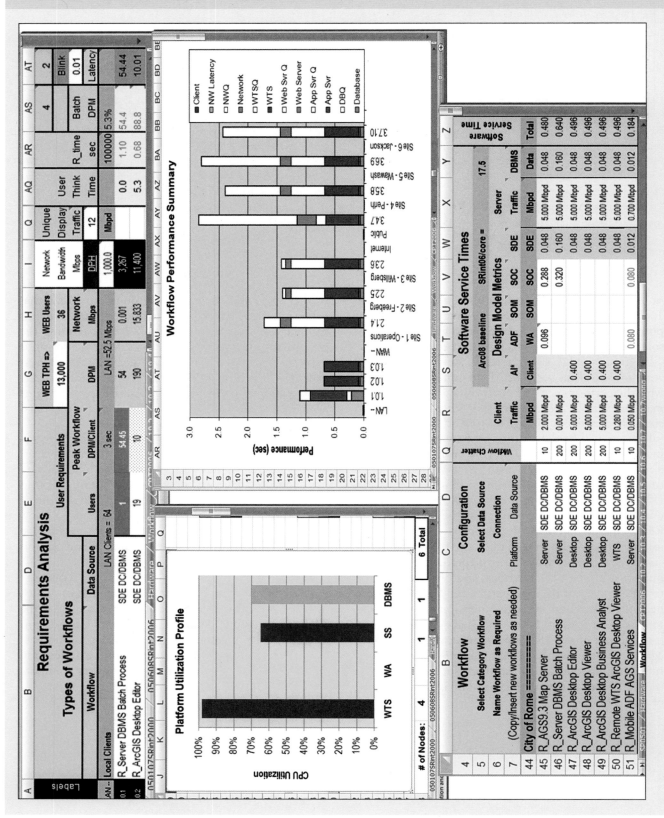

10-6 Capacity Planning Tool model representation

The models in the figure at left represent fundamental relationships that the CPT displays for you to work with. User requirements are represented by workflows (the bottom chart shows software service times established for the City of Rome case study we will discuss in chapter 11). Peak users and user productivity identify how these workflows will be used during peak work hours (the top chart shows peak ArcGIS Deskop Editor workflow loads of 19 concurrent users at 10 displays per minute—supporting a peak transaction rate of 190 displays per minute). Workflow loads are defined in terms of software component service times, and these service times are translated to platform service times based on the selected configuration strategy (displayed in the Workflow Performance Summary). System utilization during peak loads is displayed on the Platform Utilization Profile.

Requirements Analysis Module

Figure 10-7 on the following page offers a closer look at the requirements module, which represents workflows associated with a specific data server environment. Separate spreadsheet tabs can be configured for each central database server environment. Based on your workflow requirements analysis, you will select workflows from drop-down menus (column B for each workflow). The workflows are on a lookup list located by way of a separate Workflow tab.

Once you have configured your requirements analysis inputs, the CPT will update a network suitability analysis. This is updated on-the-fly, while you are entering user requirements (based on the existing platform selected) in the Platform Selection Module, which we will discuss later. As entries are made, the CPT displays several calculations: Response time shows the result of all workflow service times plus queue times. Queue times are calculated based on platform and network utilization levels. (Chapter 7 discusses queue time models; they are here in the Capacity Planning Tool to help you identify potential performance problems and translate platform service times to user response time.) Network traffic flow within the data center (server platform network interface card bandwidth requirements) will be addressed later in the Platform Selection Module. Network traffic flow within the data facility is currently not represented in the model.

10-7 Requirements Analysis Module

	Workflow	Data Source	Users	Peak Workflow DPM/Client	DPM	Network Mbps	Network Bandwidth Mbps DPH	Unique Display Traffic 12	User Think Time	R_time sec	Batch DPM	Blink 0.01 Latency
LAN --	Local Clients	LAN Clients = 64		3 sec	LAN =52.5 Mbps		1,000.0	Mbpd			100000	5.3%
1.0.1	R_Server DMS Batch Process	SDE DC/DBMS	1	60.94	61	0.001	3,656		0.0	0.98	60.9	60.95
1.0.2	R_ArcGIS Desktop Editor	SDE DC/DBMS	19	10	190	15.833	11,400		5.3	0.68	88.7	10.01
1.0.3	R_ArcGIS Desktop Viewer	SDE DC/DBMS	44	10	440	36.667	26,400		5.3	0.68	88.7	10.01
WAN	WAN Clients	WAN Clients = 102			WAN = 4.8 Mbps		6.000					0.79
Site 1	Site 1 - Operations	Site Clients = 32			Traffic =1.5 Mbps		1.500					99.6%
2.1.4	R_Remote WTS ArcGIS Desktop Viewer	SDE DC/DBMS	32	10	320	1.493	19,200		4.28	1.72	35.0	10.01
Site 2	Site 2 - Freeberg	Site Clients = 30			Traffic =1.4 Mbps		3.000					46.7%
2.2.5	R_Remote WTS ArcGIS Desktop Viewer	SDE DC/DBMS	30	10	300	1.400	18,000		4.6	1.40	42.8	10.01
Site 3	Site 3 - Willsberg	Site Clients = 40			Traffic =1.9 Mbps		3.000					62.2%
2.3.6	R_Remote WTS ArcGIS Desktop Viewer	SDE DC/DBMS	40	10	400	1.867	24,000		4.6	1.42	42.2	10.01
Interne	Internet Clients	Internet Clients = 98			Internet =10.1 Mbps		6.000					1.68
Public	Public	Site Clients = 36			Traffic =7.2 Mbps		256.000					2.8%
3.4.7	R_AGS9.3 Map Server	SDE DC/DBMS	36	6	216	7.200	12,960		7.3	2.73	22.0	6.01
Site 4	Site 4 - Perth	Site Clients = 2			Traffic =0.1 Mbps		1.500					6.2%
3.5.8	R_Remote WTS ArcGIS Desktop Viewer	SDE DC/DBMS	2	10	20	0.093	1,200		3.6	2.39	25.1	10.01
Site 5	Site 5 - Wawash	Site Clients = 40			Traffic =1.9 Mbps		1.500					124.4%
3.6.9	R_Remote WTS ArcGIS Desktop Viewer	SDE DC/DBMS	40	10	400	1.867	24,000		3.2	2.80	21.4	10.01
Site 6	Site 6 - Jackson	Site Clients = 20			Traffic =0.9 Mbps		1.500					62.2%
3.7.10	R_Remote WTS ArcGIS Desktop Viewer	SDE DC/DBMS	20	10	200	0.933	12,000		3.6	2.43	24.7	10.01
							1.500					

(Requirements Analysis — Types of Workflows; WEB TPH => 13,000; WEB Users 36)

Here is a closer look at the Requirements Analysis Module, which represents workflows associated with a specific data server environment. Separate spreadsheet tabs can be configured for each central database server environment. Based on your workflow requirements analysis, you will select workflows from drop-down menus (**column B** for each workflow). The workflows are on a lookup list located by way of a separate Workflow tab discussed on page 210.

The standard Capacity Planning Tool template presumes that there are three networks entering the computer facility (LAN—**row 7**, WAN—**row 11**, and Internet—**row 18**). Remote site network connections are identified within the WAN and Internet network segments, determined by how they connect to the data center. Three remote sites (Operations—**row 12**, Freeberg—**row 14**, and Willsberg—**row 16**) are included within the WAN environment. Three remote sites (Perth—**row 21**, Wawash—**row 23**, and Jackson—**row 25**) comprise the Internet environment. A public site (**row 19**) is included to represent Internet Web clients.

When establishing your own capacity planning model of a system, you create and label each remote network connection and adjust the client and traffic summation ranges. You also identify the existing bandwidth from a drop-down menu in column I for each network connection—you can get this information from your network administrator.

User location identifies where you will represent the workflow in the CPT requirements module. Users accessing the computer facility over the LAN connection are included as the "local client" workflows (**rows 8–10**). Users accessing the computer facility over the WAN connection are the "WAN client" workflows (**rows 12–17**). Users accessing the computer facility through the Inter-net Service Provider connection are included as the "Internet client" workflows (**rows 20-26**), for the purposes of this tool.

You need to identify the workflow name (**column B**), data source (**column D**), peak users (**column E**), and user productivity (**column F**) for each identified workflow. First you will identify the custom workflows for your implementation on the Workflow tab (discussed later in this chapter). Drop-down menus are provided for selecting the workflow name and data source. The selected data source distributes workflow software component service times to the appropriate hardware platforms for capacity planning calculations. (The variety of data source choices will be introduced later when discussing the Workflow tab.)

Once you have configured your requirements analysis inputs, the CPT will update a network suitability analysis. This is updated on-the-fly, while you are entering user requirements (based on the existing platform selected) in the Platform Selection Module, which we will discuss later. Peak displays per minute (**column G**) and peak network traffic (**column H**) will be computed for each workflow. Display response time (**column AR**) and user think time (**column AQ**) will be computed based on workflow service times (establishing workflow service times come later in this chapter, too). A minimum think time can be set in **cell F7**. **Column F** for each workflow changes to yellow if think time is less than the identified minimum think time, and turns red if there is no think time (batch process). If think time is less than zero, you will need to reduce workflow productivity until you have sufficient think time to support user requirements (less than zero is not a valid workflow).

To automatically establish a batch process, you can link **column F** to **column AT** (this will introduce a circular calculation in the Excel spreadsheet—the spreadsheet calculations are set to automatically calculate the batch process productivity value). An example batch process is configured in **row 8**. Peak users (**cell E8**) turn green when user productivity is within 0.01 of the proper batch productivity. Batch process productivity (**cell F8**) turns red, identifying zero think time. This color pattern identifies the final batch process configuration. If batch productivity (**cell F8**) does not equal batch DPM (**cell AS8**), you will need to initiate additional Excel calculations until the batch process converges to the proper productivity value.

Each site network connection row will identify the peak number of clients (centered between **column D:E**) and the total network traffic (centered on **columns G:H**). Network traffic totals are a summation of all the workflow traffic within each identified network location. Percent network traffic utilization is displayed in **column AS**. Total network traffic cell color changes to yellow when total traffic is over 50 percent network capacity, and red when total traffic is over 100 percent network capacity.

As entries are made, the CPT displays several calculations: Response time shows the result of all workflow service times plus queue times. Queue times are calculated based on platform and network utilization levels. (Chapter 7 discusses queue time models; they are here in the Capacity Planning Tool to help you identify potential performance problems and translate platform service times to user response time.) The DPM/client cell (**column F**) turns yellow if the workflow productivity does not support the minimum think time (you would need to reduce the DPM/client user productivity to establish a valid user workflow). The gray network traffic cell will turn yellow if traffic exceeds 50 percent of the bandwidth set for that network. You can specify available bandwidth for each network in the middle black column (**column I**). A cell to the right of the LAN network bandwidth setting (**cell Q6**) allows you to add a network bandwidth value that is not in the drop-down selection (the new entry will appear on the top of the network bandwidth drop-down list). Peak user totals are displayed for each network connection. Network traffic flow within the data center (server platform network interface card bandwidth requirements) will be addressed later in the Platform Selection Module. Network traffic flow within the data facility is currently not represented in the model.

Central data facility network suitability

One of the more subtle performance problems can be caused by remote client network traffic contention. Network traffic contention starts to impact performance when traffic exceeds 50 percent of the selected network bandwidth. The network traffic cell will turn yellow when this occurs. Data center network suitability (figure 10-8 on the following page) provides a network diagram showing the corresponding network bandwidth settings represented in the CPT. Traffic for the individual workflows is computed by the CPT and displayed just left of the black column. Total traffic for each network segment is displayed on the gray network header row.

Network bandwidth should be maintained at a minimum of twice the peak network traffic. Network traffic guidelines, introduced in chapter 3, follow the same performance characteristics developed in chapter 7: Traffic contention (delays) are noticed when traffic exceeds about 50 percent bandwidth capacity and grows to equal the network transport time when approaching 100 percent capacity (queuing theory discussion). These transmission delays are caused from packet random arrival times (packets arriving at the same time), accumulating transmission delays while waiting for access to the shared network segment. (You can see the affect of network performance in general in the Workflow Performance Summary, figure 10-3.)

10-8 Data center network suitability

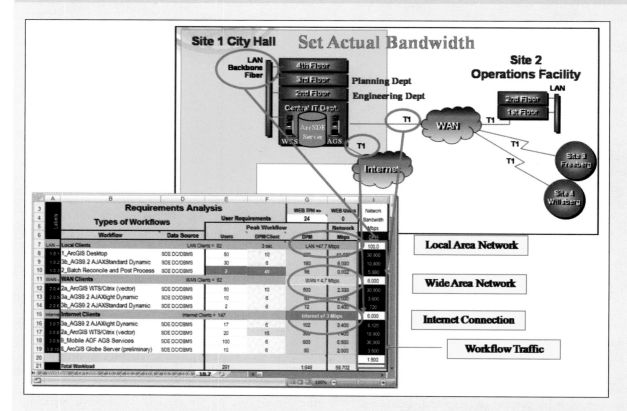

A network diagram showing the corresponding network bandwidth settings represented in the CPT. Traffic for the individual workflows is computed by the CPT and displayed just left of the black column. Total traffic for each network segment is displayed on the gray network header row. Network bandwidth should be maintained at a minimum of twice the peak network traffic.

Functions to modify the CPT

CPT Microsoft Office Excel user functions

Configuring the Requirements Analysis Module requires a basic understanding of Microsoft Office Excel functions highlighted here.

The primary functions used to modify the CPT to represent your GIS operational environment include the following:

- **Select row:** Refers to selecting a spreadsheet row for copy, insert, or delete functions. You can select a specific row by locating the mouse on the row number (mouse pointer points right) and selecting the left mouse key (selected row should be highlighted after mouse selection).

- **Copy row:** Selected row must be completed prior to copy row execution. Execute the copy function following the row selection without changing the pointer location. The right mouse button should provide a drop-down menu; select the copy function to copy row (row boundary should be highlighted with dotted line).

- **Insert row:** Select the row below where you wish to insert a copied row. Right mouse button should provide a drop-down menu that includes the Insert Copied Cells function. Select "Insert Copied Cells" to insert row above selected location.

- **Delete rows:** Select the row that you wish to delete. Right mouse button should provide a drop-down menu that includes the "Delete" function. Select "Delete" to remove selected row.

- **Update summation formula range:** The Excel SUM function is used to identify the total concurrent users and total traffic for each site location. Each of the site rows (gray rows represent the Central Data Center network connections— LAN, WAN, and Internet—and the green rows are used to represent remote site locations). Each site location row includes a concurrent user SUM function (column D) and a site traffic SUM function (column G). The LAN, WAN, and Internet SUM range covers all workflows assigned within that network (for example, WAN network traffic range includes all cells in column H, including the blank Internet cell at the end of the WAN range). When inserting a new remote site location (using a copy of the bottom green row), you need to manually expand the concurrent user and network traffic SUM ranges to calculate the site values. This can be accomplished by selecting the SUM function cell (column D or G) and then select the worksheet formula bar (colored box will display SUM range in column E or H). Carefully select the right bottom corner of the displayed range box and drag to include all of the site workflows (include the following site cell to ensure all workflows will be included). A step-by-step example of this procedure will come later in the chapter, when adding remote sites.

- **Move or copy sheet:** This is a common function used in Excel to copy contents of the existing spreadsheet to another workbook tab. Locate the mouse over the lower tab of the displayed spreadsheet, select the right mouse button (this should display a menu that includes the "Move or Copy" function), select "Move or Copy" and then the "Create a Copy" box on this menu. Identify where you would like the copy placed (usually at the end of the list) and select OK. This should result in a copy of the spreadsheet on a separate tab. You can highlight the new tab with your mouse (double-click the tab) and provide a new title as needed for your analysis. This function can be used to make copies of one design milestone, and use the new copy to update user requirements for the next user requirements milestone. This will save you a great deal of work— most of the design work is in creating the initial requirements module.

These user functions are implemented slightly differently in the Office 2003 version than they are in Office 2007. Office 2007 includes a help item that will help you identify how to complete an Office 2003 function in Office 2007.

Adding and deleting workflows

The Requirements Analysis Module is designed to adapt to any customer organization's environment. User groups can be collected together by workflow, and the number of workflows can be adjusted to support custom user workflow needs. These figures (10-9 and 10-10) describe how to add and delete workflows when using the Requirements Analysis Module.

10-9 Adding new workflows

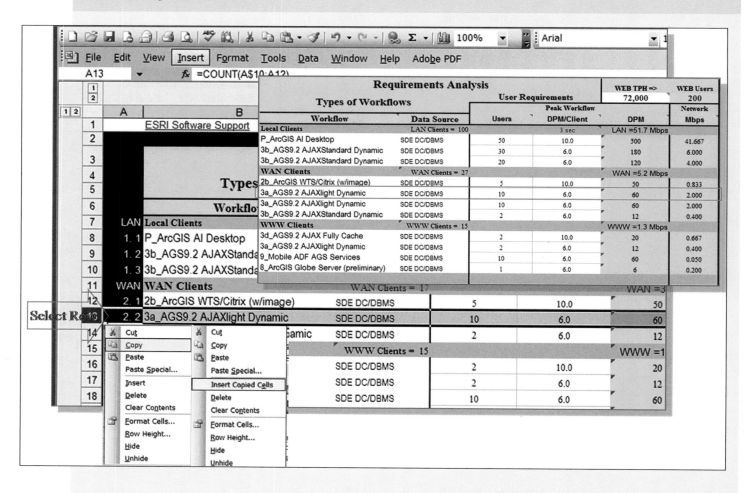

To add a workflow, simply select one of the existing workflows within the desired network area, copy the row, select the workflow row below where you want the new workflow, then "insert copied cells" to enter the new workflow. You will need to repeat the process for each new row insert. Make sure you copy a complete workflow row that includes all the required CPT formulas before each Insert Copied Cells entry (many CPT formulas are hidden in each workflow row).

10-10 Deleting workflows

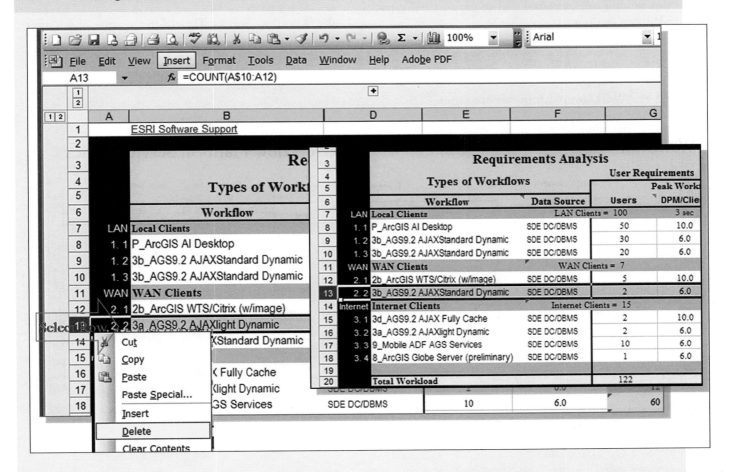

Select the complete row, right-click, and select Delete. Do not delete all the rows in a network section (you can make sure all peak clients—"users"—are empty and hide rows that you may want to open later). The hide function is on the right-click menu when the row is selected.

The platform service times are exposed in figure 10-11. The service times and display traffic identified in this module are used by Excel to support the capacity planning analysis.

The workflow labels that will be displayed in the Workflow Performance Summary (mentioned earlier) are identified in the black column A. It is good to open the Workflow Service Times for a quality check after configuring a particular performance milestone.

10-11 Workflow platform service times

There is a small plus (+) above **column P** that will open **columns J** through **O** and display the Platform Service Times. These service times are pulled from the Workflow tab based on the workflow selection.

You can see from this chart that there are performance issues with the selected application server and Internet bandwidth components. (Later in this chapter, we discuss ways to resolve the issues, using configuration tuning.)

10-12   Workflow Performance Summary—description

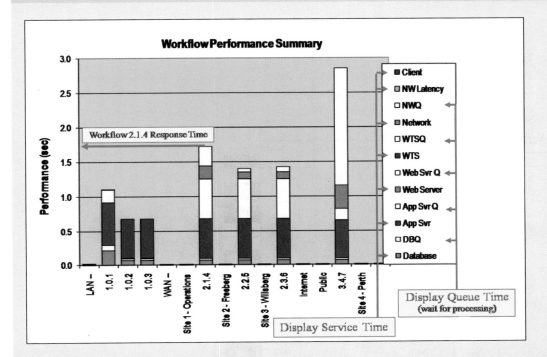

The Workflow Performance Summary graphic, located at the far right of the Requirements Module, can be moved around as needed to support your display. All of the performance metrics are shown by workflow on the Workflow Performance Summary. In figure 10-11, the workflows are labeled based on the values in **column A** to the left of each selected workflow. They are also grouped based on the supporting network (LAN, WAN, Internet). Each of the component platform service times are displayed in sequence on each workflow bar (Database, App Svr, Web Server, WTS, Network, Client). The white space above each platform component shows the calculated queue time for that component. Queue time is included with each component when utilization is over 50 percent capacity and increases to equal the service time by 100 percent capacity. Response time is the performance value shown at the top of the stacked times, and includes all of the service and queue times. The performance problems with the selected application server and Internet bandwidth components, identified here, can be remedied by nework and platform configuration tuning.

Adding remote locations

You can include remote locations on each of the networks. Figure 10-13 shows how to add remote sites to a selected network.

Once you establish the remote site locations, you can adjust the number of workflows, peak users, and DPM/client for each remote site location. You will need to adjust the new site SUM range settings when adding remote locations. It is good to check the LAN, WAN, and Internet SUM range settings to make sure they include all traffic on those networks (these ranges represent the data center network connections and should continue to include all workflows within those network environments).

This should complete the user requirements configuration resulting from the requirements analysis.

10-13   Adding remote locations

	Workflow	Data Source	Users	Peak Workflow DPM/Client	DPM	Network Mbps	DPH
3	**Requirements Analysis**				WEB TPH => 12,000	WEB Users 33	Network Bandwidth Mbps
7 LAN	**Local Clients**	LAN Clients = 101		3 sec	LAN =58.3 Mbps		100.0
8 1.0.1	1_ArcGIS Desktop	SDE DC/DBMS	50	10	500	41.667	30,000
9 1.0.2	3b_AGS9.2 AJAXStandard Dynamic	SDE DC/DBMS	50	10	500	16.667	30,000
10 1.0.3	2_Batch Reconcile and Post Process	SDE DC/DBMS	1	44.54	45	0.001	2,672
11 WAN	**WAN Clients**	WAN Clients = 57			WAN = 8.7 Mbps		6.000
12 Site 1	**Site 1**	Site Clients = 25			Traffic =4.7 Mbps		1.500
13 2.1.4	2b_ArcGIS WTS/Citrix (w/image)	SDE DC/DBMS	10	10	100	1.667	6,000
14 2.1.5	3a_AGS9.2 AJAXlight Dynamic	SDE DC/DBMS	15	6	90	3.000	5,400
15 Site 2	**Site 2**	Site Clients = 24			Traffic =3.3 Mbps		1.500
16 2.2.6	2a_ArcGIS WTS/Citrix (vector)	SDE DC/DBMS	9	6	54	0.252	3,240
17 2.2.7	3a_AGS9.2 AJAXlight Dynamic	SDE DC/DBMS	15	6	90	3.000	5,400
18 Site 3	**Remote Site 3**	Site Clients = 8			Traffic =0.7 Mbps		1.500
19 2.3.8	2a_ArcGIS WTS/Citrix (vector)	SDE DC/DBMS	5	6	30	0.140	1,800
20 2.3.9	3a_AGS9.2 AJAXlight Dynamic	SDE DC/DBMS	3	6	18	0.600	1,080
21 Interne	**Internet Clients**	Internet Clients = 147			Internet =3.5 Mbps		6.000
22 Site 4	**Site 4**	Site Clients = 147			Traffic =3.5 Mbps		256.000
23 3.4.10	2a_ArcGIS WTS/Citrix (vector)	SDE DC/DBMS	17	6	102	0.476	6,120
24 3.4.11	2a_ArcGIS WTS/Citrix (vector)	SDE DC/DBMS	20	6	120	0.560	7,200
25 3.4.12	9_Mobile ADF AGS Services	SDE DC/DBMS	100	6	600	0.500	36,000
26 3.4.13	8_ArcGIS Globe Server (preliminary)	SDE DC/DBMS	10	6	60	2.000	3,600
27							1.500

Copy the bottom green row (**row 27**) to use for identifying the remote site locations. Select the top row within the network segment (**row 12**) to insert the first remote site. You can add additional remote sites by repeating the steps above (you will need to copy a row before inserting copied cells for each insert). Once the rows are inserted, you must label the new remote sites (**column B**). Each site name will be copied to **column A**, which is used to identify workflows on the Workflow Performance Summary chart. You will also need to update the site SUM range functions for each remote site. The site client SUM range functions are located

in **column D** and the site traffic SUM functions are in **column G** on the newly inserted site rows. You can select the SUM function location (for example **cell D12**) and select the formula bar (not shown here) and you will see a colored box show up in **cell E12** (this is the SUM function range). Select the bottom right corner of the SUM range box and drag down to include all the Site 1 workflows (**range E12:E15**). Do the same for the site traffic function in **column G** (**cell G12 above**); the traffic SUM range should be extended to support all Site 1 workflow traffic (**range H12:H15**). Use the CPT on the CD to see these functions.

Remote site bandwidth suitability

Once the CPT user requirements are complete, the CPT results will provide information needed to complete the network suitability analysis. Make sure the available network bandwidth is identified properly for each site location (you can get the existing network bandwidth connections from your network administrator). Figure 10-13 identifies network traffic concerns for the data center WAN (cell G11:H11) and Internet (cell G21:H21) bandwidth connections, and also the remote network connections from remote Site 1 (G12:H12) and Site 2 (G15:H15)). The red cells indicate that network traffic exceeds the existing network bandwidth.

Figure 10-14 below shows the effects on performance of too little network bandwidth. You can see how serious this problem would be if not corrected. On the next page, bandwidth updates are shown in figure 10-15, producing the performance improvements displayed in figure 10-16 on page 197.

10-14   Bandwidth suitability—performance before upgrade

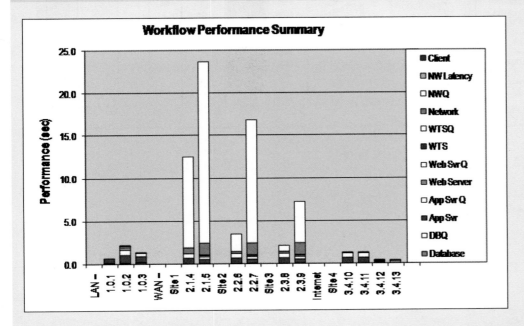

The network queue time (white) identified above the gray network service times indicates that the network traffic exceeds bandwidth capacity.

The network bandwidth bottlenecks can be resolved by upgrading network bandwidth to twice the traffic. Recommended network updates would include the following upgrades: increase WAN (**cell I11**) bandwidth to 45 Mbps, remote Site 1 (**I12**) bandwidth to 12 Mbps, and remote Site 2 (**I15**) bandwidth to 9 Mbps.

Bandwidth updates are shown in the chart used for Network performance tuning (on page 196), producing the performance improvements after upgrading bandwidth (figure 10-16). Once you make the bandwidth adjustments, the CPT will update the network requirements analysis on the fly.

Network traffic performance issues were resolved following the network upgrades.

10-15   Network performance tuning

	A	B	D	E	F	G	H	I	Q	AQ	AR	AS	AT
3		**Requirements Analysis**				WEB TPH =>	WEB Users	Network	Unique			4	2
4	Labels			User Requirements		12,000	33	Bandwidth	Display	User			Blink
5		**Types of Workflows**		Peak Workflow			Network	Mbps	Traffic	Think	R_time	Batch	0.01
6		Workflow	Data Source	Users	DPM/Client	DPM	Mbps	DPH	9	Time	sec	DPM	Latency
7	LAN –	Local Clients	LAN Clients = 101		3 sec	LAN =58.3 Mbps		100.0	Mbpd		100000	58.3%	
8	1.0.1	1_ArcGIS Desktop	SDE DC/DBMS	50	10	500	41.667	30,000		5.3	0.72	83.3	10.01
9	1.0.2	3b_AGS9.2 AJAXStandard Dynamic	SDE DC/DBMS	50	10	500	16.667	30,000		3.9	2.14	28.0	10.01
10	1.0.3	2_Batch Reconcile and Post Process	SDE DC/DBMS	1	44.54	45	0.001	2,672		0.0	1.35	44.5	44.53
11	WAN –	WAN Clients	WAN Clients = 57			WAN = 8.7 Mbps		45.000				0.19	
12	Site 1	Site 1	Site Clients = 25			Traffic =4.7 Mbps		12.000				38.9%	
13	2.1.4	2b_ArcGIS WTS/Citrix (w/image)	SDE DC/DBMS	10	10	100	1.667	6,000		4.68	1.32	45.6	10.01
14	2.1.5	3a_AGS9.2 AJAXlight Dynamic	SDE DC/DBMS	15	6	90	3.000	5,400		8.77	1.23	48.9	6.01
15	Site 2	Site 2	Site Clients = 24			Traffic =3.3 Mbps		9.000				36.1%	
16	2.2.6	2a_ArcGIS WTS/Citrix (vector)	SDE DC/DBMS	9	6	54	0.252	3,240		8.7	1.26	47.5	6.01
17	2.2.7	3a_AGS9.2 AJAXlight Dynamic	SDE DC/DBMS	15	6	90	3.000	5,400		8.72	1.28	46.8	6.01
18	Site 3	Site 3	Site Clients = 8			Traffic =0.7 Mbps		1.500				49.3%	
19	2.3.8	2a_ArcGIS WTS/Citrix (vector)	SDE DC/DBMS	5	6	30	0.140	1,800		8.6	1.42	42.3	6.01
20	2.3.9	3a_AGS9.2 AJAXlight Dynamic	SDE DC/DBMS	3	6	18	0.600	1,080		7.61	2.39	25.1	6.01
21	Interne	Internet Clients	Internet Clients = 147			Internet =3.5 Mbps		6.000				0.59	
22	Site 4	Site 4	Site Clients = 147			Traffic =3.5 Mbps		256.000				1.4%	
23	3.4.10	2a_ArcGIS WTS/Citrix (vector)	SDE DC/DBMS	17	6	102	0.476	6,120		8.7	1.29	46.6	6.01
24	3.4.11	2a_ArcGIS WTS/Citrix (vector)	SDE DC/DBMS	20	6	120	0.560	7,200		8.7	1.29	46.6	6.01
25	3.4.12	9_Mobile ADF AGS Services	SDE DC/DBMS	100	6	600	0.500	36,000		9.6	0.42	142.1	6.01
26	3.4.13	8_ArcGIS Globe Server (preliminary)	SDE DC/DBMS	10	6	60	2.000	3,600		9.6	0.40	148.6	6.01
27								1.500					

10.13-14

Calculate   100%

Recommended network updates are done now and include the following upgrades shown here: increase WAN (cell I11) bandwidth to 45 Mbps, remote Site 1 (I12) bandwidth to 12 Mbps, and remote Site 2 (I15) bandwidth to 9 Mbps. After completing these adjustments, the LAN and Internet are slightly over 50 percent of the selected bandwidth (G7:H7 and G21:H21). These performance impacts are marginal, as indicated in figure 10-16. Traffic on these networks should be monitored to ensure target performance values are met.

10-16   Bandwidth suitability—performance after upgrade

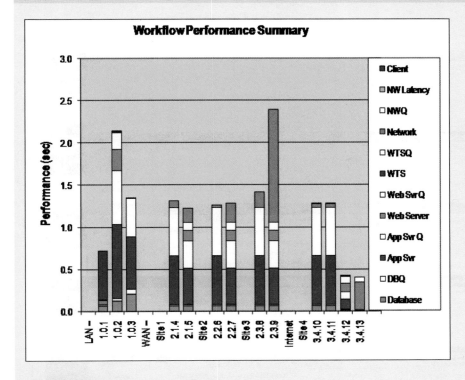

How bandwidth upgrades improved performance.

Platform Selection Module

This section displays various aspects of the CPT's Platform Selection Module. To support high availability or to meet peak system workflow requirements, you can configure a group of servers in the same way to form a platform tier. The CPT model represents several platform tiers. Each of the four primary ones pictured at the top left of the Platform Selection Module in figure 10-17 can include several server platforms, or nodes, depending on the system configuration. Figure 10-18, highlighting the center rows of the Platform Selection Module, shows the kind of information about performance that is provided on each platform tier. Performance infor-

mation is calculated based on your platform selection. All of these calculations are based on the fundamentals of performance and scalability discussed in chapter 7.

Figure 10-19 shows how hardware selections are made from a drop-down menu. Afterwards, provided for display purposes only, a spreadsheet to the right of the Platform Selection Model (see figure 10-20) projects the results of your choices in terms of platform utilization. It can be useful to see the results of the platform sizing analysis in graphic form, as you approach the next step, which is configuration performance tuning.

10-17 Platform tier layout

At left of the Platform Selection Module, shown below, are the platform tiers represented in the CPT.

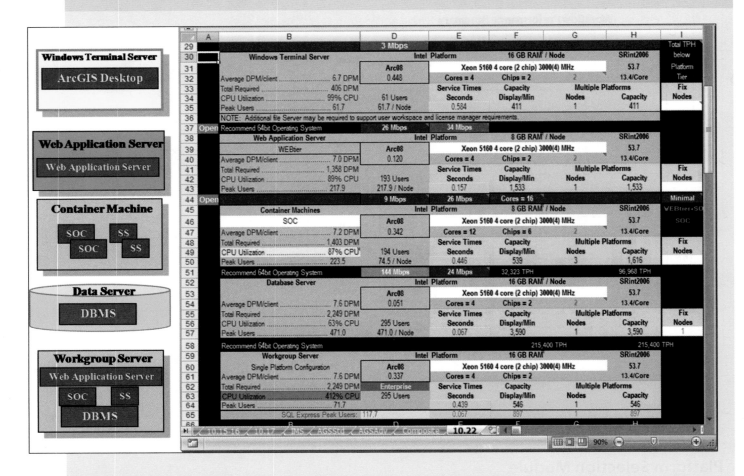

In configuring the right platform, your best solution will normally be the current recommended application servers. In this case, however, we used the Xeon 5160 4-core (2-chip) 3000(4) MHz platforms. A server platform graphic is provided on the left of the worksheet above to help you visualize the platform tiers represented by the CPT. The platform tiers identified on the graphic are configured much like what you might expect in a platform configuration drawing, although a significant amount of performance information is included in each box. Each platform tier can represent several server platforms, or nodes, depending on the system configuration. Just as a reminder, a platform tier is a group of servers configured the same way to support high availability or peak system workflow requirements. Standard GIS configurations generally include Windows Terminal Servers (desktop clients), Web or container machine application servers (two- or three-tier Web services configurations), and a data server tier (central geodatabase server configuration).

The platform configuration model includes four primary platform tiers that represent the Windows Terminal Servers, Web application servers, container machines, and data server within the central computer center. For convenience, the CPT assumes that a local client workstation platform (not part of the data center server environment) will have the same per-core performance as the Windows Terminal Server tier.

There is an additional workgroup server provided at the bottom tier. This tier can be used for sizing a server when the Web and DBMS software are all supported on a single platform. This platform tier can be used for sizing an ArcGIS Server Workgroup license with SQL Express database—if capacity exceeds licensing limitations, the red enterprise cell lights up to let you know you need an enterprise license. You can use any database on this platform if you plan to use an enterprise license.

10-18   Platform tier performance factors

In this section of the Platform Selection Module, hardware (Web application server, container machine) can be selected based on projected performance. Each platform tier display provides a lot of basic performance information.

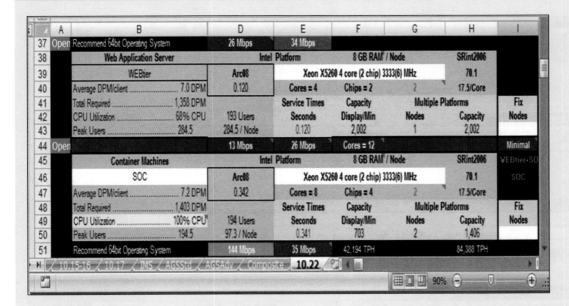

There are only two places for you to make selections in this area of the CPT: hardware platform selection (Web application server—**E39:G39**, and container machine—**E46:G46**) and a simple switch that determines if you want results for a two- or three-tier Web configuration (**B46**). If you have your favorite platform already selected, you are pretty much good to go. You have already configured the requirements module for your peak system workflows, and those loads are in the CPT.

There is lots of basic performance information provided on each platform tier display. The displayed performance information is generated based on the platform and Web tier configuration selections. The Average DPM/client (**B40**) and Total Required DPM (**B41**) are assigned to this platform tier from the user requirements. The baseline platform service time (**D40**) is provided from the workflow selections.

The rest of the performance information is calculated based on your platform selection. The platform SRint2006 performance benchmark values (**H39:H40**) come from a lookup table on the Hardware tab, along with the number of cores (**E40**) and chips (**F40**) available with the platform selection. The SPEC benchmarks are used to translate the CPT performance baseline service time to the selected platform service time (**E43**). The platform capac-

ity (**F43**) is calculated from the platform service time. The total numbers of platform nodes (**G43**) are calculated by comparing the platform capacity with the total required DPM (required capacity provided from the requirements module). The total tier capacity (**H43**) is then calculated from the number of nodes and the platform capacity. The CPU Utilization (**B42**) is calculated by comparing the Total Required DPM to the total capacity. The memory requirements (**F38**) for the Web platforms is based on the number of cores. Memory on the Windows Terminal Server and data server platforms is based on the number of peak concurrent user connections (these memory recommendations are calculated—please do not change these formulas and expect the model to do something right). Other interesting calculations include peak users per node (**D43**), calculated from the Peak Users, and number of nodes and total users, calculated from Total Required DPM divided by Average DPM/client (**D42**). You can specify a fixed number of nodes (**I43**). If you do provide a fixed node specification, the nodes will be changed to what you specify and all calculations will be made based on the set node value. All of these calculations are very simple, based on the fundamental performance and scalability functions discussed in chapter 7.

10-19   Selecting the hardware platforms

This is what platfom configuration selection looks like on the CPT.

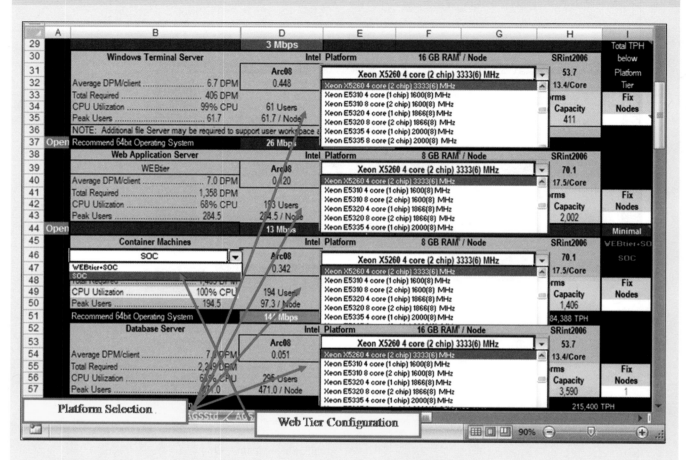

Platform configuration selection—the hardware selection—is made from a drop-down menu (the white cells on platform tier in **columns E:G**). The hardware list is provided on the Hardware tab (discussed later).

The Web tier configuration has two options (SOC or WEBtier+SOC). The SOC selection will support the Web application server on the Web tier. The WEBtier +SOC selection will support the Web applications on the container machine tier. Web platform performance calculations will follow the selected software assignment.

10-20   Platform configuration graphics

It is nice to see graphic results of the platform sizing analysis. The results are provided on the spreadsheet to the right of the Platform Selection Module. The platform configuration graphics simply display what is calculated in the platform module.

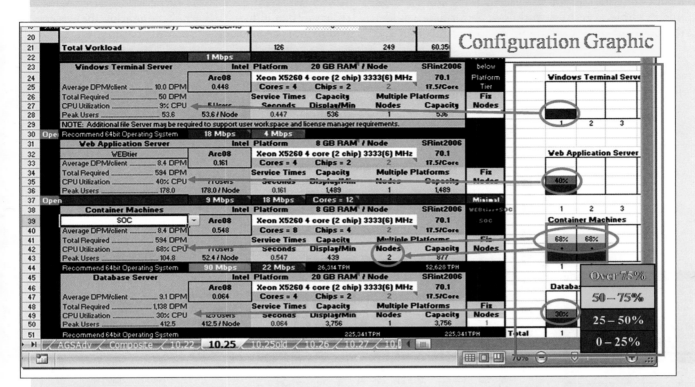

Each platform server node in the platform graphic is supported by a set of four cells. The cell colors change based on platform utilization. The bottom cell turns blue indicating that platform node is required to support the configuration. The second cell turns green when platform utilization exceeds 25 percent capacity. The third cell turns yellow when platform utilization exceeds 50 percent capacity. The fourth cell turns red when platform capacity exceeds 75 percent utilization. The percent platform utilization is displayed in the top colored cell.

The platform graphics are provided for display purposes only; there are no model calculations that depend on the values in these displays. This capacity planning analysis shows a single Windows Terminal Server at about 19 percent utilization; a single Web application server at 40 percent utilization; two container machines at 68 percent utilization, and one database server at 30 percent utilization.

Configuration performance tuning

The next series of three figures, 10-21 through 10-23, traces the process of gauging then tuning the platform configuration you've selected using the CPT. In this example, assessing workflow performance under the particular platform configuration reveals that too much queue time could reduce user display performance. Upgrading the Internet bandwidth (in figure 10-22) will resolve the issue of network performance delays . Then in figure 10-23, reconfiguring the servers will reduce the load across the container machine tier.

10-21 Configuration performance issues

Using the CPT to discover issues that may affect performance gives you a chance to do performance tuning before they happen.

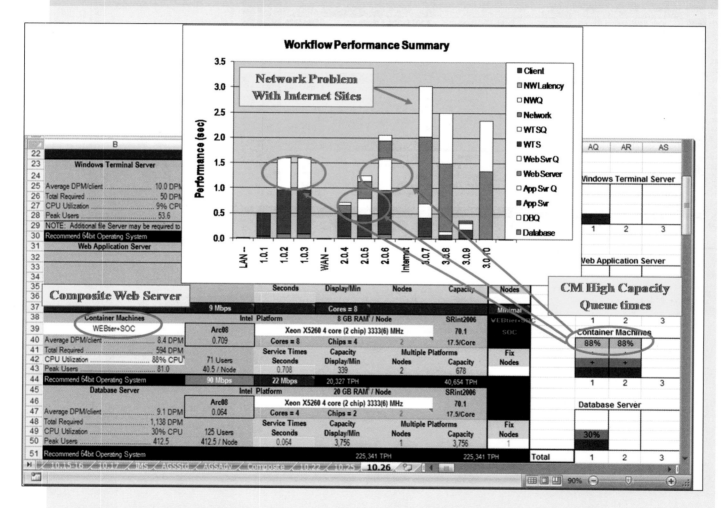

The figure above displays the platform configuration along with the Workflow Performance Summary. We have changed from the three-tier Web configuration shown in the last graphic to a two-tier Web configuration, and you can see the container machine platform utilization is now at 88 percent capacity. The CPU Utilization, **cell B42**, turns yellow when utilization exceeds 75 percent.

The Workflow Performance Summary shows a considerable amount of queue time for the Internet traffic and container machine application servers. The Workflow Performance Summary provides an estimate of what these queue times are costing us in terms of reduced user display performance.

10-22 Network performance tuning

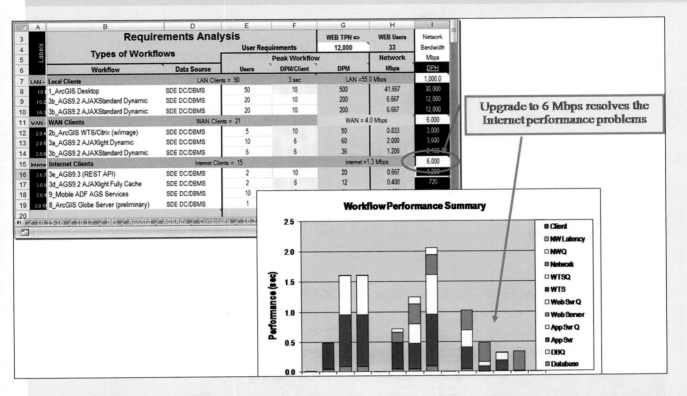

Upgrading the Internet bandwidth from 1.5 Mbps to 6 Mbps will resolve the network performance delays. The Internet bandwidth cell returns from yellow back to gray again. Notice the WAN traffic cell is still yellow, since traffic is slightly over 50 percent capacity. We are not seeing a loss in performance at this point, so no WAN upgrade is needed at this time.

10-23   Platform performance tuning

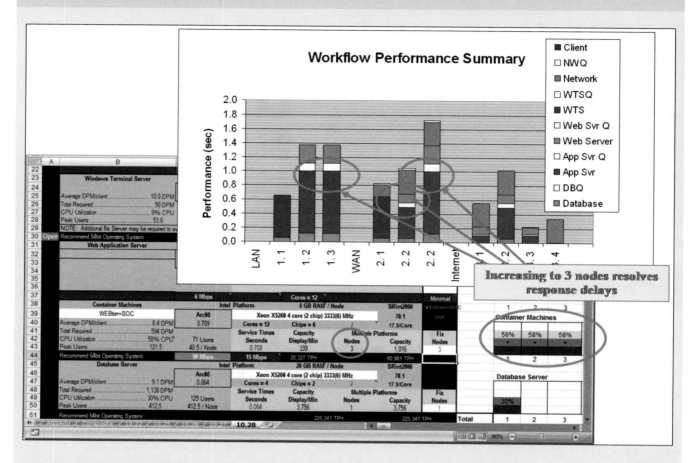

Increasing the number of container machine nodes reduces platform capacity on this tier. In this case, we decide to support the two Web servers on this same tier along with the three container machines. We will configure the servers so the ArcGIS SOM components can balance the load across the container machine tier.

The additional container machine reduces platform utilization to 58 percent capacity, which leaves only a minor performance degradation (you can still see a small amount of queue time on the Workflow Performance Summary).

Configuring Web server zones

There may be times when you want to configure separate groups of servers and the following three figures (10-24 through 10-26) display platform recommendations for each of these three configurations mentioned.

Figure 10-24 shows an example with three sets of servers (ArcIMS ArcMap Server, ArcGIS Server Standard, and ArcGIS Server Advanced). If you configured all of this software together on the same set of Web servers, you would need an Enterprise Advanced license for all Web servers; so you may want to configure on separate groups of servers, to minimize the price you pay for licensing. There could be other reasons for configuring multiple groups of servers on any one of the tiers; the procedure described in the caption below figure 10-24 addresses these types of configurations as well. You can also use this procedure to configure mixed platform server environments, although you may need to be a bit more creative.

10-24   Configuring Web server zones

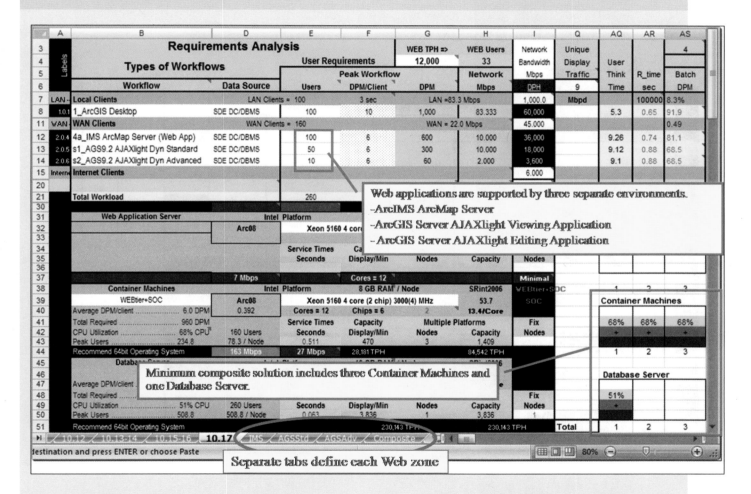

Web applications are supported by three separate environments.
- ArcIMS ArcMap Server
- ArcGIS Server AJAXlight Viewing Application
- ArcGIS Server AJAXlight Editing Application

Minimum composite solution includes three Container Machines and one Database Server.

Separate tabs define each Web zone

Using the following procedure to develop various configuration alternatives for separate groups of servers is much simpler than trying to modify the standard capacity planning tool design.

First configure the system as if everything was going to be installed on all the boxes. In this example, we have 100 desktop, 100 ArcIMS, 50 ArcGIS Server standard, and 10 ArcGIS Server Advanced users. Once you have the complete system configured with all the user workflows, copy the composite sheet (use the "Move or Copy Sheet" function) to a separate tab or folder. Make three copies of the Composite tab following the procedure described previously, and label the new spreadsheets AGS Std, AGS Adv, and IMS.

Now select the AGS Std tab and remove users from all rows except the AGS9.2AJAXlightDyn standard workflow. This tab will identify the ArcGIS Server Standard Web server configuration. See that display on figure 10-25.

Now select the AGS Adv tab and remove users from all rows except the AGS9.2AJAXlightDyn Advanced workflow. This tab will

10-24 Configuring Web server zones (continued)

identify the ArcGIS Server Advanced Web server configuration (see figure 10-26).

Finally, select the IMS tab and remove users from all rows except the ArcIMS ArcMap workflow. This tab will identify the ArcIMS Arc-Map Web Server configuration, as shown in figure 10-27.

Return to the Composite tab to complete the configuration. Add up all the platforms required to support the separate Web server

configurations and include that number in the Fix nodes cell (in this case you need four servers to support the three separate Web environments). The Composite tab can be used for database sizing and network traffic analysis. The application server on this tab will be shared, so the workflow performance may not be accurate. Figure 10-28 provides the results.

10-25 ArcGIS Server Standard Web zone

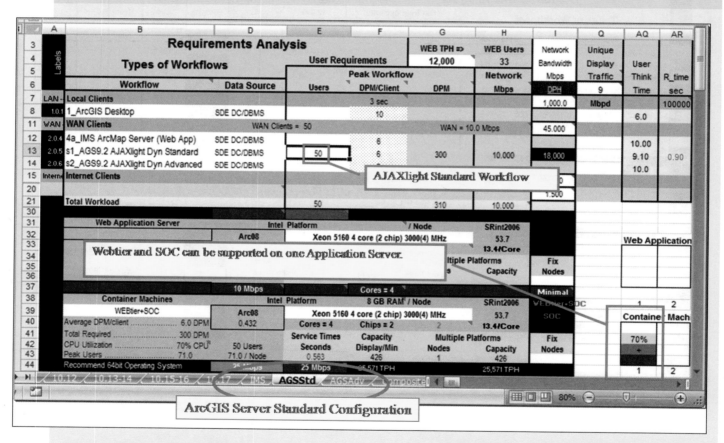

The AGSstd tab will reveal the platform recommendations for the ArcGIS Server standard Web server configuration. (You can ignore

the database configuration, since that will be identified on the Composite tab.)

10-26 ArcGIS Server Advanced Web zone

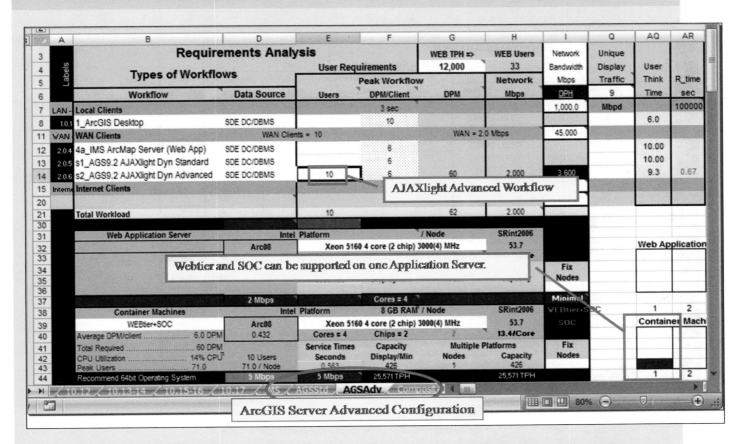

Selecting the AGS Adv tab and removing users from all rows except the AGS9.2 AJAXLightDyn Advanced workflow will reveal the platform recommendations for the ArcGIS Server Advanced configuration.

10-27 ArcIMS ArcMap Server Web zone

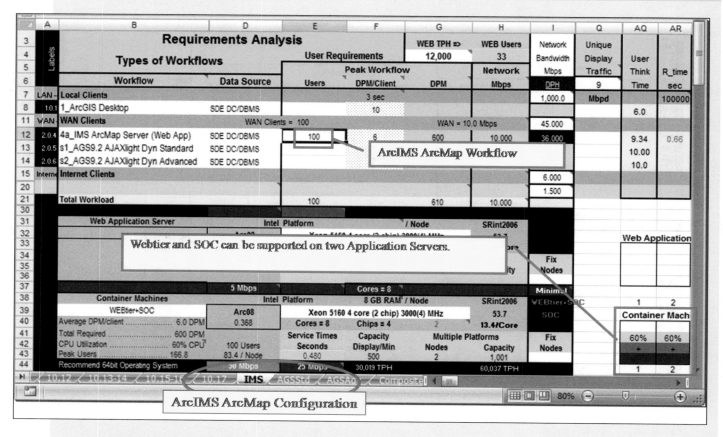

Selecting the IMS tab and removing users from all rows except the ArcIMS ArcMap workflow will reveal the platform recommendations for the ArcIMS ArcMap Web server configuration.

10-28 Composite summary configuration

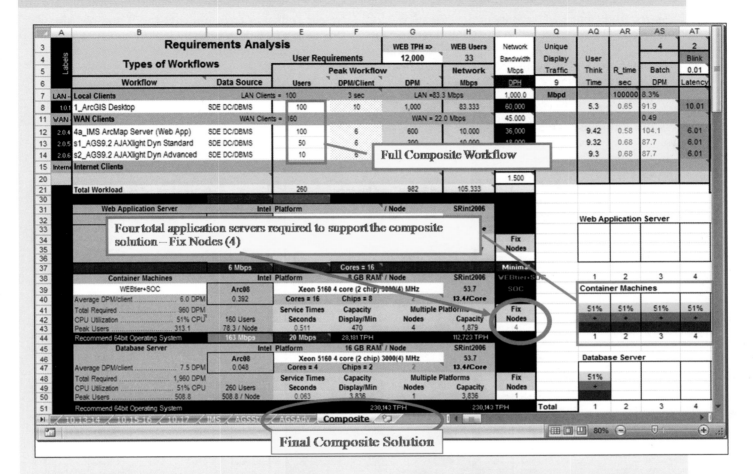

To complete the configuration, return to the Composite tab. Add up all the platforms required to support the separate Web server configurations and include that number in the Fix nodes cell (in this case you need four servers to support the three separate Web environments). The Composite tab can be used for database sizing and network traffic analysis. The application server on this tab will be shared, so the workflow performance may not be accurate. The figure above provides the results.

Workflow tab

Figures 10-29 and 10-30 represent the Workflow Module layout. Before configuring the CPT Requirements Analysis Module, you must complete your workflow assessment and identify the workflows that best represent your user environment. The Workflow Module includes all of the workflow lookup tables to support capacity planning in the CPT.

10-29 Defining the workflows
Which ones represent your user environment?

Workflow	Configuration		Platform Service Times (seconds)							Vflow Chatter	Software Service Times								Software Service Time
Select Category Workflow	Select Data Source		Arc08 baseline			SRint06/core =		17.5			Arc08 baseline		SRint06/core =			17.5			
Name Workflow as Required	Connection		Client	Desktop	Web	App Servers		Data	Data		Client	Design Model Metrics					Server		
(Copy/Insert new workflows as needed)	Platform	Data Source	Traffic	Wkstn	WTS	WA	SS	Traffic	DBMS		Traffic	AI	ADF	SOM	SOC	SDE	Traffic	DBMS	
Standard ESRI Workflows			Mbpd	Client	WTS	WA	SS	Mbpd	DBMS		Mbpd	Client	WA	SOM	SOC	SDE	Mbpd	Data	Total
1_ArcGIS Desktop	Desktop	SDE DC/DBMS	5.000	0.448				5.000	0.048	200	5.000 Mbpd	0.400				0.048	5.000 Mbpd	0.048	0.496
2a_ArcGIS WTS/Citrix (vector)	WTS	SDE DC/DBMS	0.280		0.448			5.000	0.048	10	0.280 Mbpd	0.400				0.048	5.000 Mbpd	0.048	0.496
2b_ArcGIS WTS/Citrix (w/image)	WTS	SDE DC/DBMS	1.000		0.448			5.000	0.048	10	1.000 Mbpd	0.400				0.048	5.000 Mbpd	0.048	0.496
3a_AGS9.2 AJAXlight Dynamic	Server	SDE DC/DBMS	2.000			0.096	0.336	5.000	0.048	10	2.000 Mbpd		0.096		0.288	0.048	5.000 Mbpd	0.048	0.480
3b_AGS9.2 AJAXStandard Dynamic	Server	SDE DC/DBMS	2.000			0.192	0.672	5.000	0.096	10	2.000 Mbpd		0.192		0.576	0.096	5.000 Mbpd	0.096	0.960
3c_AGS9.2 (Web Service)	Server	SDE DC/DBMS	2.000			0.032	0.336	5.000	0.048	10	2.000 Mbpd		0.032		0.288	0.048	5.000 Mbpd	0.048	0.416
3d_AGS9.2 AJAXlight Fully Cache	Server	SDE DC/DBMS	2.000			0.080	0.009	2.000	0.001	10	2.000 Mbpd		0.080		0.008	0.001	2.000 Mbpd	0.001	0.090
3e_AGS9.3 (REST API)	Server	SDE DC/DBMS	2.000			0.032	0.336	5.000	0.048	10	2.000 Mbpd		0.032		0.288	0.048	5.000 Mbpd	0.048	0.416
4a_IMS ArcMap Server (Web App)	Server	SDE DC/DBMS	1.000			0.080	0.288	5.000	0.048	10	1.000 Mbpd		0.080		0.240	0.048	5.000 Mbpd	0.048	0.416
4b_IMS ArcMap Server (Web Service)	Server	SDE DC/DBMS	1.000			0.032	0.304	5.000	0.048	10	1.000 Mbpd		0.032		0.256	0.048	5.000 Mbpd	0.048	0.384
5a_IMS Image Server (Web App)	Server	SDE DC/DBMS	1.000			0.040	0.168	5.000	0.048	10	1.000 Mbpd		0.040		0.120	0.048	5.000 Mbpd	0.048	0.256
5b_IMS Image Server (Web Service)	Server	SDE DC/DBMS	1.000			0.032	0.176	5.000	0.048	10	1.000 Mbpd		0.032		0.128	0.048	5.000 Mbpd	0.048	0.256
6_AGS9.1 Map Server	Server	SDE DC/DBMS	2.000			0.144	0.336	5.000	0.048	200	2.000 Mbpd		0.144		0.288	0.048	5.000 Mbpd	0.048	0.528
7_ArcGIS Image Server (preliminary)	Server	SDE DC/DBMS	2.000			0.016	0.336	5.000	0.016	10	2.000 Mbpd		0.016		0.336		5.000 Mbpd	0.016	0.368
8_ArcGIS Globe Server (preliminary)	Server	SDE DC/DBMS	2.000			0.001	0.004	2.000		10	2.000 Mbpd		0.001		0.004		2.000 Mbpd		0.005
9_Mobile ADF AGS Services	Server	SDE DC/DBMS	0.050					0.700		10	0.050 Mbpd		0.080		0.080	0.012	0.700 Mbpd	0.012	0.184
Batch Processing			Mbpd	Client	WTS	WA	SS		DBMS		Mbpd	Client	WA	SOM	SOC	SDE	Mbpd	Data	Total
1_Batch Map Production Process	Server	SDE DC/DBMS	2.000			0.192	0.628	5.000	0.052	10	2.000 Mbpd		0.192		0.576	0.052	5.000 Mbpd	0.052	0.872
2_Batch Reconcile and Post Process	Server	SDE DC/DBMS	0.001				0.480	5.000	0.160	200	0.001 Mbpd				0.320	0.160	5.000 Mbpd	0.160	0.640
Custom Workflows			Mbpd	Client	WTS	WA	SS		DBMS		Mbpd	Client	WA	SOM	SOC	SDE	Mbpd	Data	Total
Network Analysis Workflow	Server	SDE DC/DBMS	2.000			0.160	1.600	5.000		10	2.000 Mbpd		0.160		1.600		5.000 Mbpd		1.760
Custom ArcGIS Server Workflow	Server	SDE DC/DBMS	5.000			0.101	0.368	5.000		10	2.000 Mbpd		0.101		0.336	0.032	5.000 Mbpd	0.032	0.501
s1_AGS9.2 AJAXlight Dyn Standard	Server	SDE DC/DBMS	2.000			0.096	0.336	5.000	0.048	10	2.000 Mbpd		0.096		0.288	0.048	5.000 Mbpd	0.048	0.480
s2_AGS9.2 AJAXlight Dyn Advanced	Server	SDE DC/DBMS	2.000			0.096	0.336	5.000	0.048	10	2.000 Mbpd		0.096		0.288	0.048	5.000 Mbpd	0.048	0.480
City of Rome ==========			Mbpd	Client	WTS	WA	SS		DBMS		Mbpd	Client	WA	SOM	SOC	SDE	Mbpd	Data	Total
R_AGS9.3 Map Server	Server	SDE DC/DBMS	2.000			0.096	0.336	5.000	0.048	10	2.000 Mbpd		0.096		0.288	0.048	5.000 Mbpd	0.048	0.480
R_Server DBMS Batch Process	Server	SDE DC/DBMS	0.001				0.480	5.000	0.160	200	0.001 Mbpd				0.320	0.160	5.000 Mbpd	0.160	0.640
R_ArcGIS Desktop Editor	Desktop	SDE DC/DBMS	5.000	0.448				5.000	0.048	200	5.000 Mbpd	0.400				0.048	5.000 Mbpd	0.048	0.496
R_ArcGIS Desktop Viewer	Desktop	SDE DC/DBMS	5.000	0.448				5.000	0.048	200	5.000 Mbpd	0.400				0.048	5.000 Mbpd	0.048	0.496
R_ArcGIS Desktop Business Analyst	Desktop	SDE DC/DBMS	5.000	0.448				5.000	0.048	200	5.000 Mbpd	0.400				0.048	5.000 Mbpd	0.048	0.496
R_Remote WTS ArcGIS Desktop Viewer	WTS	SDE DC/DBMS	0.280		0.448			5.000	0.048	10	0.280 Mbpd	0.400				0.048	5.000 Mbpd	0.048	0.496
R_Mobile ADF AGS Services	Server	SDE DC/DBMS	0.050					0.700		10	0.050 Mbpd		0.080		0.080	0.012	0.700 Mbpd	0.012	0.184

* AI = ArcGIS Desktop

Data Source - Configuration Strategies:	
	File Geodatabase
SDE DC/DBMS = direct connect	Small Shape File
SDE DBMS = App Server Connect	Medium Shape File
SDE Server/DBMS = Remote Server Connect	Large Shape File

This is a high-level overview of the modules included on the Workflow tab worksheet. The Workflow tab supports several lookup tables used by the CPT Requirements Analysis Module. Before configuring the CPT Requirements Analysis Module, complete your workflow assessment and identify which workflows you need to represent your user environment.

10-30 Workflow Module layout

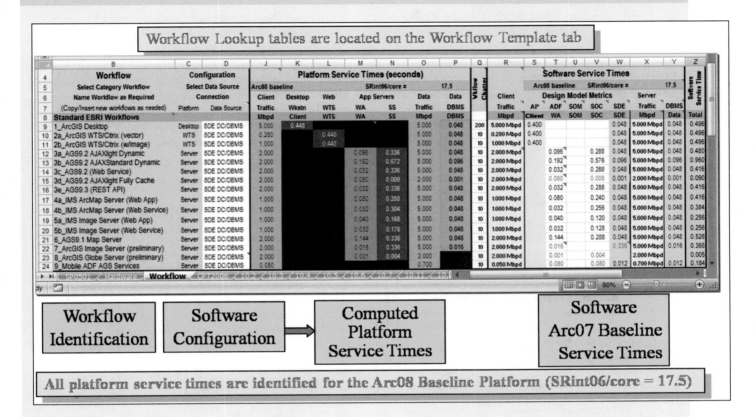

Workflow Lookup tables are located on the Workflow Template tab

All platform service times are identified for the Arc08 Baseline Platform (SRint06/core = 17.5)

The Workflow Lookup tables provide a unique workflow name and workflow software display service times, including client traffic. The data source selection for each workflow determines how the software service times will be distributed to support the platform service times. The figure above provides an overview of the Workflow Module.

The platform service times are generated on the Workflow tab and also on the Capacity Planning Tool—both are generated directly from the software service times based on the data source selection. The platform service times included on the Workflow tab are for display purposes only— the values on the CPT tab are also calculated from the software service times, and they are the ones used in the CPT sizing models.

Standard ESRI workflows

The CPT is an adaptive configuration tool that can be used with standard ESRI workflows, or if you prefer, you can define your own performance targets. Figures 10-31 through 10-35 take us from an overview of standard ESRI workflow options to comparing ArcGIS Desktop and ArcGIS Server workflows (and others), in terms of the relative performance of each software technology configuration.

Under the Workflow tab you will find the standard ESRI workflows used as a baseline reference for capacity planning. These are the same standard workflow models used in the platform sizing charts in chapter 9, where they are discussed at more length. The use of the term "standard" here refers to the fact that these workflows follow the platform sizing guidelines consistently used by ESRI in system design consulting. Developed and modified over time, they are also called "performance sizing models" and are updated annually based on feedback from real implementations.

10-31 Standard ESRI workflows

Several standard ESRI workflows are included with the Workflow tab.

These are the standard ESRI workflows used as a baseline reference for capacity planning, and can be used as a source for establishing the customer specific workflow models. For many years, we have provided standard platform sizing guidelines based on user implementation experience with ESRI core technology. These standard sizing guidelines provide consistency for system design consulting and ESRI sales. These performance sizing models are updated annually based on feedback from ESRI customer implementations. The standard ESRI workflows, along with the CPT, represent the standard platform sizing guidelines for future systems.

The number of standard workflow options increased with the ArcGIS Server 9.3 release, and are expected to continue expanding as users implement more variations of the technology. Most customers use the standard ESRI workflows to establish performance targets, since these workflow loads are evaluated against real customer experience. These standard workflows could be refined based on early performance measurements or customer experience—normally to provide a more conservative performance target.

Workflow performance models are defined based on software service times. You will see in figure 10-32 that software service times are automatically modified, based on the data source you select for the configuration. Therefore, you do not need to—and you should not—change the standard ESRI workflow service times in this section. Later on, under "custom workflows," the graphics will show how you can use a copy of the standard ESRI workflow to establish a custom workflow, one you can modify to represent your own performance targets. For planning purposes, until you can validate your own measurements, use the factors as provided.

10-32   Data source performance factors

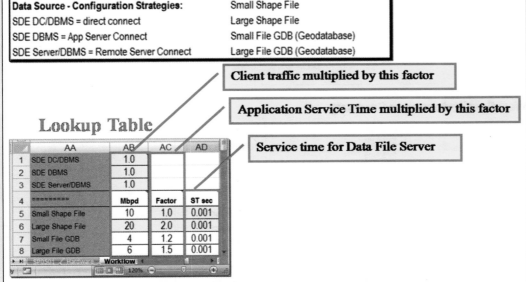

The data source selection modifies the software service times based on the selected data source. The data source options are identified under configuration options, and the performance factors are supported in a lookup table located in the Workflow tab **cell range of Y1:AB8**. The Lookup Table includes a network traffic and performance factor used to modify performance relative to the SDE performance. The factors used in the model are conservative; ArcGIS 9.3 test results with current technology suggest these factors can be reduced. Using these factors for planning purposes, until you can validate your own measurements, is recommended. (To get a sense of the size difference between small and large shapefiles, see next page.)

Processing performance will vary depending on the size of the file data source and the complexity of the map display. Baseline models are provided for both small and large file types for use in establishing appropriate performance targets—service times that can be adjusted based on a specific workflow environment. For a basic idea of what we're talking about when we say "small" (or "large") shapefiles and file GDBs—and the presumed difference between them—here are some generic definitions to work with:

- Small shapefile: Neighborhood parcel layer using ESRI_Optimized feature symbols (20 MB file with 20,000 features).
- Large shapefile: County parcel layer using ESRI_Optimized feature symbols (130 MB file with 130,000 features).
- Small file GDB: Neighborhood geodatabase using ESRI_Optimized feature symbols (140 MB file with 150,000 features).
- Large file GDB: County geodatabase using ESRI_Optimized feature symbols (1.17 GB file with 1,000,000 features).

10-33 ArcGIS Desktop workflows

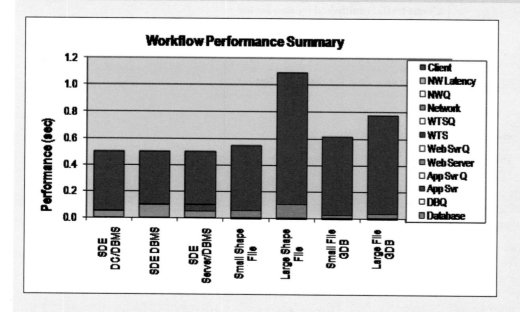

This is a Workflow Performance Summary of the ArcGIS Desktop workflows, which shows the relative performance of each software technology configuration, based on generating the same map display output. The Workflow Performance Summary provides a stacked performance profile of platform service times and computed queue times. Platform service times are based on the standard ESRI workflows discussed earlier. Platform queue times (identified after each hardware component in the key to the right) are included based on the random arrival delays discussed in chapter 7. These standard ESRI workflows are shown with capacity less than 50 percent, so there are no queue times generated for these workflows (this is the ideal performance situation).

10-34 ArcGIS Server workflows

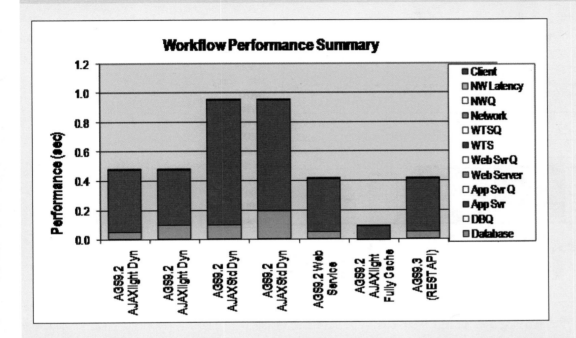

This figure provides the Workflow Performance Summary of the ArcGIS Server workflows. This chart shows the relative performance for these standard ESRI workflows, with all supporting the same peak workload and producing the same map display output. (Note that "ArcGIS 9.2 StdDyn" in the figure represents a higher quality map display—twice the processing load of the "ArcGIS 9.2 AJAX-light" display.)

10-35   Additional standard workflows

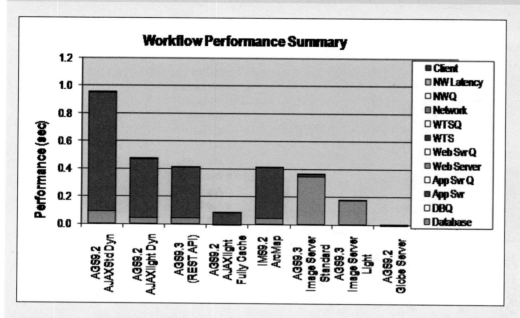

This is a collection of the remaining standard ESRI workflows—some ArcGIS Server, ArcIMS, Image Server, and Globe Server workflows—each supporting the same single-user workload and producing the same map display output.

You can generate your own custom Workflow Performance Summary by configuring the workflows you want to compare and including a single user for each workflow in the CPT Requirements Analysis Module. I changed the x-axis labels to show the data source; you can do this by right-clicking the Workflow Summary Chart, selecting source data and the Series tab, then redefining the category x-axis range to show the data source column (this should only be done by advanced Excel users).

Custom workflows

A workflow requirements analysis should be accomplished before completing a capacity planning exercise. The workflow requirements analysis should include an evaluation of the workflow loads, and how they compare with the standard ESRI workflows. Chapter 8 discusses software performance and some of the factors that affect software processing times, including examples of their impact on software performance. That's the place to start your thinking when doing your workflow analysis. The best way to compare your map display to a standard ESRI workflow is to do some simple testing. A relative performance difference between your custom map display (MXD) and the standard workflow display (MXD) will provide the information you need to modify your workflow models. For example, if a measure of the processing time to support your average map display is twice as long as what is identified in the CPT standard workflow (after correcting for any platform performance differences), you may want to establish a custom workflow with double the map services times identified in the CPT standard ESRI workflow. Custom workflows are defined from the standard ESRI workflows by copying the standard workflows to a separate area within the lookup range on the Workflow tab.

10-36 Sample workflow section

A custom set of workflows should be defined before configuring a solution on the Capacity Planning Tool.

Sample Portland Workflow Section

Here is a sample set of custom workflows included in the Workflow tab for a city.

10-37 Building a custom workflow section

Workflow (Name Workflow as Required)	Platform	Data Source	Vkflow Chatter	Client Traffic (Mbpd)	AI* (Client)	ADF (WA)	SOM (SOM)	SOC (SOC)	SDE (SDE)	Server Traffic (Mbpd)	DBMS Data	Software Service Time (Total)
Standard ESRI Workflows				Mbpd	Client	WA	SOM	SOC	SDE	Mbpd	Data	Total
1_ArcGIS Desktop	Desktop	SDE DC/DBMS	200	5.000 Mbpd	0.400				0.048	5.000 Mbpd	0.048	0.495
2a_ArcGIS WTS/Citrix (vector)	WTS	SDE DC/DBMS	10	0.280 Mbpd	0.400				0.048	5.000 Mbpd	0.048	0.495
2b_ArcGIS WTS/Citrix (w/image)	WTS	SDE DC/DBMS	10	1.000 Mbpd	0.400				0.048	5.000 Mbpd	0.048	0.495
3a_AGS9.2 AJAXlight Dynamic	Server	SDE DC/DBMS	10	2.000 Mbpd		0.096		0.288	0.048	5.000 Mbpd	0.048	0.480
3b_AGS9.2 AJAXStandard Dynamic	Server	SDE DC/DBMS	10	2.000 Mbpd		0.192		0.576	0.096	5.000 Mbpd	0.096	0.960
3c_AGS9.2 (Web Service)	Server	SDE DC/DBMS	10	2.000 Mbpd		0.032		0.288	0.048	5.000 Mbpd	0.048	0.416
3d_AGS9.2 AJAXlight Fully Cache	Server	SDE DC/DBMS	10	2.000 Mbpd		0.080		0.008	0.001	2.000 Mbpd	0.001	0.090
3e_AGS9.3 (REST API)	Server	SDE DC/DBMS	10	2.000 Mbpd		0.032		0.288	0.048	5.000 Mbpd	0.048	0.416
4a_IMS ArcMap Server (Web App)	Server	SDE DC/DBMS	10	1.000 Mbpd		0.080		0.240	0.048	5.000 Mbpd	0.048	0.416
4b_IMS ArcMap Server (Web Service)	Server	SDE DC/DBMS	10	1.000 Mbpd		0.032		0.256	0.048	5.000 Mbpd	0.048	0.384
5a_IMS Image Server (Web App)	Server	SDE DC/DBMS	10	1.000 Mbpd		0.040		0.120	0.048	5.000 Mbpd	0.048	0.256
5b_IMS Image Server (Web Service)	Server	SDE DC/DBMS	10	1.000 Mbpd		0.032		0.128	0.048	5.000 Mbpd	0.048	0.256
6_AGS9.1 Map Server	Server	SDE DC/DBMS	10	2.000 Mbpd		0.144		0.288	0.048	5.000 Mbpd	0.048	0.528
7a_ArcGIS Image Server Light (prelimina	Server	SDE DC/DBMS	10	2.000 Mbpd		0.008			0.168	5.000 Mbpd	0.008	0.184
7b_ArcGIS Image Server Standard (preli	Server	SDE DC/DBMS	10	2.000 Mbpd		0.016			0.336	5.000 Mbpd	0.016	0.368
8_ArcGIS Globe Server (preliminary)	Server	SDE DC/DBMS	10	2.000 Mbpd		0.001		0.004		2.000 Mbpd		0.005
9_Mobile ADF AGS Services	Server	SDE DC/DBMS	10	0.050 Mbpd		0.080		0.080	0.012	0.700 Mbpd	0.012	0.184
Batch Processing				Mbpd	Client	WA	SOM	SOC	SDE	Mbpd	Data	Total
1_Batch Map Production Process	Server	SDE DC/DBMS	10	2.000 Mbpd		0.192		0.576	0.052	5.000 Mbpd	0.052	0.872
2_Batch Reconcile and Post Process	Server	SDE DC/DBMS	200	0.001 Mbpd				0.320	0.160	5.000 Mbpd	0.160	0.640
Customer Workflow				Mbpd	Client	WA	SOM	SOC	SDE	Mbpd	Data	Total
New ArcGIS Desktop Light	Desktop	SDE DC/DBMS	200	5.000 Mbpd	0.200				0.024	5.000 Mbpd	0.024	0.248
Portland Workflows ==========				Mbpd	Client	WA	SOM	SOC	SDE	Mbpd	Data	Total
P_ArcGIS AI Desktop	Desktop	SDE DC/DBMS	200	5.000 Mbpd	0.400				0.048	5.000 Mbpd	0.048	0.495
P_ArcGIS AE Desktop	Desktop	SDE DC/DBMS	200	5.000 Mbpd	0.400				0.048	5.000 Mbpd	0.048	0.495
P_ArcGIS AV Desktop	Desktop	SDE DC/DBMS	200	5.000 Mbpd	0.400				0.048	5.000 Mbpd	0.048	0.495
P_ArcGIS AI WTS/Citrix (vector)	WTS	SDE DC/DBMS	10	0.280 Mbpd	0.400				0.048	5.000 Mbpd	0.048	0.495
P_ArcGIS AE WTS/Citrix (vector)	WTS	SDE DC/DBMS	10	0.280 Mbpd	0.400				0.048	5.000 Mbpd	0.048	0.495
P_ArcGIS AV WTS/Citrix (vector)	WTS	SDE DC/DBMS	10	0.280 Mbpd	0.400				0.048	5.000 Mbpd	0.048	0.495
P_AGS9.2 Map Server (AJAX App)	Server	SDE DC/DBMS	10	2.000 Mbpd		0.096		0.288	0.048	5.000 Mbpd	0.048	0.480

Software Service Times — Arc08 baseline SRint06/core = 17.5

* AI = ArcGIS Desktop

Sheet tabs: SP0501 | 050107SRint2000 | 050608SRint2006 | Hardware

Status bar: Average: 12.98494118 Workstation Count: 21 Sum: 220.744 Workstation 80%

This figure shows how to build custom workflows—ones to represent customer services, for example—to use in your capacity planning analysis. Try establishing a new workflow zone by using the gray "city" row and relabeling the row Customer Workflow, as shown above. Then you can copy a similar standard ESRI workflow from above and insert into the New Customer zone. You must then provide a new name for the custom workflow.

The new workflow can now be selected on the CPT requirements analysis (CPT2006 tab). You should define custom workflows for each design implementation, particularly if you plan to use the CPT as a performance tool during system implementation. This allows you to identify workflow names that match your specific user requirements—workflows that you would recognize in your enterprise design solution.

If you are not sure about the workflow service times, use the standard ESRI workflow models. You can change the service times for your workflows to represent your specific solution—larger service times are more conservative, and smaller service times are more aggressive. I have been setting the standard workflow model service times for the last 15 years, and I caution you about being aggressive. You will look very good in the budget meeting, but if your customized solution doesn't support the peak operational workflow requirements during implementation, you are toast.

Building composite workflows

Many of the Web applications today are supported by multiple component services. Many implementations with Image Server or cached image services may also include some dynamic operational layer overlays. Workflows that are supported by multiple services or data sources can be combined into a composite workflow used by the CPT requirements analysis. Composite workflows will provide a simpler system design analysis, and it will be much easier to explain when you present your system design recommendations.

10-38 Building composite workflows

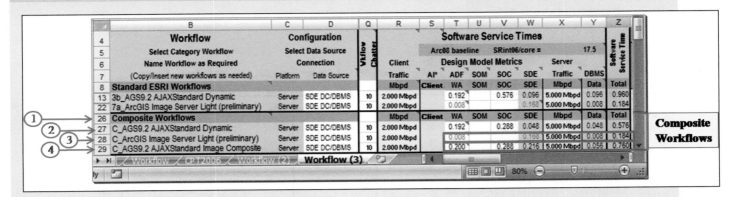

This is an example of a composite workflow. The composite workflow uses ArcGIS Server 9.2 to access an image background layer service provided by ArcGIS Image Server with a vector overlay of business layers to complete the published display.

The following steps were used in creating the composite workflow:

1. A separate composite workflow section is established on the CPT Workflow tab. Make sure the section is located above the bottom pink row so it will appear in the CPT Workflow Lookup Table. See figure 10-37, **row 39**.

2. Copy the 7_ArcGIS Image Server (preliminary) workflow row from the Standard ESRI Workflows. Insert into the Composite Workflow section and provide a unique name (C_ArcGIS Image Server).

3. Copy the 3b_AGS9.2 AJAXStandard Dynamic workflow row from the Standard ESRI Workflows. Insert into the Composite Workflow section and provide a unique name (C_AGS9.2 AJAXStandard Dynamic (vector only).

a. The ArcGIS Server vector only service times are estimated to be about 50 percent of the standard service times.

b. WA, SOC, SDE, and DBMS service times are reduced by 50 percent to provide a modified loads profile.

4. A new composite workflow is identified by adding the individual service times of the supporting services. This assumes that there is no extra overhead processing required to combine (blend) these services. The new composite workflow is provided a unique name (AGIS9.2 AJAXstandard Image Composite), which will be used to support the CPT design analysis.

The analysis above can be used to present the logic for building the composite workflow. Once the logic is understood, the new composite workflow can be used to support the CPT system design analysis. If someone wants you to explain the logic for your custom workflow, you can go to the workflow tab and track your basic assumptions and logic back to the standard ESRI workflow models.

Hardware tab

The Hardware tab takes you to the platform performance lookup tables used in the CPT. The hardware vendors provide the platform benchmark information on the SPEC Web site. We use their performance numbers in the model. Having used these SPEC benchmarks to help with thousands of customer hardware selections over two decades, we find they do quite well in representing relative platform compute performance for the purpose of supporting ESRI customer's capacity planning needs.

10-39   Hardware tab

	A	B	C	D	E	F	G	H	I
		# of	Core /	SPECint_Rate2000		SPECint_Rate2006		Total	Processor
8	==== **Server Candidates** ====	Core	Chip	Rate 2000	Per Core	Rate 2006	Per Core	Chips	Platform
9									
54	Xeon 5160 2 core (1 chip) 3000(4) MHz	2	2	60.0	30.0	28.6	14.3	1	Intel
55	Xeon 5160 4 core (2 chip) 3000(4) MHz	4	2	120.0	30.0	53.7	13.4	2	Intel
56	Xeon X5260 4 core (2 chip) 3333(6) MHz	4	2	147.2	36.8	70.1	17.5	2	Intel
65	Xeon X5355 4 core (1 chip) 2666(8) MHz	4	4	109.0	27.3	44.4	11.1	1	Intel
66	Xeon X5355 8 core (2 chip) 2666(8) MHz	8	4	200.0	25.0	80.9	10.1	2	Intel
67	Xeon X5365 4 core (1 chip) 3000(8) MHz	4	4	121.0	30.2	57.6	14.4	1	Intel
68	Xeon X5365 8 core (2 chip) 3000(8) MHz	8	4	206.0	25.8	98.1	12.3	2	Intel
71	Xeon X5450 8 core (2 chip) 3000(12) MHz	8	4	226.8	28.4	108.0	13.5	2	Intel
72	Xeon X5460 4 core (1 chip) 3166(12) MHz	4	4	130.4	32.6	62.1	15.5	1	Intel
73	Xeon X5460 8 core (2 chip) 3166(12) MHz	8	4	239.4	29.9	114.0	14.3	2	Intel
74	Xeon X7350 16 core (4 chip) 2933(8) MHz	16	4	373.8	23.4	178.0	11.1	4	Intel
92	AMD 8220 8 core (4 chips) 2800 MHz	8	2	186.9	23.4	89	11.1	4	AMD
93	AMD 8220SE 8 core (4 chips) 2800 MHz	8	2	183.1	22.9	87.2	10.9	4	AMD
94	AMD 8222SE 4 core (2 chips) 3000 MHz	4	2	105.8	26.5	50.4	12.6	2	AMD
95	AMD 8222SE 8 core (4 chips) 3000 MHz	8	2	201.8	25.2	96.1	12	4	AMD
96	AMD 8356 8 core (2 chips) 2300 MHz	8	4	186.5	23.3	88.8	11.1	2	AMD
97	AMD 8356 16 core (4 chips) 2300 MHz	16	4	336.0	21.0	160	10	4	AMD
98	==== **Exercise Candidates** ======								
99	Arc08 2 core (1 chip) Baseline	2	2	73.5	36.8	35.0	17.5	1	Intel
100	Arc08 4 core (2 chip) Baseline	4	2	147.0	36.8	70.0	17.5	2	Intel
101	Arc08 6 core (3 chip) Baseline	6	2	220.5	36.8	105.0	17.5	3	Intel
102	Arc08 8 core (4 chip) Baseline	8	2	294.0	36.8	140.0	17.5	4	Intel
128	==== **Sun Candidates** ======								
129	Sun EM4000 8 core (4 chip) 2150 MHz	8	2	144.1	18.0	68.6	8.6	4	Sun SPARC
130	Sun EM5000 16 core (8 chip) 2150 MHz	16	2	281.4	17.6	134.0	8.4	8	Sun SPARC
131	Sun EM8000 32 core (16 chip) 2400 MHz	32	2	625.8	19.6	298.0	9.3	16	Sun SPARC
132	Sun EM9000 48 core (24 chip) 2400 MHz	48	2	893.6	18.6	425.5	8.9	24	Sun SPARC
133	Sun EM9000 64 core (32 chip) 2400 MHz	64	2	1,161.3	18.1	553.0	8.6	32	Sun SPARC
134	Sun EM9000 96 core (48 chip) 2400 MHz	96	2	1,747.2	18.2	832.0	8.7	48	Sun SPARC
135	Sun EM9000 124 core (64 chip) 2400 MHz	128	2	2,333.1	18.2	1,111.0	8.7	64	Sun SPARC
160	==== **HP Candidates** ======								
161	Itanium 2 core (2 chip) 1600 MHz	3	1	35.5	11.8	16.9	5.6	3	Itanium
162	Itanium 4 core (4 chip) 1600 MHz	4	1	72.5	18.1	34.5	8.6	4	Itanium
163	Itanium 8 core (8 chip) 1600 MHz	8	1	134.0	16.8	63.8	8.0	8	Itanium
174	==== **IBM Candidates** ======								
175	IBM p6 2 core (1 chip) 4700 MHz	2	2	111.7	55.9	53.2	26.6	1	IBM pSeries
176	IBM p6 4 core (2 chip) 4700 MHz	4	2	222.6	55.7	106.0	26.5	2	IBM pSeries
177	IBM p6 8 core (4 chip) 4700 MHz	8	2	432.6	54.1	206.0	25.8	4	IBM pSeries
178	IBM p6 16 core (8 chip) 4700 MHz	16	2	861.0	53.8	410.0	25.6	8	IBM pSeries
190									

SP0050 | 050107SRint2000 | 050608SRint2006 | **Hardware**

Ready   100%

10-39 Hardware tab

The platforms that have been tested and certified for ESRI software are listed here under the Hardware tab. You will find a long list of platforms we have used in our analyses over the past several years. The CPT does not know if these are supported platforms, so you must use some judgment on selecting platforms that have been tested and certified to support your software requirements. You can find a list of supported platforms for each of ESRI's software products at the ESRI Support Center, http://support.esri.com/index.cfm?fa=software.gateway.

The Hardware tab does not include all the available vendor platforms; there are thousands of potential platform configurations, any one of which you may be interested in. I have included separate tabs with the CPT that show the SPEC published benchmarks, to be discussed later.

Hardware lookup column layout

All of the performance information needed by the CPT is found in the hardware lookup tables. There are two sets of columns for each hardware platform. One set provides the SPECrate_int2000 benchmark values and the other set of columns gives the SPECrate_int2006 results. You can see those two sets boxed in the figure below.

10-40   Hardware lookup column layout

Platforms	Core	Core/Chip	SRint2000		SRint2006		Chips	Vendor
A	B	C	D	E	F	G	H	I
7	# of	Core /	SPECint_Rate2000		SPECint_Rate2006		Total	Processor
8 ==== Server Candidates ====	Core	Chip	Rate 2000	Per Core	Rate 2006	Per Core	Chips	Platform
54 Xeon 5160 2 core (1 chip) 3000(4) MHz	2	2	60.0	30.0	28.6	14.3	1	Intel
55 Xeon 5160 4 core (2 chip) 3000(4) MHz	4	2	120.0	30.0	53.7	13.4	2	Intel
56 Xeon X5260 4 core (2 chip) 3333(6) MHz	4	2	147.2	36.8	70.1	17.5	2	Intel
65 Xeon X5355 4 core (1 chip) 2666(8) MHz	4	4	109.0	27.3	44.4	11.1	1	Intel
66 Xeon X5355 8 core (2 chip) 2666(8) MHz	8	4	200.0	25.0	80.9	10.1	2	Intel
67 Xeon X5365 4 core (1 chip) 3000(8) MHz	4	4	121.0	30.2	57.6	14.4	1	Intel
68 Xeon X5365 8 core (2 chip) 3000(8) MHz	8	4	206.0	25.8	98.1	12.3	2	Intel
71 Xeon X5450 8 core (2 chip) 3000(12) MHz	8	4	226.8	28.4	108.0	13.5	2	Intel
72 Xeon X5460 4 core (1 chip) 3166(12) MHz	4	4	130.4	32.6	62.1	15.5	1	Intel
73 Xeon X5460 8 core (2 chip) 3166(12) MHz	8	4	239.4	29.9	114.0	14.3	2	Intel
74 Xeon X7350 16 core (4 chip) 2933(8) MHz	16	4	373.8	23.4	178.0	11.1	4	Intel
92 AMD 8220 8 core (4 chips) 2800 MHz								AMD
93 AMD 8220SE 8 core (4 chips) 2800 MHz								
94 AMD 8222SE 4 core (2 chips) 3000 MHz								
95 AMD 8222SE 8 core (4 chips) 3000 MHz								
96 AMD 8356 8 core (2 chips) 2300 MHz								
97 AMD 8356 16 core (4 chips) 2300 MHz								
98 ==== Exercise Candidates ====								
99 Arc08 2 core (1 chip) Baseline	2	2	73.5	36.8	35.0	17.5	1	Intel
100 Arc08 4 core (2 chip) Baseline	4	2	147.0	36.8	70.0	17.5	2	Intel
101 Arc08 6 core (3 chip) Baseline	6	2	220.5	36.8	105.0	17.5	3	Intel
102 Arc08 8 core (4 chip) Baseline	8	2	294.0	36.8	140.0	17.5	4	Intel
128 ==== Sun Candidates ====								
129 Sun EM4000 8 core (4 chip) 2150 MHz	8	2	144.1	18.0	68.6	8.6	4	Sun SPARC
130 Sun EM5000 16 core (8 chip) 2150 MHz	16	2	281.4	17.6	134.0	8.4	8	Sun SPARC
131 Sun EM8000 32 core (16 chip) 2400 MHz	32	2	625.8	19.6	298.0	9.3	16	Sun SPARC
132 Sun EM9000 48 core (24 chip) 2400 MHz	48	2	893.6	18.6	425.5	8.9	24	Sun SPARC
133 Sun EM9000 64 core (32 chip) 2400 MHz	64	2	1,161.3	18.1	553.0	8.6	32	Sun SPARC
134 Sun EM9000 96 core (48 chip) 2400 MHz	96	2	1,747.2	18.2	832.0	8.7	48	Sun SPARC
135 Sun EM9000 124 core (64 chip) 2400 MHz	128	2	2,333.1	18.2	1,111.0	8.7	64	Sun SPARC
160 ==== HP Candidates ====								
161 Itanium 2 core (2 chip) 1600 MHz	3	1	35.5	11.8	16.9	5.6	3	Itanium
162 Itanium 4 core (4 chip) 1600 MHz	4	1	72.5	18.1	34.5	8.6	4	Itanium
163 Itanium 8 core (8 chip) 1600 MHz	8	1	134.0	16.8	63.8	8.0	8	Itanium
174 ==== IBM Candidates ====								
175 IBM p6 2 core (1 chip) 4700 MHz	2	2	111.7	55.9	53.2	26.6	1	IBM pSeries
176 IBM p6 4 core (2 chip) 4700 MHz	4	2	222.6	55.7	106.0	26.5	2	IBM pSeries
177 IBM p6 8 core (4 chip) 4700 MHz	8	2	432.6	54.1	206.0	25.8	4	IBM pSeries
178 IBM p6 16 core (8 chip) 4700 MHz	16	2	861.0	53.8	410.0	25.6	8	IBM pSeries

General Benchmark Translation

SRint2006 = SRint2000 / 2.1

050107SRint2000   050608SRint2006   **Hardware**

Ready   100%

These are lookup tabs from figure 10-5. The CPT2006 tab uses the SRint2006 columns. The SRint2000 benchmarks provide relative performance values for the older server platforms. Performance for all the new platforms are published on the SRint2006 benchmarks (www.spec.org). Some of the new hardware configured form the old processor technology (dual-core 3.2GHz platforms) may only have performance benchmarks published on the SRint2000 site, so you may need to convert these values for use in your CPT2006 analysis. You can divide the published baseline SPECrate_int2000 values by a factor of 2.1 to estimate an equivalent SPECrate_int2006 value for use in the Capacity Planning Tool. Both values should be included in the Hardware tab.

10-41   Hardware lookup platform listing

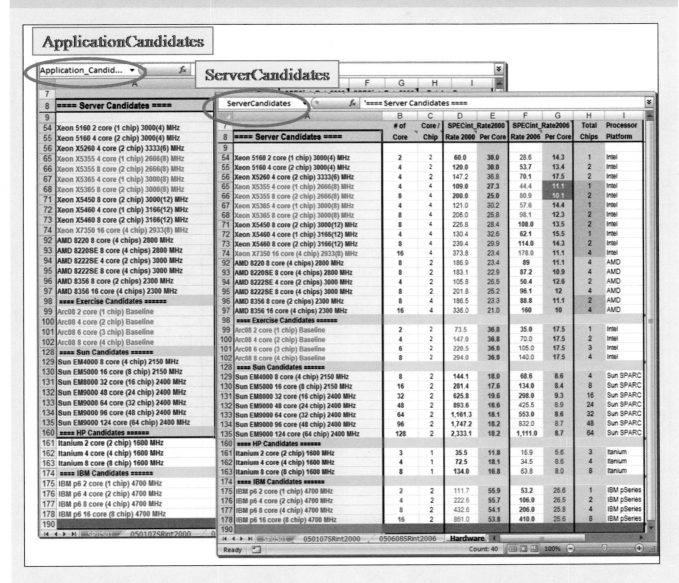

This figure zeroes in on the CPT2006 platform selection lookup tables located on the Hardware tab. The application server candidates (ASC) include all the platforms above the yellow HP candidates line on the table (this keeps the application server lookup shorter). The ASC are used for configuring Windows Terminal Server and the Web Servers. Sun Solaris platforms are included in this list. If you need to include HP or IBM Unix platforms in this list, you may have to insert the platforms you are interested in above this line. The database server candidates column (DSC) includes the full list.

The performance lookups used in the CPT will review the complete list; only the ASC platform selection drop-downs provide less than the full list. If you have a platform selected that is identified in the lookup table, the CPT will find the performance benchmark numbers.

The CPT uses the SPEC benchmark, number of cores, number of chips, and the benchmark/core values to enable the model calculations.

Vendor-published hardware performance tabs

The SRint2000 (CINT2000 Rates) and the SRint2006 (CINT2006 Rates) from SPEC are downloaded into separate tabs in the CPT. These tabs are for reference only, so you have easy access to what you could find on the Web. A date is included with the tab to identify the last update. You can access the SPEC site and create your own worksheet update if you need more current information. The figure below provides an overview of the SPEC reference tabs (www.specbench.org).

10-42 Vendor-published hardware performance benchmark tabs

Hardware Vendor	System	# Cores	# Chips	# Cores Per Chip	CPU	MHz	2nd Level Cache	Operating System	Baseline	Baseline/ core
Dell Inc.	Dell	2	1	2	Intel Core 2	2933	4 MB I+D on chip	Windows XP	29.7	14.9
Dell Inc.	Dell	2	1	2	Intel Xeon 5160	3000	4 MB I+D on chip	Windows XP	29.2	14.6
Dell Inc.	PowerEd	4	2	2	Intel Xeon 5160	3000	4 MB I+D on chip	Microsoft	53.7	13.4
Dell Inc.	PowerEd	4	2	2	Intel Xeon 5160	3000	4 MB I+D on chip	Microsoft	53.7	13.4
Dell Inc.	PowerEd	4	2	2	Intel Xeon 5160	3000	4 MB I+D on chip	Microsoft	53.7	13.4
Dell Inc.	Dell	4	2	2	Intel Xeon 5160	3000	4 MB I+D on chip	Windows XP	53.2	13.3
Dell Inc.	PowerEd	4	2	2	Intel Xeon			Microsoft	50	12.5
Dell Inc.	PowerEd	4	2	2	AMD Opteron	3000	2 MB I+D on chip	SuSE Linux	48.5	12.1
Dell Inc.	PowerEd	8	2	4	Intel Xeon E5355	2666	8 MB I+D on chip	Microsoft	69	8.6
Dell Inc.	PowerEd	8	2	4	Intel Xeon E5345	2333	8 MB I+D on chip	Microsoft	75	9.4
Dell Inc.	PowerEd	8	2	4	Intel Xeon X5355	2666	8 MB I+D on chip	Windows	80.8	10.1
Dell Inc.	PowerEd	8	2	4	Intel Xeon X5355	2666	8 MB I+D on chip	Windows	80.4	10.1
Dell Inc.	PowerEd	8	2	4	Intel Xeon E5335	2000	8 MB I+D on chip	Microsoft	69	8.6
Dell Inc.	PowerEd	8	2	4	Intel Xeon E5345			Microsoft	74.9	9.4
Dell Inc.	PowerEd	8	2	4	Intel Xeon X5355			Windows	80.9	10.1
Dell Inc.	PowerEd	8	2	4	Intel Xeon E5335			Microsoft	68.8	8.6
Dell Inc.	PowerEd	8	2	4	Intel Xeon E5345			Microsoft	74.8	9.4
Dell Inc.	PowerEd	8	2	4	Intel Xeon X5355			Windows	80.9	10.1
Dell Inc.	Dell	2	1	2	Intel Pentium			Windows XP	23.1	11.6
Dell Inc.	PowerEd	4	1	4	Intel Xeon X3210			Windows	39.8	10.0
Dell Inc.	PowerEd	4	1	4	Intel Xeon X3220			Windows	43	10.8
Dell Inc.	PowerEd	4	1	4	Intel Xeon X3210			Windows	39.5	9.9
Dell Inc.	PowerEd	4	1	4	Intel Xeon X3220			Windows	43.2	10.8
Dell Inc.	Dell	2	1	2	Intel Core 2 Duo	2333	4 MB I+D on chip	Windows XP	23	11.5
Dell Inc.	PowerEd	4	2	2	Intel Xeon 5140	2333	4 MB I+D on chip	Microsoft	45.9	11.5
Dell Inc.	PowerEd	4	2	2	Intel Xeon 5140	2333	4 MB I+D on chip	Microsoft	45.9	11.5
Dell Inc.	PowerEd	4	2	2	AMD Opteron 2220	2800	1 MB I+D on chip	SuSE Linux	45.4	11.4
Dell Inc.	PowerEd	4	2	2	Intel Xeon 5130	2000	4 MB I+D on chip	Microsoft	41.6	10.4
Dell Inc.	PowerEd	4	2	2	Intel Xeon 5130	2000	4 MB I+D on chip	Microsoft	41.6	10.4
Dell Inc.	PowerEd	4	2	2	AMD Opteron 2216	2400	1 MB I+D on chip	SuSE Linux	40.3	10.1
Dell Inc.	PowerEd	4	2	2	Intel Xeon 5120	1866	4 MB I+D on chip	Microsoft	37.9	9.5

AMD Opteron 3000 MHz 4 core, 2 Core/Chip SRint2006 = 48.5

Use the SPEC tabs to identify new platform performance benchmarks. The example above identifies an AMD Opteron 3000 MHz 4-core, 2-core/chip platform with SRint2006 = 48.5.

Inserting a new hardware platform

Your existing platform environment may not be included on the Hardware tab, or you may want to include a new vendor platform release as a candidate for your system design. In either case, you need to know how you can insert a new hardware platform into the Hardware tab.

10-43   Inserting a new hardware platform

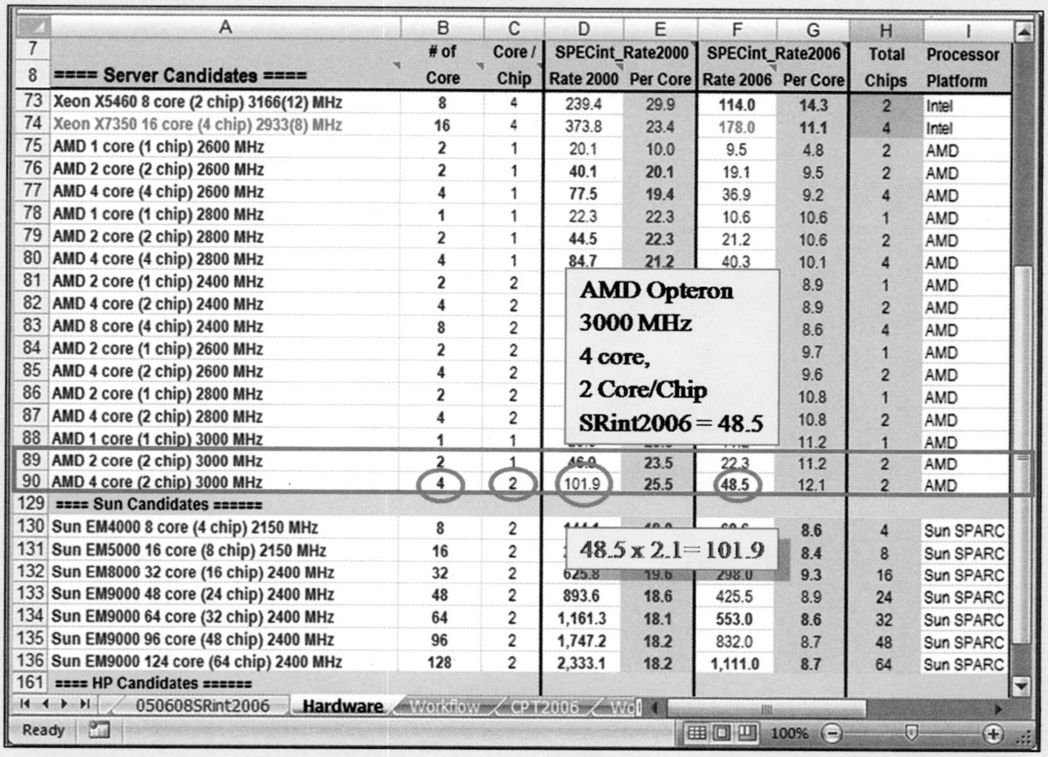

	A	B	C	D	E	F	G	H	I
7		# of	Core /	SPECint_Rate2000		SPECint_Rate2006		Total	Processor
8	==== Server Candidates ====	Core	Chip	Rate 2000	Per Core	Rate 2006	Per Core	Chips	Platform
73	Xeon X5460 8 core (2 chip) 3166(12) MHz	8	4	239.4	29.9	114.0	14.3	2	Intel
74	Xeon X7350 16 core (4 chip) 2933(8) MHz	16	4	373.8	23.4	178.0	11.1	4	Intel
75	AMD 1 core (1 chip) 2600 MHz	2	1	20.1	10.0	9.5	4.8	2	AMD
76	AMD 2 core (2 chip) 2600 MHz	2	1	40.1	20.1	19.1	9.5	2	AMD
77	AMD 4 core (4 chip) 2600 MHz	4	1	77.5	19.4	36.9	9.2	4	AMD
78	AMD 1 core (1 chip) 2800 MHz	1	1	22.3	22.3	10.6	10.6	1	AMD
79	AMD 2 core (2 chip) 2800 MHz	2	1	44.5	22.3	21.2	10.6	2	AMD
80	AMD 4 core (4 chip) 2800 MHz	4	1	84.7	21.2	40.3	10.1	4	AMD
81	AMD 2 core (1 chip) 2400 MHz	2	2				8.9	1	AMD
82	AMD 4 core (2 chip) 2400 MHz	4	2				8.9	2	AMD
83	AMD 8 core (4 chip) 2400 MHz	8	2				8.6	4	AMD
84	AMD 2 core (1 chip) 2600 MHz	2	2				9.7	1	AMD
85	AMD 4 core (2 chip) 2600 MHz	4	2				9.6	2	AMD
86	AMD 2 core (1 chip) 2800 MHz	2	2				10.8	1	AMD
87	AMD 4 core (2 chip) 2800 MHz	4	2				10.8	2	AMD
88	AMD 1 core (1 chip) 3000 MHz	1	1				11.2	1	AMD
89	AMD 2 core (2 chip) 3000 MHz	2	1	46.0		23.5	11.2	2	AMD
90	AMD 4 core (2 chip) 3000 MHz	4	2	101.9	25.5	48.5	12.1	2	AMD
129	==== Sun Candidates ======								
130	Sun EM4000 8 core (4 chip) 2150 MHz	8	2				8.6	4	Sun SPARC
131	Sun EM5000 16 core (8 chip) 2150 MHz	16	2				8.4	8	Sun SPARC
132	Sun EM8000 32 core (16 chip) 2400 MHz	32	2	625.8	19.6	298.0	9.3	16	Sun SPARC
133	Sun EM9000 48 core (24 chip) 2400 MHz	48	2	893.6	18.6	425.5	8.9	24	Sun SPARC
134	Sun EM9000 64 core (32 chip) 2400 MHz	64	2	1,161.3	18.1	553.0	8.6	32	Sun SPARC
135	Sun EM9000 96 core (48 chip) 2400 MHz	96	2	1,747.2	18.2	832.0	8.7	48	Sun SPARC
136	Sun EM9000 124 core (64 chip) 2400 MHz	128	2	2,333.1	18.2	1,111.0	8.7	64	Sun SPARC
161	==== HP Candidates ======								

Annotations within the figure:

AMD Opteron 3000 MHz 4 core, 2 Core/Chip SRint2006 = 48.5

48.5 x 2.1 = 101.9

Tabs: 050608SRint2006 / **Hardware** / Workflow / CPT 2006 / Wo
Ready — 100%

First identify the new platform on the SPEC benchmark tabs (new platforms are on the CPT SRint2006 tabs or on the SPEC Web site). If the benchmark information is not available, you will need to work with the vendor to estimate relative performance from a published benchmark value. For our insert example, we use the AMD Opteron 3000 MHz 4-core, 2-core/chip platform with SRint2006 = 48.5, identified on the SRint2006 tab (figure 10-38). Find a platform similar to the new selected platform (in this case an AMD 2-core (2-chip) 3000 MHz platform was found in the CPT Hardware tab). Select row, copy, and insert copied cells. Our example produces a copy of **row 86** that can be updated with the new platform specifications. Provide a unique platform name (AMD 4-core (2-chip) 3000 MHz platform). Update the row columns in white with the new benchmark information. Insert the SRint2006 benchmark baseline value in the SRint2006 column (**F90**). Enter an estimated value for the SRint2000 benchmark by multiplying the SRint2006 value by a factor of 2.1 (**D90**). All columns should be checked to be sure you have identified the proper number of cores and performance/cores (**B90 = 4, C90 = 2**). These are the numbers that will be used by the CPT when using this platform for your design analysis.

The CPT has come a long way since January 2006. All of the programming is supported by standard Microsoft Office Excel 2003 functions. The CPT has been tested on Microsoft Office Excel 2007 and works fine—there is more functionality in the 2007 edition that will be helpful in making future versions of the tool even more powerful. In providing the flexibility to adapt to technology change, the CPT should help many people better understand and manage their enterprise GIS operations. For an extended example of how that might be done, the next chapter takes you through the system design process—using the CPT and its underlying fundamentals—for the fictional City of Rome.

11 Completing the system design

OVERVIEW

This chapter describes the actual process of system design, using a case study of a fictional organization, City of Rome. Now that you have an understanding of the technology and performance models, you can follow along and use the Capacity Planning Tool (CPT) to design your own system. This chapter provides a step-by-step guide to the system design process itself: this is how the bits and pieces of the puzzle are actually brought together. Everything before this has been a part of the puzzle. Chapters 1 through 6 gave you an overview of the technology (what the parts of the elephant look like). The next section told you how these components act together, and some of the principles underlying why. All the Capacity Planning Tool functionality is based on the performance fundamentals in chapter 7, the software performance in chapter 8, and then more specifically on the verities of platform performance in chapter 9. Chapter 10 explained how to use Excel functions to modify the CPT—information you need to be able to customize the CPT for your own system design process.

The CPT is a tool, a software product that can be used to support any system design project. City of Rome is one example of how it might be used. City of Rome is also the name of the case study provided to demonstrate the planning process presented in Roger Tomlinson's book *Thinking About GIS*. Both his book's chapter 10 and this chapter show standard templates that can be used for most enterprise design studies. In this book, however, we will use the CPT to model the user requirements for three planned years of expansion and growth. With this tool, we'll be able to make network and platform tuning adjustments based on our understanding of the system performance and scalability theory explained in chapter 7. Here, you will use the CPT to establish minimum platform and network requirements, complete a network and platform performance tuning process, and generate final design specifications.

Another way to describe the system design process is as an integrated business needs assessment. I would be remiss if I didn't reiterate that the prerequisite for engaging in this system architecture design methodology is a user needs assessment. The purpose of the system design process in this chapter is for you to establish what you require in the way of hardware and network bandwidth in order to fulfill the performance and communication needs of GIS users. If you don't specifically know the needs to be met and have not quantified them, how can you possibly come to any sound conclusion about the hardware specifications required to do the job? And yet many organizations launch into procurement for GIS implementation without having identified their user needs. This can be a very costly and risky way to go, and it is not system architecture design.

The system design process allows you to figure out hardware requirements based on identified business needs. By the time you begin the process described in this chapter, your team will have planned for GIS, ideally by following the methodology Roger Tomlinson lays out in his book, *Thinking About GIS* (now in its third edition and better than ever). If you have done so, not only will you have your organization's real needs in mind but also you will have identified the specific GIS applications required to create these needed information products, right down to the details of how they will function as workflows. These details, after you take into account some principles of physics and engineering that apply to them, become the system requirements. They, in turn, become the hardware specifications and the network bandwidth describing the right system for your GIS. The CPT can automate this process for you, but the results will be only as good as your own input. In other words, you have to do your homework; otherwise, the weakest link in the chain will be you.

An application needs assessment should be completed by the business units that benefit from the GIS information products. The needs assessment should be led by in-house GIS staff with support from an executive sponsor. You may have used professional GIS consultants to facilitate the planning process, but all the decision making will come from your end—from the organization's managers and decision makers. Planning is critical in justifying the required GIS investments, in providing a framework for enterprise GIS implementation, and in ensuring upper management support throughout the process. In fact, a well-documented system design, the product of a thorough integrated business needs assessment, is prerequisite to winning final approval for the GIS project and the go-ahead for implementation scheduling and procurement. It is also prerequisite for successful GIS operations from thereon, because system design (and the user needs assessment at its heart) is an ongoing process. Before every upgrade or deployment of new technology—whenever there's a change—everything must be reassessed.

Most GIS deployments evolve over several years of incremental technology improvements, and implementation plans normally address a two- or three-year schedule to ensure that the budget is in place for the anticipated deployment needs. So in this case study, we will use the CPT to prepare for year 1, year 2, and year 3.

In case you jumped to this chapter, rather than read what is before it, just know I have anticipated that quite a few of you might do this. What do you need to have done before moving right into this chapter's case study? We haven't talked about that since chapter 1—and we won't talk about it much here because you have Roger Tomlinson's book to thoroughly show you how to do it. But do it you must: you must take the time to complete a user needs assessment, in order to identify what your organization intends to get out of the system; and you must spend some time establishing your workflow performance targets.

The official presumption is that you have already done a user needs assessment before launching into system design, but in case you haven't, this chapter begins with what you must do to be successful. We review the information collected during the user requirements analysis, then identify templates to use for data collection of peak user loads. To further guide you, examples of the standard design templates used during the system architecture design process are included along with the interfaces you will see when using the Capacity Planning Tool. You can use them both or you can use the CPT alone.

GIS user needs assessment

There are a few basic user requirements that must be understood before launching into the system design process. In order to design an effective system for GIS, you must come to grips with the three basic factors that are the focus of the system architecture needs assessment (aka integrated business needs assessment, or simply user needs assessment). These requirements comprise the system architecture needs assessment: identifying where the GIS users are located in relation to the associated data resources (site locations); what network communications are available to connect user sites with the GIS data sources; and what are the peak user workflow requirements for each user location. Figure 11-1 provides an overview of the system architecture needs assessment.

GIS user locations
All user locations requiring access to GIS applications and data resources must be identified. You want to include everyone who might need access to the system during peak work periods. Therefore, the term "user locations" encompasses local users, remote users on the wide area network (WAN), and Internet users (internal and public).

The enterprise infrastructure must be able to support the peak user workflows. Knowing where users are located, understanding what applications they will

need to do their work, and identifying the location of the required data resources provide the basis for system design analysis.

Network communications
In the system design assessment, you must identify the network communication bandwidth between the different user locations and the data center. Network bandwidth may establish communication constraints to consider when developing the software technology solution. The final technology solution may require upgrades to the network communication infrastructure. Considering that the GIS traffic must be supported with all the other traffic (e-mail, video conferencing, document management, data replication services, etc.), what bandwidth will be available for GIS operations? Is it enough for everybody?

User workflows
The peak system workflow loads are the peak users and workflow platform service times discussed in chapter 10. These will determine computer processing and network capacity requirements. Traditional capacity planning was done with simple workflow models, using separate models for GIS desktop workflows, Web services, and batch processing. Platform sizing models were established to help select the right server platform technology. General network design guidelines were used to address communication suitability. These simple platform sizing models and network design guidelines demonstrated the value of a proper design in ensuring successful enterprise GIS implementations.

The CPT introduced in 2006 (chapter 10) now provides an opportunity for more granular workflow

GIS User Locations
- User departments
- Site locations

Network Communications
- Local area network standards
- Wide area network bandwidth

GIS User Types
- ArcGIS (ArcInfo, ArcEditor, ArcView)
- Web services (Web information products, project reporting)
- Concurrent batch processes (Examples: reconcile/post, online compress, data loading, online backup, replication, and other heavy geoprocessing during peak production workflow)

Figure 11-1

System architecture needs assessment.

models. The CPT incorporates all of the platform sizing models we have developed over the past 16 years; only what we have added to it is new. Now it puts a broader range of fundamental performance and scalability relationships to the task of translating peak user workflow service time loads to specific hardware platform and network infrastructure solutions. The tool's ability to dynamically model the design alternatives provides an adaptive, integrated, management view of the complete technical solution.

There is a trade-off between simplicity and complexity in building such a system performance model. Simplicity is easier to understand, simpler to quantify, enables broader validation, helps quantify business risk, and provides valuable information on which to base business decisions. This is the 10,000-foot model of the business solution. Complexity may be more accurate, may provide a closer representation of the final implementation, and may lead to more detailed results. Yet complex models can be much more difficult to understand, harder to quantify, more difficult to validate, and may include hidden risk. This is the 1,000-foot model of the business solution. During the planning phase, it is best to provide a simple model to represent the technical solution and lead to the right business decisions. A simple model is best: it highlights the relationship between what the organization wants to do with GIS (what you want out of the system) and the technology you need to do it (software and hardware procurement decisions).

Planning should establish performance targets that you can manage throughout the implementation phase. The CPT models used during planning are based on what other people are able to do with the technology. You can use the CPT models to establish your own system performance targets, based on your specific infrastructure limitations and operational needs. This chapter will show you how to build your own solution—one that can achieve your performance goals—according to the guidelines described in chapter 8. Afterwards, you must monitor and validate that these goals are met throughout the GIS implementation (see chapter 12).

City of Rome user requirements analysis

The fictional City of Rome represents a typical organization, just right as a case study to demonstrate how you can use the Capacity Planning Tool in your system design process. In planning a GIS for this city, we're going to look ahead to year 2 and year 3 even as we plan for the year 1 implementation. A city government is indeed a typical organization to use as an example

of an enterprise GIS, particularly because no one will have any trouble imagining that it will grow. And what we are designing when we design a system infrastructure for the first launch of GIS is a scalable system and one that will grow as the organizations needs grow. Fortunately, component hardware now lends itself to growing in increments—the day of purchasing one big machine, and hoping it will last for five or ten years, is over. There is nothing to be gained by purchasing what you don't yet need. By looking ahead for two or three years or even five years, however, you can plan for future purchases and upgrades that will meet the need when you have a use for them. So our system design will take into account year 2 and year 3, in addition to year 1.

Let's begin by taking stock of the city government's current situation and how exactly the organization and its employees are looking forward to using GIS. This city has more than 580 employees who require GIS information to help in their normal work processes (information products). These employees are located in the Planning, Engineering, Police, and Operations departments throughout the city. The public will also benefit from deployment of standard GIS information products through published Web applications (services). Each department provides a set of these Web services, which it shares with the public on the city's Web site.

Figure 11-2 provides a sample format for identifying user locations throughout the operations environment. This is an overview of the facility locations and network communications to be addressed in the system design study. The point is to show how each user location is connected to the data center (LAN, WAN, Internet) along with the available network capacity (T1 here, a 1.54 Mbps bandwidth).

GIS user types

The types of GIS users can be divided into three basic categories. The ArcGIS Desktop user type will require desktop applications for GIS processing. Web users will be supported by ArcGIS Server Web applications. An additional batch process to support replication services and standard administrative batch processing (reconcile and post, online backups, etc.) is the third user type.

ArcGIS Desktop: This category includes ArcGIS Desktop specialists doing general spatial query and analysis studies, simple map production, and general-purpose query and analysis operations, including all ArcEditor and ArcView clients. GIS applications for custom business solutions and any custom ArcGIS Engine clients that support specific business needs should also be included in this category.

For this study, separate workflows are identified for ArcGIS Desktop editors and viewers. A separate

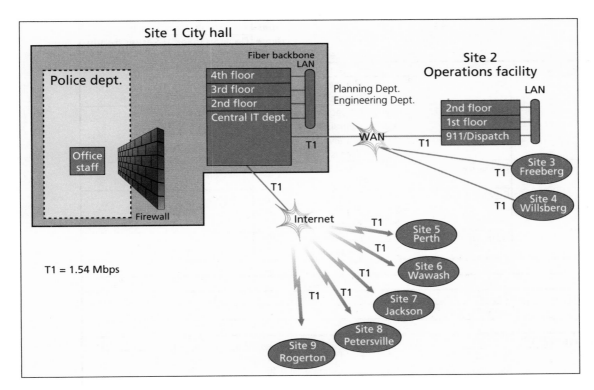

Figure 11-2

Network communications overview: A sample format for identifying
user locations throughout the operations environment.

workflow is also included for ArcGIS Desktop Business Analyst. The initial Desktop workflows use the published standard ESRI workflow models. These workflows can be modified as required following initial applications testing, which is performed to validate that capacity planning performance targets are being achieved.

Web services: ESRI ArcGIS Server software provides transaction-based map products to Internet browser clients and supports synchronization with remote mobile clients. Web dynamic mapping services are supported by the ArcGIS Server 9.2 AJAX light workflow, while mobile clients are supported by the ArcGIS Server Mobile ADF workflow. The standard ESRI workflow loads are used for initial planning purposes. These workflows can be modified as required following initial applications testing, to validate that capacity planning performance targets are being achieved.

Batch processing: A batch process load profile will be included in the design to account for the administrative batch process workflows planned during peak processing periods. These may include online backups, replication services, reconcile-and-post services, and so forth. A platform utilization profile can be established to represent the batch processing model loads (this

profile would depend on potential batch processing needs). It will be a system administration responsibility to make sure the number of concurrent batch processes are managed within these limits during peak processing loads.

The capacity planning framework introduced in the last chapter provides the flexibility to include different desktop and server workflow models in a single system design assessment. Workflow models are defined in terms of component service time metrics. Component service time target performance loads are provided for ESRI standard COTS (commercial off-the-shelf) software based on performance validation testing and customer feedback. These standard ESRI workflow models, which are used as the ESRI baseline for system architecture design consulting, have proven to be adequate to the task of helping ESRI customers find successful design solutions. New technology options introduced with the ArcGIS 9.3 software release may require different workflow models, and these new models can be represented in the future by appropriate component service times identified for these workflows in the capacity planning framework. Careful selection of appropriate service time metrics is an important part of the system architecture design study.

City of Rome - Year 1				Total Users	Peak Workflow Loads		
Department	Workflow	IPD	User type		Desktop (users) ArcInfo	ArcView	Server (req/hr) AGS
Site 1 - City Hall							
Planning	Zoning	1.0	Planner	20		8	
		1.1	Web services				2,600
	Permits	1.2	Inspector	20		10	
		1.3	Appraiser	15	8		
		1.4	Supervisor	2		2	
		1.5	Web services				600
Engineering	Sewer Backup	2.1	Engineer	4		3	
		2.2	Web services				800
	Electrical Breaks	2.3	Electrician	13	6		
		2.4	Supervisor	2	1		
		2.5	Web services				600
	Hwy. repair	2.6	Field engineer	10		4	
		2.7	Contracts	4		4	
City hall totals				90	15	31	
IT Department	Public		Web Services				**5,800**
Site 2 - Operations							
Operations	Clean-up prog.	3.1	Ops. Staff	4		2	
Operations totals				4	0	2	0
Remote field offices (WAN)							
Site 3 - Freeberg	Inspection	4.1	Field engineer	40		30	
Site 4 - Willsberg	Inspection	4.1	Field engineer	30		20	
Field Offices	Inspection	4.2	Web Services				1,200
Remote totals				70	0	50	
City Totals				164	15	83	0

Note: Peak Web Users = Peak Requests per minute / 6 displays per minute

Figure 11-3

City of Rome planning workflow requirements: Year 1. A simple spreadsheet identifies user workflow requirements (peak workflow loads) at each user location at the department level.

User workflow requirements

The user-needs template of figure 11-3, used here to document the year 1 user application requirements for the City of Rome, was designed to integrate the requirements analysis provided in Tomlinson's *Thinking About GIS* with what is needed to complete the system architecture design. A simple spreadsheet, it identifies user workflow requirements (peak workflow loads) at each user location at the department level.

Peak workflow loads (identified during the requirements definition—or user needs assessment—stage) establish processing and traffic requirements used by the Capacity Planning Tool to generate hardware specifications. Each department manager should be called upon during the design process to validate that these peak user loads are accurate estimates. Final negotiations on hardware selection should focus on these peak user workflow requirements. Peak user requirements are used by the Capacity Planning Tool to generate system loads (the processing and traffic requirements referred to above) and determine the final hardware and network solution.

The following information is included in the columns on the chart:

Department: This is the lowest organizational level addressed in the case study.

Workflow: This is the type of work the department does or is involved in.

Information Product Description (IPD): This provides a key for relating workflow to information products identified in the user needs assessment. (An information product is GIS-generated information rendered in a way that is useful—usually a map, but sometimes a table or a list. Information products are made by GIS applications—applications which, when used together, become a workflow.)

User Type: This identifies the job title or type of user requiring specific GIS applications to support their workflow.

Total Users: This is the total number of GIS application users in the department.

Peak Workflow Loads: This represents the peak number of users on the system at one time. ArcGIS

Desktop users are identified by product level (ArcInfo, ArcEditor, ArcView). Web services are represented by estimated peak users (peak users were calculated from estimated peak map displays per hour). The peak user loads will be applied in the design analysis to identify system capacity requirements.

The Web services are published from the IT data center to the public Internet site (in this case study,

departments did not access Web services from internal locations).

Again, each department manager is responsible for validating the final workflow requirements. Workflow requirements are identified for each implementation phase. Figure 11-4 identifies the City of Rome year 2 workflow requirements, and figure 11-5 (next page) identifies workflow requirements for year 3.

City of Rome - Year 2				Total Users	Peak Workflow Loads			
					Desktop (Users)		Server (req/hr)	
Department	Workflow	IPD	User type		ArcInfo	ArcView	Map	Mobile
Site 1 - City Hall								
Planning	Zoning	1.0	Planner	25		15		
		1.1	Web services				2,600	
	Permits	1.2	Inspector	25		15		
		1.3	Appraiser	20	10			
		1.4	Supervisor	5	2			
		1.5	Web services				900	
Engineering	Sewer Backup	2.1	Engineer	5		3		
		2.2	Web services				1,000	
	Electrical Breaks	2.3	Electrician	13	6			
		2.4	Supervisor	2	1			
		2.5	Web services				1,900	
	Hwy. repair	2.6	Field engineer	11		7		
		2.7	Contracts	4		4		
City Hall LAN Totals				110	19	44		
IT Department	Public		Web Services				**12,900**	
Police (Firewall)	Patrol sched.	5.1	Admin.	10		3		
		5.4	Web services					100
	Crime analysis	5.2	Detectives	10	5			
	Spec. events	5.3	Traffic	10		3		**100**
Remote Patrols	Patrols	5.5	Patrol officers	20				
Police Network Totals				50	5	6		100
Site 2 - Operations								
Operations	Clean-up prog.	3.1	Ops. staff	4		2		
911	Response	3.2	Call takers	50		30		
		3.3	Web services				4,000	
Remote vehicles	Dispatch	3.4	Drivers	30				
Operations totals				84	0	32		
Remote field offices (WAN)								
Site 3 - Freeberg	Inspection	4.1	Field engineer	45		30		
Site 4 - Willsberg	Inspection	4.1	Field engineer	60		40		
WAN Field Offices	Inspection	4.2	Web Services				1,200	
Remote field office (WAN) totals				105	0	70		
Remote field offices (Internet)								
Site 5 - Perth	Inspection	4.3	Field engineer	10		2		
Site 6 - Wawash	Inspection	4.3	Field engineer	50		40		
Site 7 - Jackson	Inspection	4.3	Field engineer	60		20		
WAN Field Offices	Inspection	4.2	Web Services				1,300	
Remote field office (Internet) totals				120	0	62		
City totals (excluding Police private network)				419	19	208	12,900	

Note: Peak Web Users = Peak Requests per minute / 6 displays per minute

Figure 11-4

City of Rome planning workflow requirements: Year 2. This user needs assessment identifies information you will eventually transfer to the CPT (see pages 241–242).

City of Rome - Year 3				Total Users	Peak User Workflow				
					Desktop (Users)			Server (req/hr)	
Department	Workflow	IPD	User type		ArcInfo	ArcView	Bus Anal	Map	Mobile
Site 1 - City Hall									
Planning	Zoning	1.0	Planner	25		15			
		1.1	Web services					2,600	
	Permits	1.2	Inspector	25		15			
		1.3	Appraiser	20	10				
		1.4	Supervisor	5	2				
		1.5	Web services					900	
Engineering	Sewer Backup	2.1	Engineer	5		3			
		2.2	Web services					1,000	
	Electrical Breaks	2.3	Electrician	13	6				
		2.4	Supervisor	2	1				
		2.5	Web services					1,900	
	Hwy. repair	2.6	Field engineer	11		7			
		2.7	Contracts	4		4			
	Work Orders	2.8	Managers	4		3			
	Work Reports	2.9	Field Units	10		2			
Business	Site Sel. and Natureserve	6.1	Planners	10			8		
Development	FMA Flood Zone, Serviced Land	6.2	Planners	10			8		
	Emergency Response, Time, etc.	6.3	Planners	2			2		
	Web Services	6.4						2,000	
City Hall LAN Totals				146	19	49	18		
IT Department	Public		Web Services					17,600	
Police (Firewall)	Patrol sched.	5.1	Admin.	10		3			
		5.5	Web services						200
	Crime analysis	5.2	Detectives	20	15				
	Spec. events	5.3	Traffic	10		3			
Remote Patrols	Police Dispatch	5.4	Traffic	10		3			
	Patrols/Routing	5.6	Patrol officers	20					200
Police Network Totals				70	15	9			200
Site 2 - Operations									
Operations	Clean-up prog.	3.1	Ops. staff	4		2			
911	Response	3.2	Call takers	50		30			
		3.3	Web services					4,000	
Remote vehicles	Fire and Ambalance Dispatch	3.4	Schedulers	30		30			
	Routing	3.5	Drivers	30					
Snow Clearing	Scheduling	3.6	Engineers	4		4			
	Snow Plows	3.7	Drivers	100					
Operations totals				218	0	66			
Remote field offices (WAN)									
Site 3 - Freeberg	Inspection	4.1	Field engineer	45		30			
Site 4 - Willsberg	Inspection	4.1	Field engineer	60		40			
WAN Field Offices	Inspection	4.2	Web Services					1,200	
Remote field office (WAN) totals				105	0	70			
Remote field offices (Internet)									
Site 5 - Perth	Inspection	4.3	Field engineer	10		2			
Site 6 - Wawash	Inspection	4.3	Field engineer	50		40			
Site 7 - Jackson	Inspection	4.3	Field engineer	60		20			
Site 8 - Petersville	Inspection	4.3	Field engineer	80		60			
Site 9 - Rogerton	Inspection	4.3	Field engineer	80		60			
Internet Field Offices	Inspection	4.2	Web Services					4,000	
Remote field office (Internet) totals				120	0	182			
City totals (excluding Police private network)				589	19	367	18	0	

Note: Peak Web Users = Peak Requests per minute / 6 displays per minute

---

Figure 11-5

City of Rome planning workflow requirements: Year 3.

Year 2 includes implementation of a separate secure network to support police operations. The police network will be a separate design, and geodatabase replication services will provide communication between the city network and the police network. A new Mobile ADF application will support the police patrols, using background communications over dial-up connections to synchronize with the ArcGIS Server mobile application.

Year 2 city network deployment adds 911 services within the Operations Department along with a new dispatch operation and implementation of three additional field offices (Perth, Wawash, and Jackson). (See figure 11-15 for the CPT analysis for how all these changes will impact the network.)

Year 3 includes deployment of new Business Analyst and Job Tracking (JTX) work order management applications. A tracking server implementation is deployed to facilitate snowplow scheduling. Two new remote field offices are added to the system. The police department adds a police dispatch and implements tracking server solution with the 20 police patrols.

Now that you've got it all down on the template, your overall task is to figure out what infrastructure you'll need to support all these new workflows. In other words, to design the system you must do a GIS user requirements analysis. Performing a proper one is hard work. The organization staff must work together to identify workflow processes and agree on business needs. Estimating the peak number of users for each user workflow is a fundamental part of doing business; once identified, this number will affect decisions on staffing and software licensing, as well as the hardware and infrastructure cost to support these workflows. Understanding and getting this right will make a big difference in user productivity and the success of the system.

City of Rome system architecture design review

People skills and experience in maintaining distributed computer system solutions are important considerations when selecting a system design. Maintenance of the distributed computer environment is a critical factor in selecting appropriate vendor solutions. What experience and training in maintaining specific computer environments already exists in the organization? The answer to this may in itself identify a particular design solution as the best fit for an organization. This and the other considerations listed in figure 11-6 must be analyzed and understood before you can develop a proper design solution.

Current technology enables distribution of GIS solutions to clients throughout an enterprise environment, but there are limitations that apply to any distributed computer system design. It is important to clearly understand real GIS user needs and discuss alternative options for meeting those needs with system support staff to identify the most cost-effective solution. It may be necessary to review multiple software workflows and a variety of system deployment alternatives to identify and establish the best implementation strategy.

Platform and network environments

Whether on your own or in concert with a design consultant, you should review the vendor platforms and network environments currently maintained by the organization. Hardware experience, maintenance relationships, and staff training represent a considerable amount of investment for any organization. Proposed GIS design solutions should take advantage of corporate experience gained from working with the established platform and network environment.

Hardware policies and standards

Organizations develop policies and standards that support their hardware investment decisions. Understanding management preferences and associated vendor relationships will provide insight into a design solution that can be supported best by the organization.

Operational constraints and priorities

Understanding the type of operations supported by the GIS solution will identify requirements for fault tolerance, security, application performance, and the type of client/server architecture that would be appropriate to support these operations.

System administration experience

The skills and experience of the system support staff provide a foundation for the final design solution. Understanding network administration and hardware support experience, in conjunction with support staff preference for future administration strategies, will help guide the consultant to compatible hardware vendor solutions.

Financial considerations

The final design must be affordable. An organization will not implement a solution that is beyond its financial resources. With system design, cost is a function of performance and reliability. If cost is an issue, the system design must facilitate a compromise between user application performance, system reliability, and cost. The design consultant must identify a hardware solution that provides optimum performance and reliability within identified budget constraints.

Figure 11-6

General system design considerations.

Year 1 capacity planning

Figure 11-7 provides an overview of City of Rome year 1 implementation strategy. A server-based architecture will be deployed from the central IT data center for the year 1 implementation. Server platforms will include a Windows Terminal Server farm to support the remote ArcGIS Desktop users, ArcGIS Server to support the public Web services, and a central GIS data server to support the enterprise geodatabase.

Year 1 workflow requirements analysis
The system architecture design process is supported by the Capacity Planning Tool. The planning workflow

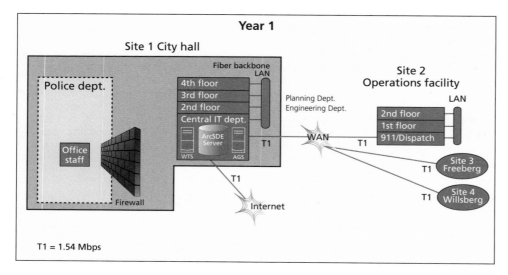

Figure 11-7

User locations and network: Year 1.

requirements identified earlier in figure 11-13 are used for the workflow analysis. For some organizations, the CPT may be used to collect workflow requirements and complete the analysis without using such separate workflow requirements templates. Organizations should use the methodology appropriate to their design needs. The templates pictured earlier, with planning workflow requirements inserted as on a spreadsheet, provide the most complete user needs representation and establish appropriate documentation for validating the simpler workflow representations displayed in the CPT, as follows.

The first step in the system design process is to create custom GIS workflows for City of Rome in the CPT Workflow tab. Procedures for creating custom workflows were discussed in chapter 10, wherein standard ESRI

workflow templates (included in the CPT Workflow tab) were used to establish such custom workflows. Workflow consolidation simplifies the look of the requirements analysis, but user workflows can only be consolidated when they are supported by the same software technology and have the same software service times. You can use a large number of separate workflows in the CPT, yet keeping workflows to a reasonable number will clarify the presentation. The City of Rome example simplifies the analysis display by assuming all ArcGIS Desktop workflows are similar and can be represented by common productivity. If a more detailed analysis is needed, this can be completed on more detailed tabs with a summary provided for presentation. Figure 11-8 shows the custom workflows established for the City of Rome system design.

	B		C	D	Q	R	S	T	U	V	W	X	Y	Z
4	**Workflow**		**Configuration**		Wkflow Chatter	**Software Service Times**								Software Service Time
5	Select Category Workflow		Select Data Source				Arc08 baseline		SRint06/core =			17.5		
6	Name Workflow as Required		Connection			Client	**Design Model Metrics**				Server			
7	(Copy/Insert new workflows as needed)		Platform	Data Source		Traffic	AI*	ADF	SOM	SOC	SDE	Traffic	DBMS	
31	City of Rome Workflows ==========					Mbpd	Client	WA	SOM	SOC	SDE	Mbpd	Data	Total
32	R_ArcGIS9.2 AJAX Map Server		Server	SDE DC/DBMS	200	2.000 Mbpd		0.096		0.288	0.048	5.000 Mbpd	0.048	0.480
33	R_Server DBMS Batch Process		Server	SDE DC/DBMS	200	5.000 Mbpd		0.101		0.336	0.032	5.000 Mbpd	0.032	0.501
34	R_ArcGIS Desktop Editor		Desktop	SDE DC/DBMS	200	5.000 Mbpd	0.400				0.048	5.000 Mbpd	0.048	0.496
35	R_ArcGIS Desktop Viewer		Desktop	SDE DC/DBMS	10	0.280 Mbpd	0.400				0.048	5.000 Mbpd	0.048	0.496
36	R_ArcGIS Desktop Business Analyst		Desktop	SDE DC/DBMS	10	0.280 Mbpd	0.400				0.048	5.000 Mbpd	0.048	0.496
37	R_Remote WTS ArcGIS Desktop Viewer		WTS	SDE DC/DBMS	10	0.280 Mbpd	0.400				0.048	5.000 Mbpd	0.048	0.496
38	R_Mobile ADF AGS Services		Server	SDE DC/DBMS	10	2.000 Mbpd		0.080		0.080	0.012	0.700 Mbpd	0.012	0.184
39							* AI = ArcGIS Desktop							

Figure 11-8

City of Rome custom workflows.

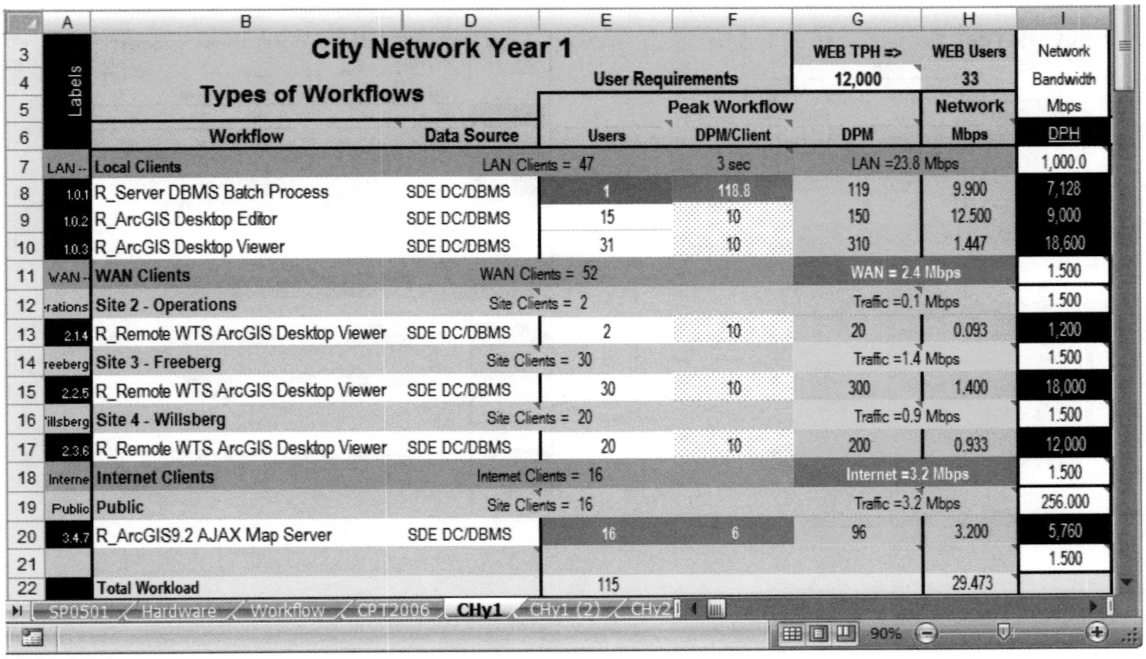

Figure 11-9

Workflow requirements analysis: Year 1. The CPT computes all site workflow traffic going over the network connections to complete the performance analysis (DPM x Mbpd/60 sec.).

Each organization's solution will be different. Several decisions must be made during the design process before a final representation is collected in the CPT. The process and discussion leading up to the final design should be documented to buttress the design process. Design documentation should clearly define the basis for the final workflow representation.

Once the custom workflows are defined, we are ready to complete the phase 1 workflow requirements analysis. Figure 11-9 shows a CPT representation of the year 1 workflow requirements analysis for City of Rome.

The CPT workflow requirements analysis includes all of the workflows identified during planning. Remember that common workflows can be consolidated at each site location to simplify the display. The CPT workflows must track back to represent the individual user workflow requirements (figure 11-3). While you configure the user requirements and site locations in the CPT, and update the site traffic summation (SUM) ranges to include all site workflow traffic going over the network connections (these CPT configuration procedures were discussed in chapter 10), the CPT completes the performance analysis. Workflow network traffic is computed from total workflow requirements (DPM x Mbpd/60 sec.).

Year 1 workflow component service time validation

In the CPT, ESRI's standard workflow service times are provided as a starting point for configuration planning. These workflow templates represent target performance milestones for the application software. By opening CPT columns J through P you can find the workflow platform components service times that will be used by the configuration tool for the capacity planning analysis. It is a good idea to verify that the workflow service times represented in the CPT are correct and that you agree with the custom workflows established to support the workflow analysis. Figure 11-10 (next page) provides an overview of the workflow component service times used for the City of Rome analysis.

Once the user workflow requirements are configured in the CPT, it will provide a quick assessment of network bandwidth requirements. The network bandwidth connections represented on the network diagram (figure 11-7) should be entered on the CPT (figure 11-9) .

Data Center Network Connections (represented by the gray rows)

    LAN: Cell I7
    WAN: Cell I11
    Internet: Cell I18

Remote Site Network Connections

    Site 2–Operations: Cell I12

	B	D	E	F	G	H	I	J	K	L	M	N	O	P
3	**City Network Year 1**				WEB TPH =>	WEB Users	Network	**Platform Service Times (seconds)**						
4	**Types of Workflows**		User Requirements		12,000	33	Bandwidth	Arc08 4 core (2 chip) Baseline					BM/Core	17.5
5			Peak Workflow				Mbps	Client	Desktop	Desktop	Web	Server	DB	Data
6	Workflow	Data Source	Users	DPM/Client	DPM	Network Mbps	DPH	Mbpd	Client	WTS	WA	SS	Mbpd	DBMS
7	Local Clients	LAN Clients = 47		3 sec	LAN =23.8 Mbps		1,000.0	9	11	12	13	14	15	16
8	R_Server DBMS Batch Process	SDE DC/DBMS	1	118.8	119	9.900	7,128	5.000			0.101	0.368	5.000	0.032
9	R_ArcGIS Desktop Editor	SDE DC/DBMS	15	10	150	12.500	9,000	5.000	0.448				5.000	0.048
10	R_ArcGIS Desktop Viewer	SDE DC/DBMS	31	10	310	1.447	18,600	0.280	0.448				5.000	0.048
11	WAN Clients	WAN Clients = 52			WAN = 2.4 Mbps		1.500							
12	Site 2 - Operations	Site Clients = 2			Traffic =0.1 Mbps		1.500							
13	R_Remote WTS ArcGIS Desktop Viewer	SDE DC/DBMS	2	10	20	0.093	1,200	0.280		0.448			5.000	0.048
14	Site 3 - Freeberg	Site Clients = 30			Traffic =1.4 Mbps		1.500							
15	R_Remote WTS ArcGIS Desktop Viewer	SDE DC/DBMS	30	10	300	1.400	18,000	0.280		0.448			5.000	0.048
16	Site 4 - Willsberg	Site Clients = 20			Traffic =0.9 Mbps		1.500							
17	R_Remote WTS ArcGIS Desktop Viewer	SDE DC/DBMS	20	10	200	0.933	12,000	0.280		0.448			5.000	0.048
18	Internet Clients	Internet Clients = 16			Internet =3.2 Mbps		1.500							
19	Public	Site Clients = 16			Traffic =3.2 Mbps		256.000							
20	R_ArcGIS9.2 AJAX Map Server	SDE DC/DBMS	16	6	96	3.200	5,760	2.000			0.096	0.336	5.000	0.048
21							1.500							
22	Total Workload		115			29.473			0.448	0.448	0.099	0.354		0.046

Figure 11-10

Workflow component service time validation: Year 1. A way to verify that the workflow service times represented in the CPT are correct and that you agree with the custom workflows established to support the workflow analysis.

Site 3–Freeberg: Cell I14
Site 4–Willsberg: Cell I16
Public Internet Clients: Cell I19 (This represents multiple sites, and should be set at a high bandwidth—more than twice the total Internet traffic—so it will not affect the analysis.)

For WANs, the service provider (i.e., telephone company) lends its services based on the peak bandwidth requirements at each connection to the WAN cloud. The central data center connection is often more than each of the remote office connections. The Internet Service Provider provides bandwidth access to the Internet based on peak bandwidth requirements as well.

The network traffic totals should be checked to make sure the SUM range is correct. Data center site ranges should include upper and lower gray rows bounding each network workflow (LAN, WAN, and Internet). Remote site ranges should include all workflows within the site and can include the upper and lower green rows. Validating that these ranges are correct will ensure that the CPT does the proper calculations.

Year 1 network bandwidth suitability

Once the network connections are defined and the summary ranges validated, the CPT will identify potential traffic bottlenecks. The bottlenecks show up both as colored network traffic summary cells and as network queue time on the Workflow Performance Summary. You can see the latter in figure 11-11, showing the

performance problems identified by the CPT during the workflow requirements analysis in figure 11-9.

In figure 11-10, the network traffic summary cells turn yellow when traffic is more than 50 percent of existing network bandwidth, and turn red when traffic exceeds existing network bandwidth. The network queue times (NWQ) are identified on the Workflow Performance Summary above the network service time for each workflow.

For year 1, both the WAN and Internet traffic are well above the current bandwidth capacity. The Workflow Performance Summary shows the expected display response time due to network contention—the Internet application display response time (workflow 3.4.7) is over 12 seconds during peak loads. The Workflow Performance Summary is a powerful information product generated by the Capacity Planning Tool, and is described in detail in chapter 10 (figure 10-12). Each of the component platform and network services times are identified in the key, with the associated queue time for each platform and network component identified by a white bar above each. Each workflow is displayed (identified by an identification key from the CPT—in column A) located left of each workflow in the User Requirements Module. For each workflow, you have a visual display of the service times and the overall response time (sum of all service times and queue times). The Workflow Performance Summary is dynamically updated (based on the selected hardware in the Plat-

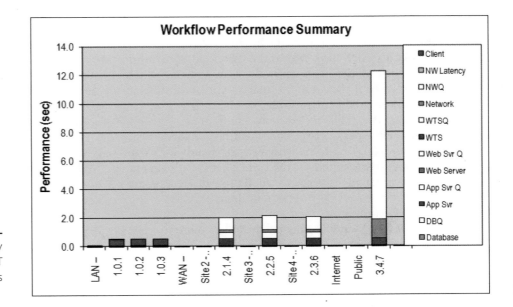

Figure 11-11

City network suitability: Year 1. Shows any performance problems identified by the CPT during the workflow requirements analysis in figure 11-9.

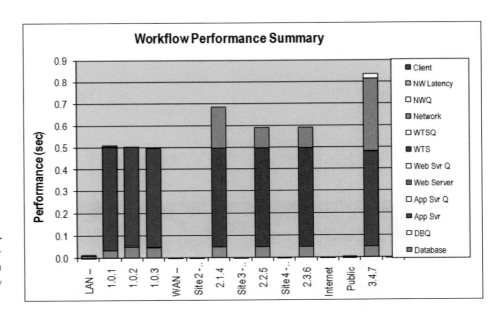

Figure 11-12

Network performance tuning: Year 1. Following the recommended network bandwidth upgrades, the public workflow 3.0.7 display response time is now less than 1 second.

form Selection Module) as you complete the user requirements configuration.

Year 1 network performance tuning

After identifying the network performance bottlenecks, the Capacity Planning Tool can be adjusted to resolve the particular performance problems. The standard recommendation is to configure network bandwidth to at least double the projected peak traffic requirements.

Remember, we only show the GIS traffic in this analysis; the WAN and Internet connections are shared by other users throughout the organization, and their needs should also be represented in the analysis. It is

possible to include an additional workflow representing other projected network traffic—this would certainly be advisable when performance is important, which it usually is.

City of Rome decided to upgrade the data center WAN and Internet connections to 6 Mbps, and the Site 3 (Freeberg) and Site 4 (Willsberg) connections to 3 Mbps. Figure 11-12 shows the Workflow Performance Summary following the recommended network bandwidth upgrades. The public workflow 3.4.7 display response time is now less than 1 second.

Appropriate network bandwidth upgrades can be represented in the CPT, and the network traffic cell colors and Workflow Performance Summary will respond with the proper adjustments. This is a discussion you should have with the network administrator; he or she can identify other traffic requirements, which can easily be included in the analysis and confirm what network upgrades are possible.

Once the CPT is properly configured, you can compare the peak network bandwidth (provided by the workflow analysis tool) with existing bandwidth connections (represented on the earlier network diagram). Then you can identify the required upgrades, if any, and recommend them in your system design.

The City of Rome year 1 implementation encompasses users at four locations. City hall includes an intranet connection to enable WAN communications with the remote offices and an Internet connection over which the published public Web mapping services can be transported. Each remote office includes a router connection to the city WAN. Traffic requirements for each user workflow are represented in the workflow analysis network traffic (column H), as seen in figures 11-9 and 11-10.

Year 1 hardware platform configuration

After you have input the year 1 peak user workflow requirements in the CPT User Requirements Module, the CPT Platform Configuration Module can be used to make the final platform selection and complete the system configuration recommendations. The CPT incorporates the performance models (chapter 7) and the vendor platform performance benchmarks

(chapter 9) discussed earlier. The CPT uses the established peak user workflow requirements, along with vendor-relative performance benchmarks, to generate the number of platform nodes required to handle the peak workflow loads. (The system architect can select from a variety of vendor platforms for the final configuration.)

The hardware platform configuration in figure 11-13 is supported by Xeon X5260 4 core (2 chip) 3333(6) MHz hardware platforms (as of mid-2008, our favorite application server platform). These platforms represent the ESRI 2008 platform performance baseline.

Peak capacity workflows projected for the year 1 deployment can be supported with two Windows Terminal Servers; a single ArcGIS Server container machine supporting the ADF, SOC, and ArcSDE direct-connect software components; and a Windows database platform. All hardware platforms are supported with Xeon X5260 4-core (2-chip) 3333(6) MHz platforms configured with 20 GB physical memory (RAM) (Note that 8 GB RAM would be sufficient for the Web server.) Projected CPU utilization rates are shown for each tier of the recommended platform solution.

Year 2 capacity planning

A similar analysis can be completed for the year 2 City of Rome implementation. Most GIS deployments evolve over several years of incremental technology

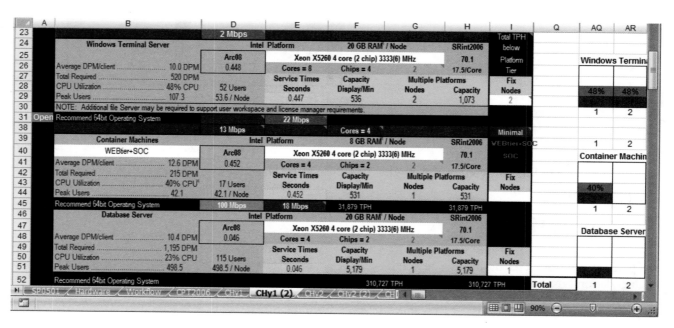

Figure 11-13

Hardware platform solution: Year 1. The favored application server platform as of mid-2008—Intel Xeon X5260 4-core (2-chip) 3333 (6) MHz platforms—supports this configuration and represents the ESRI 2008 performance baseline.

improvements, and implementation plans normally address a two- or three-year schedule, to ensure that the budget is in place for the anticipated deployment needs. Figure 11-14 identifies user locations for the year 2 implementation.

Several additional Internet remote sites will be included in year 2, along with deployment of the 911 dispatch and the initial police network. The police network will be supported by a separate server environment (ArcSDE DBMS server for the geodatabase and an ArcGIS Server for the Mobile ADF police patrol application. Geodatabase replication will be used for data updates to the police geodatabase). City network operations will continue to be administered from the IT data center.

Year 2 workflow analysis

See figure 11-15 for results of the year 2 city network workflow analysis, as they appear in the CPT. You will need to transfer the workflow requirements identified

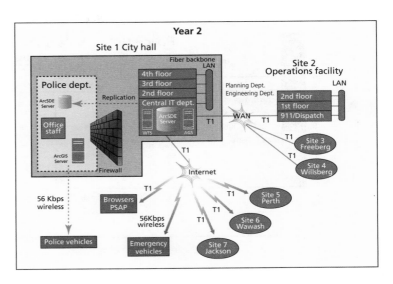

Figure 11-14

User locations and network: Year 2.

	A	B	D	E	F	G	H	I
3	Labels	**City Network Year 2**				WEB TPH =>	WEB Users	Network
4		**Types of Workflows**		User Requirements		12,000	33	Bandwidth
5				Peak Workflow			Network	Mbps
6		Workflow	Data Source	Users	DPM/Client	DPM	Mbps	DPH
7	LAN -	Local Clients	LAN Clients = 64		3 sec	LAN =26.4 Mbps		1,000.0
8	1.0.1	R_Server DBMS Batch Process	SDE DC/DBMS	1	102.11	102	8.509	6,127
9	1.0.2	R_ArcGIS Desktop Editor	SDE DC/DBMS	19	10	190	15.833	11,400
10	1.0.3	R_ArcGIS Desktop Viewer	SDE DC/DBMS	44	10	440	2.053	26,400
11	WAN -	WAN Clients	WAN Clients = 102			WAN = 4.8 Mbps		6.000
12	ations	Site 2 - Operations	Site Clients = 32			Traffic =1.5 Mbps		1.500
13	2.1.4	R_Remote WTS ArcGIS Desktop Viewer	SDE DC/DBMS	32	10	320	1.493	19,200
14	eeberg	Site 3 - Freeberg	Site Clients = 30			Traffic =1.4 Mbps		3.000
15	2.2.5	R_Remote WTS ArcGIS Desktop Viewer	SDE DC/DBMS	30	10	300	1.400	18,000
16	llsberg	Site 4 - Willsberg	Site Clients = 40			Traffic =1.9 Mbps		3.000
17	2.3.6	R_Remote WTS ArcGIS Desktop Viewer	SDE DC/DBMS	40	10	400	1.867	24,000
18	Interne	Internet Clients	Internet Clients = 98			Internet =10.1 Mbps		6.000
19	Public	Public	Site Clients = 36			Traffic =7.2 Mbps		256.000
20	3.4.7	R_ArcGIS9.2 AJAX Map Server	SDE DC/DBMS	36	6	216	7.200	12,960
21	- Perth	Site 5 - Perth	Site Clients = 2			Traffic =0.1 Mbps		1.500
22	3.5.8	R_Remote WTS ArcGIS Desktop Viewer	SDE DC/DBMS	2	10	20	0.093	1,200
23	awash	Site 6 - Wawash	Site Clients = 40			Traffic =1.9 Mbps		1.500
24	3.6.9	R_Remote WTS ArcGIS Desktop Viewer	SDE DC/DBMS	40	10	400	1.867	24,000
25	ackson	Site 7 - Jackson	Site Clients = 20			Traffic =0.9 Mbps		1.500
26	3.7.10	R_Remote WTS ArcGIS Desktop Viewer	SDE DC/DBMS	20	10	200	0.933	12,000
27								1.500
28		Total Workload		264			41.249	

CPT2006 / CHy1 / CHy1 (2) / **CHy2** / CHy2 (2) / CHy2 (3) — 80%

Figure 11-15

City network workflow requirements analysis: Year 2. Results as they appear in the CPT, after transferring information from the user needs assessment chart on page 219 (figure 11-4).

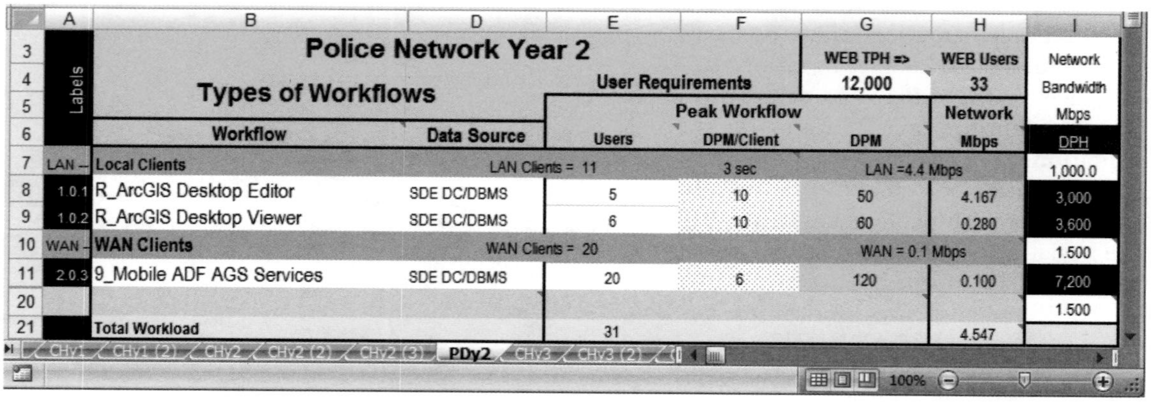

Figure 11-16

Police network workflow requirements analysis: Year 2. The CPT interface using requirements identified in the user needs assessment charted in figure 11-4.

in the user needs assessment (figure 11-4 on page 233) to the workflow analysis module in the configuration tool. Two separate CPT tabs must be completed, one for the city network and a separate CPT tab for the police network. The year 2 worksheet for the city network was updated in the CPT by copying the year 1 worksheet to a separate tab and inserting the additional locations and user workflows, to complete the user requirements. Year 1 network bandwidth upgrades were included as a starting point for the year 2 analysis.

The year 2 implementation includes three additional remote sites with access over the data center Internet network. The new remote site network bandwidth connections represented on the network diagram (figure 11-14) are identified on the CPT workflow requirements analysis display. The new remote site bandwidth connections identified from figure 11-14 are as follows:

New remote site network connections
Site 5—Perth: 1.5 Mbps (Cell I21)
Site 6—Wawash: 1.5 Mbps (Cell I23)
Site 7—Jackson: 1.5 Mbps (Cell I25)
Internet Web browser PSAP and emergency vehicles were included with the public Web services.

The network traffic totals should be checked to make sure the SUM ranges are updated (remote site ranges should include upper and lower green lines bounding the site workflows, and the central data center ranges should include the total range of each gray network—LAN, WAN, Internet). Validating that these ranges are correct will ensure the proper calculations are supported by the CPT.

Figure 11-16 provides the results of the year 2 police network workflow requirements analysis. The workflow requirements identified in the user needs assessment (figure 11-4) were transferred to the Configuration Tool Workflow Analysis Module. You will need to com-

plete the year 2 police network user requirements on a separate CPT worksheet tab, configured as inputs to the police network workflow requirements analysis.

The year 2 police implementation includes local network clients and remote mobile police patrol cars. Each police patrol communications unit is supported over dedicated dial-up phone connections (bandwidth is set at 1.5 Mbps, well beyond twice the total connection bandwidth, since there is no need to evaluate network suitability).

Year 2 network suitability

With the network connections defined and the summary ranges validated, the CPT will identify potential traffic bottlenecks. The bottlenecks show up both as colored network traffic summary cells and as network queue time on the Workflow Performance Summary. The city network Workflow Requirement Analysis (figure 11-15) shows four site connections over 50 percent capacity (City WAN, Operations, Willsberg, and Jackson) and two site connections exceeding bandwidth capacity (City Internet and Wawash). Figure 11-17 provides an overview of the city network year 2 Workflow Performance Summary. Within it, the network queue times (NWQ) are identified above the (brown) network service time for each workflow. Both the WAN and Internet traffic are well above the current bandwidth capacity. The Workflow Performance Summary shows the expected display response time due to network contention—the Wawash (site 4 workflow 3.6.9) browser display response time is over 2 seconds.

Year 2 network performance tuning

Once the network performance bottlenecks are identified, the CPT can be adjusted to resolve the identified performance problems. The standard guideline is to

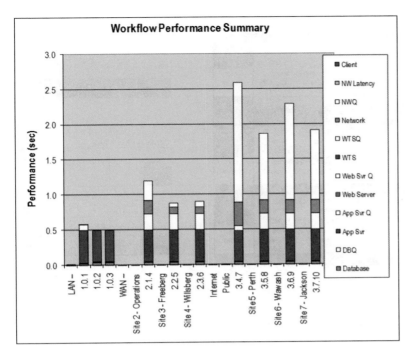

Figure 11-17

City network suitability: Year 2. Identifying the bottlenecks.

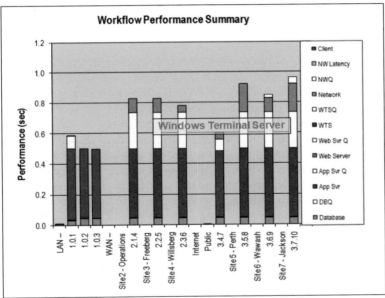

Figure 11-18

Network performance tuning: Year 2.

configure the network bandwidth to at least double the projected peak traffic requirements. The network administrator may have other network traffic loads to consider, so final recommendations should address enterprise traffic requirements overall.

City of Rome decided to upgrade network connections for Operations, Freeberg, and Wawash to 3 Mbps, Willsberg to 6 Mbps, city WAN to 12, and the city Internet connection to 45 Mbps. Figure 11-18 shows the results of the recommended network bandwidth adjustments.

Appropriate network bandwidth upgrades can be entered in the Capacity Planning Tool, and the network

traffic cell colors and Workflow Performance Summary will respond with the proper adjustments. Other traffic requirements can easily be included in the analysis, so have a discussion with the network administrator to see what they are and confirm that your recommended network upgrades can be supported

Year 2 hardware selection, performance review

The CPT can be used to select the year 2 platform configuration. The CPT display (figure 11-19) shows the minimum recommended platform solution for the year 2 city network workflows when using the

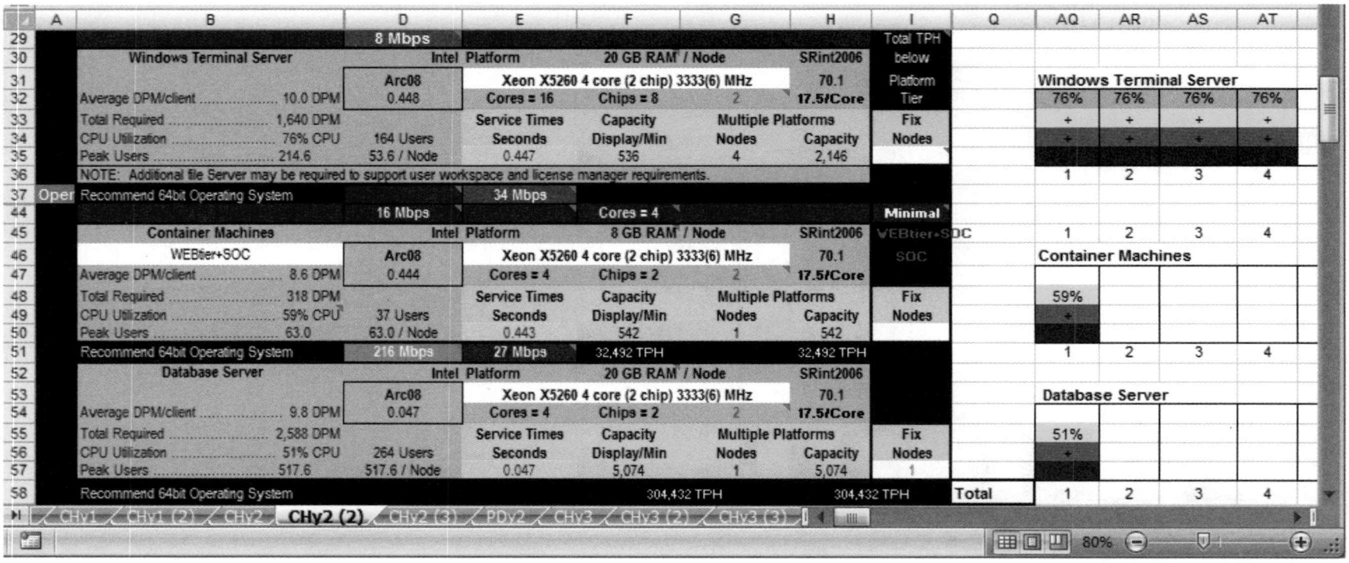

Figure 11-19

Hardware selection and performance review: Year 2. The CPT can be used to select the year 2 platform configuration. The CPT display shows the minimum recommended platform solution for the year 2 city network workflows when using the Intel Xeon 5260 4-core (2-chip) 3333(6) MHz platforms. It also provides an overview of the CPT platform configurations.

Xeon X5260 4 core (2 chip) 3333(6) MHz platforms. Platform configuration requirements are provided by the CPT after the user requirements and platform selections are complete. Figure 11-19 also provides an overview of the CPT platform configurations.

The phase 2 city network workflow requirements can be supported with four Windows Terminal Servers, one Web server/container machine, and one database server with the selected platform configurations.

The Workflow Performance Summary can be used to review user response times for the selected configuration. Process queue times located above each hardware component represent platform processing wait times (queue-time models are discussed in chapter 7). User performance will decrease when platform capacity exceeds 50 percent utilization. Figure 11-18 identifies workflow response times with the current high-capacity configuration: Windows Terminal Servers are at 76 percent CPU utilization during peak loads, which causes the high workflow queue times and reduced user display performance. Remote terminal client display response times are taking more than 0.9 seconds during peak loads.

The level of acceptable performance depends on customer needs. In this example, performance degradation was determined to be unacceptable for the Windows Terminal Server configuration. The Windows Terminal Server farm will be increased to improve user productivity.

City network year 2 hardware platform solution

For year 2, an additional server node will be included with the Windows Terminal Server tier to improve peak user performance levels. Figure 11-20 identifies the workflow performance improvement to expect from that addition. Remote terminal client display response times improved to under 0.8 seconds (0.1 second gain for each display). For some organizations, this small performance gain is not significant. For others, including an additional server is a wise investment.

Police network year 2 platform solution

The police network platform configuration requirements are shown in figure 11-21, which includes the CPT tool interface at the top bolstered by the Workflow Performance Summary at the bottom (you can drag the graphs located in the CPT to locations that are convenient for display purposes). The number of ArcGIS Desktop users required to support the peak workflow loads was 11 in City of Rome year 2. The new ESRI workgroup server can support up to 10 concurrent ArcGIS Desktop connections and could be used to support the police network needs, if the peak workflow requirements could be reduced by one desktop user. This should be discussed during the design evaluation as a potential cost reduction—it is important that these peak loads were not overestimated. The ESRI workgroup server is provided as a separate platform tier

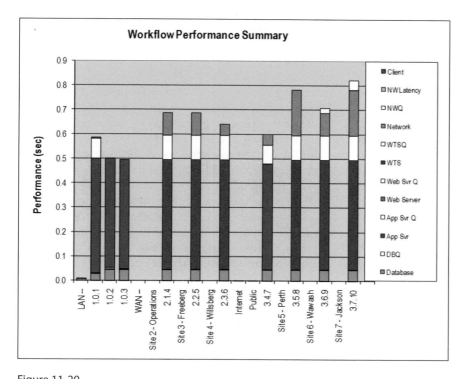

Figure 11-20

Hardware performance tuning: Year 2.

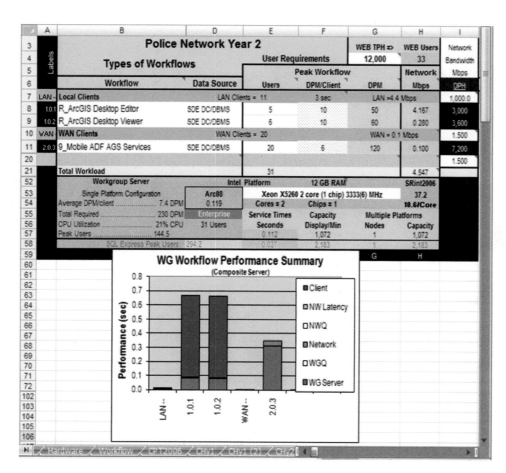

Figure 11-21

Police network hardware platform requirements: Year 2.

below the database tier on the CPT tab. This platform tier is provided to support small workgroup environments (like the police network) and includes sizing for a SQL Server Express database with the ESRI software license. The peak police network capacity requirements could be satisfied with one single-chip server (2-core) handling its peak workflow needs.

The Capacity Planning Tool workgroup server can also be used to configure enterprise server configurations that support the Web and database all on a single platform. A review of the Workflow Performance Summary suggests there are no identified performance bottlenecks.

Year 3 capacity planning

Year 3 includes deployment of two new regional office sites, increase in the number of city hall departments and workflows using GIS technology, and introduction of tracking analysis to monitor mobile vehicle operations, including up to 100 snowplows during winter storm operations. Here are the user locations for year 3 (figure 11-22).

Year 3 workflow analysis

Figure 11-23 provides the results of the year 3 workflow analysis. The workflow requirements identified in the user needs assessment (figure 11-5) were transferred to the configuration tool Workflow Requirements Module. The workflow templates for the city network and the police network were generated from the capacity planning workbook in Excel, to complete the site configurations: by copying the year 2 worksheets to separate tabs and inserting the additional remote site locations and workflow upgrades.

Year 3 includes adding a new Business Analyst workflow to the city LAN, and adding two new remote sites (Petersville and Rogerton) to the data center Internet connections. The remote site network bandwidth connections are shown on the far right of the CPT display (column I):

New remote site network connections
   Site 8—Petersville: 1.5 Mbps (Cell I28)
   Site 9—Rogerton: 1.5 Mbps (Cell I30)

The network traffic totals should be checked to make sure the SUM ranges are correct (remote site ranges should include upper and lower green lines bounding the site workflows, and the central data center ranges

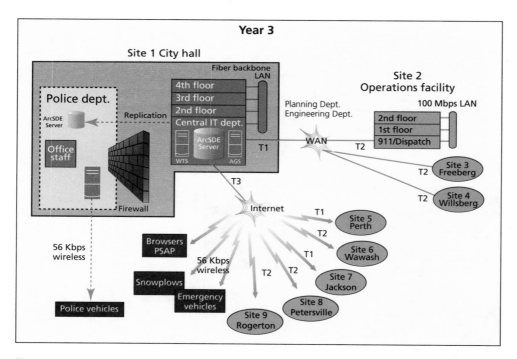

Figure 11-22

User locations and network: Year 3

	A	B	D	E	F	G	H	I
3		**City Network Year 3**			**User Requirements**	**WEB TPH =>**	**WEB Users**	Network
4	Labels	**Types of Workflows**				**12,000**	**33**	Bandwidth
5					**Peak Workflow**		**Network**	Mbps
6		**Workflow**	**Data Source**	**Users**	**DPM/Client**	**DPM**	**Mbps**	DPH
7	LAN--	**Local Clients**	LAN Clients = 87		3 sec	LAN =27.5 Mbps		1,000.0
8	1.0.1	R_Server DBMS Batch Process	SDE DC/DBMS	1	102.1	102	8.508	6,126
9	1.0.2	R_ArcGIS Desktop Editor	SDE DC/DBMS	19	10	190	15.833	11,400
10	1.0.3	R_ArcGIS Desktop Viewer	SDE DC/DBMS	49	10	490	2.287	29,400
11	1.0.4	R_ArcGIS Desktop Business Analyst	SDE DC/DBMS	18	10	180	0.840	10,800
12	WAN--	**WAN Clients**	WAN Clients = 136			WAN = 6.3 Mbps		12.000
13	rations	Site 2 - Operations	Site Clients = 66			Traffic =3.1 Mbps		3.000
14	2.1.5	R_Remote WTS ArcGIS Desktop Viewer	SDE DC/DBMS	66	10	660	3.080	39,600
15	reeberg	Site 3 - Freeberg	Site Clients = 30			Traffic =1.4 Mbps		3.000
16	2.2.6	R_Remote WTS ArcGIS Desktop Viewer	SDE DC/DBMS	30	10	300	1.400	18,000
17	illsberg	Site 4 - Willsberg	Site Clients = 40			Traffic =1.9 Mbps		6.000
18	2.3.7	R_Remote WTS ArcGIS Desktop Viewer	SDE DC/DBMS	40	10	400	1.867	24,000
19	Interne	**Internet Clients**	Internet Clients = 261			Internet =19.7 Mbps		45.000
20	Public	**Public**	Site Clients = 49			Traffic =9.8 Mbps		256.000
21	3.4.8	R_ArcGIS9.2 AJAX Map Server	SDE DC/DBMS	48.9	6	293	9.780	17,604
22	-Perth	Site 5 - Perth	Site Clients = 2			Traffic =0.1 Mbps		1.500
23	3.5.9	R_Remote WTS ArcGIS Desktop Viewer	SDE DC/DBMS	2	10	20	0.093	1,200
24	Vawash	Site 6 - Wawash	Site Clients = 40			Traffic =1.9 Mbps		3.000
25	3.6.10	R_Remote WTS ArcGIS Desktop Viewer	SDE DC/DBMS	40	10	400	1.867	24,000
26	ackson	Site 7 - Jackson	Site Clients = 170			Traffic =0.9 Mbps		1.500
27	3.7.11	R_Remote WTS ArcGIS Desktop Viewer	SDE DC/DBMS	20	10	200	0.933	12,000
28	ersville	Site 8 - Petersville	Site Clients = 60			Traffic =2.8 Mbps		1.500
29	3.8.12	R_Remote WTS ArcGIS Desktop Viewer	SDE DC/DBMS	60	10	600	2.800	36,000
30	ogerton	Site 9 - Rogerton	Site Clients = 574			Traffic =4.2 Mbps		1.500
31	3.9.13	R_Remote WTS ArcGIS Desktop Viewer	SDE DC/DBMS	90	10	900	4.200	54,000
32								1.500
33		**Total Workload**		484			53.488	

Figure 11-23

City network workflow loads analysis: Year 3.

should include the total range of each network—LAN, WAN, Internet). Validating that these ranges are correct will ensure that the Capacity Planning Tool makes the proper calculations.

Year 3 network suitability

Once the network connections are defined and the summary ranges are validated, the CPT will identify potential traffic bottlenecks. The bottlenecks show up both as colored network traffic summary cells and as network queue time on the Workflow Performance Summary. The CPT (figure 11-23) shows network traffic over 50 percent bandwidth capacity for Wawash, Jackson, and the city WAN and over 100 percent capacity for, Operations, Petersville, and Rogerton. Figure 11-24 provides

an overview of the city network year 3 Workflow Performance Summary.

The network queue times (NWQ) are identified on the Workflow Performance Summary above the network service time for each workflow. Both the WAN and Internet traffic are well above the current bandwidth capacity. The terminal client display response time for ArcGIS users at Rogerton (Workflow 3.9.13) is more than 3 seconds. With productivity of 10 displays per minute, this leaves less than 3 seconds user think time.

Year 3 network performance tuning

Having identified the network performance bottlenecks, the Capacity Planning Tool can be adjusted to resolve the identified performance problem. The standard

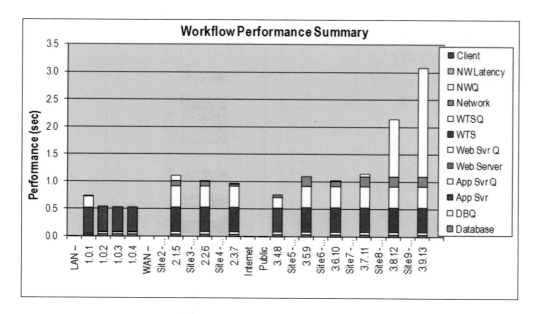

Figure 11-24

City network suitability: Year 3.

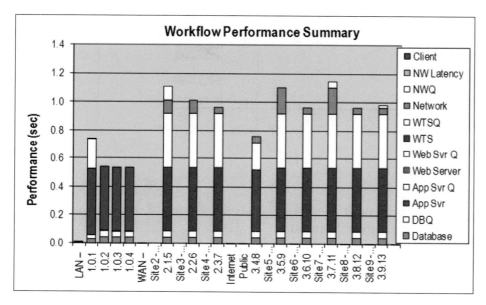

Figure 11-25

Network performance tuning: Year 3.

recommendation is to configure network bandwidth to at least double the projected peak traffic requirements. The City of Rome's network administrator agrees to the following network upgrades for year 3:

    City WAN (Cell I12)—45 Mbps
    Operations (Cell I13), Wawash (Cell I24),
    Petersville (Cell I28), and Rogerton (Cell
    I30)—6 Mbps

Figure 11-25 shows the results of the expanded network bandwidth adjustments.

Appropriate network bandwidth upgrades can be represented in the Capacity Planning Tool, and the Workflow Performance Summary will respond with the proper adjustments. Again, this analysis considers only the GIS traffic, so there will come a time to conduct the discussion with the network administrator about including the other traffic requirements. They can be estimated using a custom workflow representing existing traffic loads, and thereby easily included in the analysis.

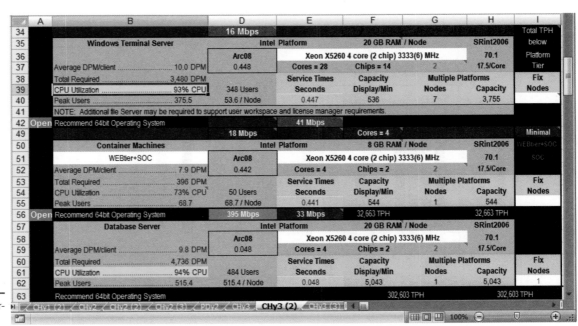

Figure 11-26

Hardware selection and performance review: Year 3.

Year 3 hardware selection, performance review

The CPT can be used to select the year 3 platform configuration. The CPT interface in figure 11-26 shows the minimum recommended platform solution for the year 3 city network workflows when using the Xeon X5260 4 core (2 chip) 3333(6) MHz platforms. Platform configuration requirements are provided by the CPT once the user requirements and platform selections are complete.

You can use the Windows Performance Summary we just looked at (figure 11-25) to review user response times for the selected configuration. Process queue times are represented above each hardware component to identify platform processing wait times (queue time models are discussed in chapter 7). Response times will increase when platform capacity exceeds 50 percent utilization. For this solution, Windows Terminal Server platforms are at 93 percent capacity (cell B39) and the ArcGIS Server container machines at 73 percent capacity (cell B54), both well above 50 percent utilization.

Again, the level of acceptable performance does depend on the particular organization's needs, however in this example, performance degradation was declared unacceptable for the Windows Terminal Server and ArcGIS Server configurations. The Windows Terminal Server farm, therefore, will be increased to 10 servers to

improve user productivity, and an another server will be added to improve Web client performance.

City network year 3 hardware platform solution

An additional three server nodes have been included with the Windows Terminal Server tier to improve peak user performance levels. One additional server was included to support the container machine tier. Figure 11-27 (next page) identifies the workflow performance improvement (display response times for workflow 2.2.6 improved from 1.0 sec to about 0.75 seconds). We've found that for some customers, this small performance gain is not important enough to justify the additional hardware and software licensing cost. For others, including the extra hardware to improve user productivity turns out to be a wise investment.

Police network year 3 platform solution

The police network year 3 configuration requirements are provided in figure 11-28. Capacity requirements can continue to be satisfied with a single-socket server configuration. Peak desktop user connections have increased to a total of 24 ArcGIS Desktop users, which would require an enterprise ArcGIS Server license, although peak loads could continue to be supported by the SQL Server Express database.

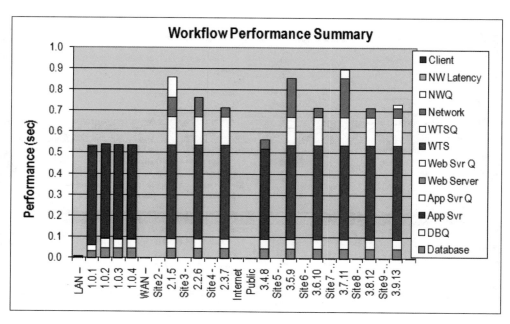

Figure 11-27

City network hardware performance tuning: Year 3.

The Workflow Performance Summary shows excellent performance with the proposed system design solution.

Choosing a system configuration

The best solution for a given organization depends on the distribution of the user community and the type of operational data in use. User requirements determine the number of machines necessary (to support the operational environment), the amount of memory required (to support the applications), and the amount of disk space needed (to support the system solution). The system design models provide target metrics to aid in capacity planning. The Capacity Planning Tool incorporates standard templates representing the sizing models and provides a manageable interface to help in enterprise-level capacity planning efforts. The CPT can be a big help in applying the results of the user needs assessment.

There's understanding and then there's applying—a basic ingredient in both, when you're looking at the results of the user needs assessment, is identifying the type of users and the platform performance needed to support required business functions. You need to know a lot of information about an organization—its intentions and its realities—in order to derive truly useful results from a user needs assessment. Information required includes the number of users of the system, the percent of time each user will be using his or her GIS application, size of the user directories (workspace), size and type of other applications on the system, and user performance requirements. Additional considerations include where data files will be located on the system, how users will access the data, and how much

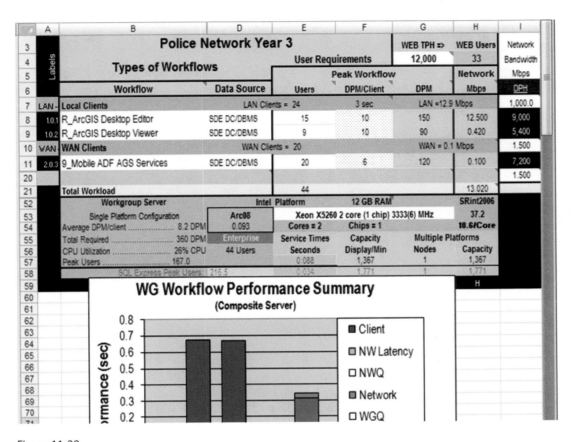

Figure 11-28

Police network hardware platform requirements: Year 3.

disk space will be needed to store the data. Also, it is important to understand the facility layout and available network communications and to evaluate the environment for potential performance bottlenecks. Other factors include accounting for existing equipment, organizational policies, preference toward available system design strategies, future growth plans, and available budget.

All of the above should be elements of interest to you when assessing user needs, the keystone of planning for success in GIS.

User needs change as an organization changes, so this assessment not only precedes the system design process, it is part of the ongoing process. Upgrades, deploying new solutions, tuning and optimizing performance—every implementation or change is like a new launch, insofar as you need to plan for it. Planning provides an opportunity to establish performance milestones that can be used to manage a successful GIS implementation. Performance targets used in capacity planning can provide target milestones to validate performance and scalability of the final system. The next and last chapter is about what to do with a system design once you have it: we will discuss the fundamentals of systems integration and some best practices that lead you to implementation success. Whether it's a first launch or a periodic deployment, basically the same principles apply.

12 | System implementation

OVERVIEW

A system architecture design provides the framework for establishing a system implementation plan. In chapter 11 we took user needs and used the Capacity Planning Tool (CPT) to figure out what's necessary to meet those needs (platform size to support peak workflow requirements). That completes the system design. Once the final design is approved, you must implement the design within established performance targets. To complete the system deployment, you can use the methodology described in this chapter along with its standard recommendations for project management, staffing, scheduling, and testing. Final recommendations on system tuning and planning for business continuance completes the chapter—and the book, which closes, appropriately, with an emphasis on managing technology change by integrating the GIS planning process with your annual business plan. Technology is changing faster each year, and every indication suggests that this trend will continue and possibly accelerate. You need to follow a strategic plan, maintain a realistic understanding of your hardware needs, and make smart decisions to stay ahead.

With ArcGIS Server, GIS is moving from department-level to enterprise-level operations at a rapid pace. Most of the implementations we have supported over the years have involved integrating GIS into existing systems, geographic or otherwise. Many existing work processes have been around for a long time, and GIS becomes the relatively new technology helping them run better. Often it's an inside team that manages the system design process, does the user needs assessment, hires the GIS staff, and implements the applications and solutions. They also end up maintaining the final system, integrating new technology on a periodic upgrade schedule. Even though GIS is a unique technology, organizations think of GIS implementation as part of an ongoing process, and rightfully so.

Enterprise IT environments involve integration of a variety of vendor technologies (software, hardware, networks, etc.). Interoperability standards within IT environments are voluntary, meaning that there is no guarantee that they will work together (integration problems are often "the other guy's fault"). Even the most simple integration must demonstrate—at each step of the integration process—that all the technologies involved will work together.

Integrating the pieces of computer technology has become easier over the years as interface standards have matured. At the same time, distributed environments have become larger and more complex, which has not made implementing these pieces any easier. The complexity and risk associated with an enterprise system deployment are directly proportional to the variety of vendor components required to support the final integrated solution. Centralized computing solutions with a single database are the easiest environments to implement and support.

Distributed computer systems, on the other hand, can be very complex and difficult to deploy and support, in large part because of what's required to integrate and maintain all the distributed copies of the same database. An example might be a local city government, with not only a central geodatabase but also a number of separate department-level database environments. Preserving the integrity of all the information products coming from these different data sources can be difficult—hundreds of changes made daily in several databases do not play well together. Many such enterprise organizations are consolidating their data resources and application processing environments to reduce implementation risk and simplify administrative support. For example, many government organizations are consolidating their GIS operations at state or federal data centers, rather than continuing to rely on distributed database environments in the field offices.

Regardless of the size or type of environment, a good understanding of every component of the system is critical in putting together an implementation strategy. Successful system implementation requires good leadership and careful planning.

GIS staffing

Good leadership begins with proper staffing. Successful GIS enterprise deployments are usually supported by an executive business sponsor, and a GIS manager that reports to senior management. Ideally a team will already be in place, having been formed to plan for the GIS well before the time for implementation. Some members of these teams will carry on through implementation or become part of GIS management.

Figure 12-1 shows the traditional GIS organizational structure. At the top of enterprise GIS operations is an executive committee with the influence and power to make financial and policy decisions on behalf of the user community. A technical coordinating committee is responsible for providing technical direction and leadership. Working groups are assigned, usually with each group aligned with a technical discipline, to address organizational issues and report on system status. The GIS user community, end users as well as power users (developers), should be represented during the review meetings.

A formal organizational structure provides a framework for establishing and maintaining the executive support and sponsorship required for successful enterprise GIS operations. The fundamental management concepts underpinning this structure apply to managing small or large organizations; and the same type of structure can be just as useful in coordinating community GIS operations.

Several technical disciplines are involved in achieving successful GIS operations. GIS staff roles might include a GIS manager, GIS analyst, database manager, application and Web programmers, along with GIS users located throughout the organization. The complexity of their responsibilities will vary with the size and extent of each individual GIS implementation, although every organization will need some level of support and expertise in each of these areas.

Building qualified staff

Training is available to develop qualified staff and promote GIS user productivity, and organizations should make sure their teams receive it. Even experts need training from time to time if only to stay up to date. The ESRI Training and Education Web site (http://training.esri.com/gateway/index.cfm) provides recommendations for GIS staff training, and tools to develop individual training plans. Roger Tomlinson's book, *Thinking About GIS*, contains an appendix (A) with information about GIS jobs and training as well.

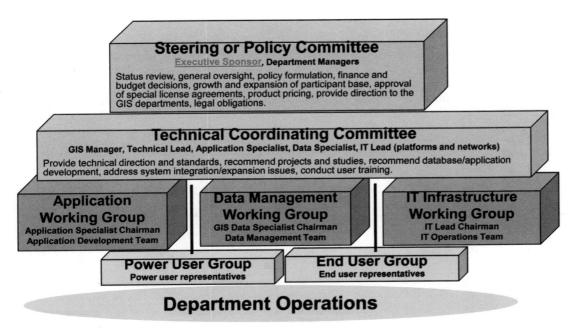

Figure 12-1

Traditional GIS organizational structure.

Job titles and descriptions vary, yet from understanding the skills your organization requires and the training required to get them, you can build the qualified staff you need to be successful.

System architecture deployment strategy

Planning is the first step in achieving a successful system deployment. A system design team has reviewed current GIS and hardware system technology and user requirements, and established a system architecture design based on user workflow needs. Once the system design is approved, the design staff enters the process of developing in earnest the implementation strategy. A deployment schedule, as shown in figure 12-2, is developed to establish overall implementation milestones. Such milestones are part of a phased implementation strategy, which is the best way to approach technology deployment and the risks that go with it.

Risk comes with the territory these days and must be managed: Computer technology continues to evolve at a remarkable pace. New ideas are introduced into the marketplace every day, and a relatively small number of these ideas develop into dependable long-term product solutions. Integration standards are constantly changing with technology and, at times, may not be ready to support immediate system deployment needs. A phased implementation strategy can significantly reduce implementation risk. The frequency of change is accelerating at a rapid pace, and today the traditional waterfall approach operates more like a spiral, with a

series of incremental upgrades planned throughout the year. It is more important than ever before to retain the fundamental discipline required to ensure success.

Best practices for implementing enterprise GIS
The following best practices are recommended to support a successful enterprise GIS implementation.

Pilot phase
- Represent all critical hardware components planned for the final system solution.
- Use proven low-risk technical solutions to support full implementation.
- Include test efforts to reduce uncertainty and implementation risk.
- Qualify hardware solutions for initial production phase.

Initial production phase
- Do not begin until final acceptance of pilot phase.
- Deploy initial production environment.
- Use technical solutions qualified during the pilot phase.
- Demonstrate early success and payoff of the GIS solution.
- Validate organizational readiness and support capabilities.
- Validate initial training programs and user operations.
- Qualify advanced solutions for final implementation.

Final implementation phase
- Do not begin until final acceptance of initial production phase.
- Plan a phased rollout with reasonable slack for resolving problems.
- Use technical solutions qualified during previous phases.
- Prioritize rollout timelines to support early success.

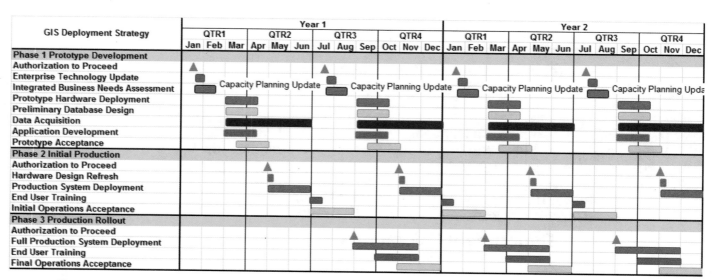

Figure 12-2

System deployment strategy for GIS.

The strategic approach that follows (along with best practices) puts into place separate steps for testing and validation before final launch.

The three implementation phases may be called by different names, but the basic concept remains the same. First you need to make sure the technology will work together, then you need to make sure you can deploy and support the new technology in a real environment. Finally you need to roll out the new technology to support full production operations. Some best practices on how to do this right are provided in the following discussion.

Data center architecture

Information technology (IT) is an integral part of most business organizations now, and it's the "ITs" (e.g., system administrators, et al.) who are responsible for building the infrastructure for the enterprise. GIS applications are becoming more and more common within the enterprise architecture, but protecting and supporting enterprise operations is traditionally the IT's primary mission. Managing technology across an enterprise can be an overwhelming responsibility, and with growing dependence on Internet access and Web technologies, the challenges are expanding. At the same time, business is more competitive and organizations must find ways to improve productivity *and* reduce cost in this rapidly changing environment. It's no wonder then that the

trend is toward centralizing: Centralized operations provide the lowest cost operational environment. The perceived need to simplify data center administration and reduce enterprise security risks is driving organizations toward a common strategy for supporting a secure data center architecture. The platform tiers that compose this evolving common or standard enterprise IT architecture are shown in figure 12-3.

This layered design is evolving into a standard architecture strategy for all business operations. Within its multitiered structure, each single tier addresses specific functional responsibilities and the different security administration needs that come with them—Web access, enterprise applications, component applications (business functions), data, and storage, as described here.

- **Web access tier.** Commercial Web access products are entering the marketplace in growing numbers. These solutions include traffic accelerators, reverse proxy servers, network traffic load balancing functions, data caching, and a variety of security enhancements to enhance and protect Internet access. Maintaining high-performance Internet access and lock-tight enterprise security is an increasing challenge. Addressing this challenge with enterprise-wide security standards and common Web-tier solutions simplifies administration and reduces complexity by providing a common enterprise access strategy.

- **Enterprise application tier.** A service-oriented architecture provides the highest level of security

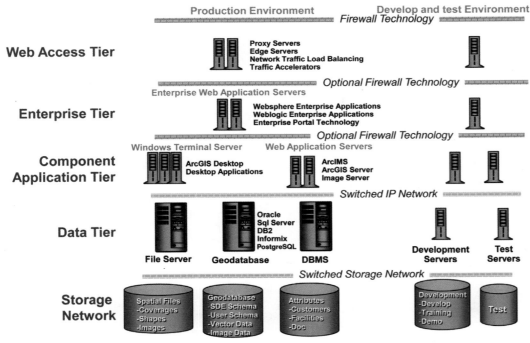

Figure 12-3

Data center architecture.

and adaptability for building enterprise business applications. Abstracting enterprise applications from functional business service components provides a more adaptive and secure enterprise solution. Several enterprise-level Web application server solutions are on the market, including the IBM WebSphere and the BEA WebLogic Web application-development frameworks.

- **Component application tier.** High-performance business applications provide the fundamental technology underlying business operations. These are more tightly coupled technical solutions that enable core business functions—heavy lifting core applications and services that interface directly with the database tier. More tightly coupled than the Web tier applications—these are the applications and services that interface directly with the database tier. These solutions include desktop applications and software such as ArcGIS Desktop and ArcEngine on Windows Terminal Server, ArcGIS Server Web applications and geospatial services, and ArcGIS Image Server.

- **Data server tier.** This broad range of data management solutions includes the geodatabase, facility management systems, accounting systems, work management systems, document management systems, cached image data sources, and many more.

- **Storage network tier.** Data resources are growing at an ever-increasing rate, as data collection technology matures and user demand for rich visualization and geographic data display becomes the norm. The demand for temporal data and analysis of data over time requires that large volumes of data be maintained online to provide real-time access and data sharing. Maintaining security and availability of such large volumes of data resources requires the administrative advantages of a dedicated networked storage architecture. A full range of focused vendor solutions are entering the marketplace to address the need.

These technology trends seem to be pushing data center design in a common direction. Understanding these trends can provide insight into what we can expect to see in the marketplace as we integrate new enterprise technology solutions—as this technology matures.

System testing

Conducting proper testing can contribute to implementation success. For new technology, conduct functional-component and system-integration testing during the initial pilot phase of the system deployment, but delay performance testing until the initial *production* phase. Best practices for planning and conducting functional-

system testing are highlighted below. Make a special note to complete prototype integration testing before production deployment. Test in a production environment (configuration control). Be sure to list functional requirements and the test procedures required to demonstrate success.

Functional-system testing should be completed for all new technology that will be integrated into the production system. Develop a test plan that identifies test requirements, establishes configuration control (software versions, operating system environment), and spells out test procedures. Conduct the testing using the software versions and operating system that will be deployed in the actual production environment. Testing should be completed before production deployment.

Be aware of the pitfalls associated with system performance testing (below) in general. These pitfalls and the best practices to avoid them are highlighted here. Performance testing can be expensive and the results misleading. Test environments seldom replicate normal enterprise operations—real-world user workflow environments are difficult to simulate. That's why you can't count on initial system deployments to achieve final performance goals. But you can set some performance targets early on, based on your workflow service time assumptions used in the Capacity Planning Tool. You can also measure early on if you are meeting those performance targets. During and after

---

Functional-system testing best practices

Test planning:
- Complete a risk analysis: identify functionality that requires testing.
- Identify test objectives and establish configuration control plan.
- Identify test hardware/software configuration.
- Develop test procedures.

Test implementation:
- Identify implementation team and establish implementation schedule.
- Order hardware and software and publish installation plan.
- Conduct test plan and validate functional acceptance.
- Collect test performance parameters (CPU, memory, network traffic, etc.).

Test results and documentation:
- Document the results of the testing.
  - Include specific hardware/software/network components that were tested.
  - Include installation and test procedures that were followed, test anomalies, and final resolution.
  - Complete test compliance matrix identifying validation of functional requirements.

- Publish the test results for reference during system implementation.

Performance testing	
Performance testing pitfalls	**Performance testing best practices**
False sense of security: • Tendency to accept test results over analysis. • Problem 1: Test results are a function of input parameters often not understood. • Problem 2: System bottlenecks in testing can generate false conclusions. Simulated load testing may not represent the real world: • Load generation seldom represents actual user environments. • Relationships between load generation and the real world are seldom understood. • Several system configuration variables can contribute to test anomalies.	• Model system components and response parameters. • Validate models based on real-world user loads testing. • Predict test results from models (hypothesis) before conducting testing. • Evaluate test results against models and original hypothesis. • Update models and hypothesis, and repeat testing until you reach consensus. A strong mindset with which to approach performance testing and employ the best practices associated with it is one that keeps awareness of these three related truisms at the forefront: 1. Test only when you think you know the answer. 2. Testing only confirms what you already know. 3. Testing does not teach you what you don't know.

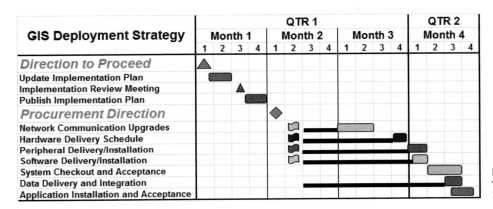

Figure 12-4

Systems integration management.

the initial deployment, you'll be tuning and optimizing the system. Early application development, however, will focus primarily on functional requirements, and performance tuning is often not complete until the final release.

For the best practices that directly apply to system performance testing, simply hark back to the scientific method introduced with grade school science fair projects. These are the fundamentals:

- Performance testing should only be conducted to validate a hypothesis (something you think you know).
- The primary objective of a performance test is to validate the hypothesis (confirm what you know).
- The test is a success only if it proves the hypothesis (testing does not teach you what you don't know).
- If the results of the test do not confirm your hypothesis, it is important to ask why and answer that question before accepting the test results.

Initial performance testing results often fail to support the test hypothesis. With further analysis and investigation, you can identify configuration bottlenecks or improper assumptions that may change your hypothesis and possibly change the results. Performance test results should be trusted only if they validate your test hypothesis.

The best performance validation can be collected during the initial production deployment. During this phase, real users doing real workflows with the production software and data on the actual system will generate the most realistic user environment. Critical system components should be monitored during the initial deployment, to identify processing bottlenecks and resolve system conflicts. Initial deployment acceptance should include validation that workflow performance targets are satisfied and user performance needs are met.

Management

As mentioned earlier, following basic project management practices promotes implementation success:

- Establish project teams.
- Assign individuals specific responsibilities.
- Break down implementation planning into a task plan.
- Develop a configuration control plan and a change control process.
- Publish the implementation schedule to define the project deployment milestones for everyone.

After selecting the hardware vendor solution for your final system design, you are ready to document your implementation plan. Figure 12-4 illustrates a typical system deployment schedule—decision milestones with the major task efforts that follow them. The project manager responsible for implementation should make sure all these tasks are well defined and that participants understand their responsibilities. To ensure that integration issues are identified and resolved at the earliest opportunity, develop a clear set of acceptance criteria for each implementation task, and follow a formal acceptance process.

System tuning

System performance tuning is a critical part of final system integration and deployment. System performance has never been far from your thoughts in figuring out the design puzzle; now it comes to the forefront. You can use the performance targets you set in your initial system design (represented as workflows in your capacity planning model) as the final system performance goals. Check your progress toward these goals at your earliest opportunity to begin performance tuning. If you are not meeting your performance goals, now is the time to make adjustments. These are the variables you have to bring your performance targets in line with:

- Separate heavy batch-processing efforts from interactive user workflows through a separate batch process queue.

- Plan to do system backups and heavy processing workloads during off-peak workflow periods.
- Monitor system component performance metrics on a periodic basis, particularly during peak workflow periods.
- Use this monitoring process to identify performance bottlenecks and address system deficiencies.

Efforts on behalf of optimal performance begin during the earliest planning stages and continue on as part of managing a GIS. Following best practices in planning for and implementing all the components of a system reduces cost and improves productivity. But your efforts don't stop there, not by a long shot. You must maintain a productive environment by attending to every factor influencing each component, especially those listed under the component headings in figure 12.5. Remember: in an enterprise environment, any component carries the potential of introducing a weak link in overall system performance. The primary point here is a simple one—you saw the basic outlines of figure 12-5 in chapter 1 (figure 1-2) when it (and your design) was empty of detail. Since then, the system design process (chapter 11) has shown you how to select the right software and hardware technology, using the CPT to help identify specifications for each system component. That's not the end of the story. Even after you have purchased and installed the system, there is more work to do: you have to make sure each component supports the established performance targets. In terms of performance tuning and optimization, the performance factors now identi-

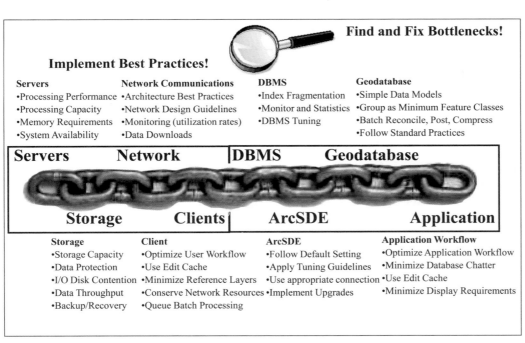

Figure 12-5

System performance tuning.

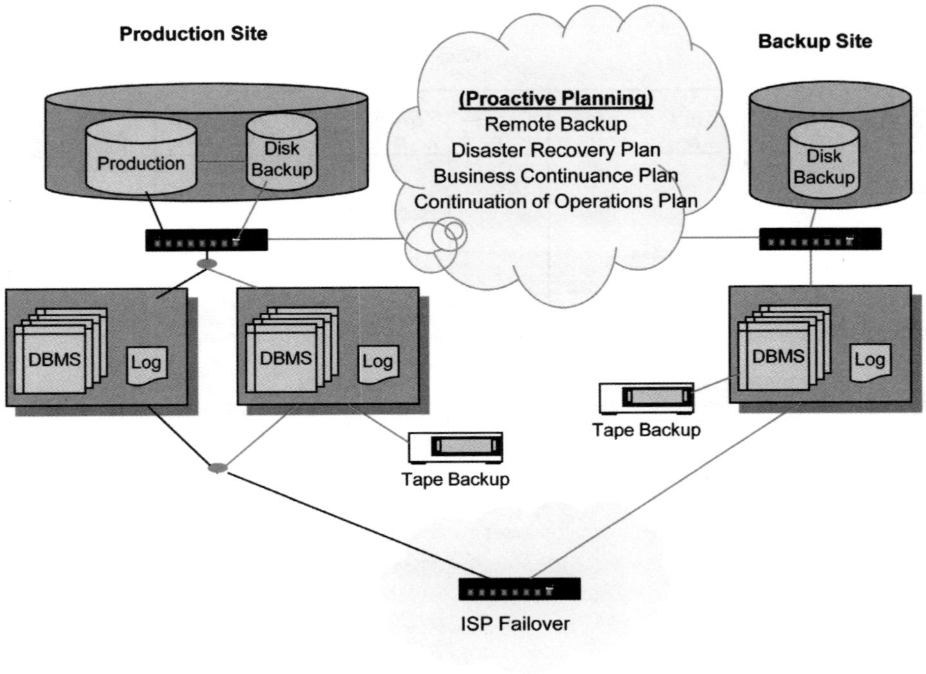

Figure 12-6

Plan for business continuance.

fied in figure 12-5 remain the standard things to look at once the system comes together as well.

Business continuance plan

Every organization should carefully assess the potential failure scenarios within its system environment and protect critical business resources against such failures. Enterprise GIS environments require a significant investment in GIS data resources. These data resources must be protected in the event of a system failure or physical disaster. Business recovery plans should be developed to cover all potential failure scenarios. Figure 12-6 provides an overview of the different system backup strategies. Be sure to develop a business continuance plan to specifically address what would be required to keep your organization going in the event of a system failure or a need for disaster recovery.

Managing technology change

Enterprise GIS operations require a combination of strategic planning and a continued investment in technology. Technology is changing very rapidly, and organizations that fail to manage this change fall behind in productivity and operational cost management. Managing technology change is a major IT challenge.

Enterprise operations should include a periodic cycle of coordinated system deployment. The planning and technology evaluation should occur one periodic cycle ahead of technology deployment, and these efforts should be coordinated to support operational deployment needs. Figure 12-7 identifies a conceptual system architecture planning and deployment strategy for technology change management.

Planning and evaluation: Establish planning activities within a periodic cycle, coordinated with the organization's operational and budget planning. Regularly update strategic plans according to a multiyear deployment strategy, and publish them periodically (i.e., annually).

The planning and evaluation process includes a requirements evaluation (strategic plan update), technology refresh (training and research), requirements analysis (process and requirements review), test and evaluation (evaluate new technology alternatives), and prototype validation (pilot test programs). The planning and evaluation program should also include scheduling for new technology procurement and deployment as well as for system upgrades.

System deployment: System upgrades should be planned on a periodic cycle (usually annually); they are scheduled in order to implement enhancements validated during the planning and evaluation process. All production system upgrades should be managed so as not to interrupt ongoing operations. You can achieve

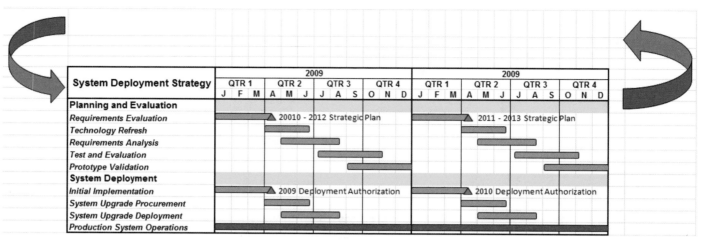

System Deployment Strategy	2009 QTR 1 J F M	QTR 2 A M J	QTR 3 J A S	QTR 4 O N D	2009 QTR 1 J F M	QTR 2 A M J	QTR 3 J A S	QTR 4 O N D
Planning and Evaluation								
Requirements Evaluation		▲ 20010 - 2012 Strategic Plan				▲ 2011 - 2013 Strategic Plan		
Technology Refresh								
Requirements Analysis								
Test and Evaluation								
Prototype Validation								
System Deployment								
Initial Implementation		▲ 2009 Deployment Authorization				▲ 2010 Deployment Authorization		
System Upgrade Procurement								
System Upgrade Deployment								
Production System Operations								

Figure 12-7

System architecture design strategic planning: Managing technology change is one of the biggest challenges facing organizations today. Successful organizations maintain an up-to-date strategic plan that includes scheduled periodic updates and an incremental deployment schedule. Enterprise business evolution is supported through a spiral development cycle, carefully maintaining implementation discipline and configuration control for the waterfall tasks of each deployment stage. The difference today from 10 years ago is the frequency of change.

that by including in any system deployment the phase of initial implementation, thereby first implementing changes only in an operational test environment. Including that testing phase is your best assurance of winning authorization for deployment.

Successful implementation depends on a good solid design, appropriate hardware and software product selection, successful systems integration, and careful incremental evaluation during installation. A phased approach to implementation reduces project risk and promotes success. With this opportunity for early success comes the flexibility to incorporate new technology at low risk, prior to final system delivery.

Yet even with the best laid plans, the challenges are so unique that if there are problems the project manager is not able to resolve, the system architect will need to address those problems. Keep in mind that the success of a system architecture design is demonstrated in the final implementation, after all the performance milestones are met, which sometimes can be as much as three years later. I have several clients I have supported over and over again; where there are performance issues,

we get a call to help. Guidelines are also available to help in devising a successful system design (see chapter 8), even for large, complex systems.

A primary objective of this book is to provide the project manager with tools—and most importantly, the fundamental concepts underlying those tools—to support a successful system deployment. Understanding the technology, knowing your performance needs, and proper systems implementation management truly can take you far in ensuring success.

The methodology in this book has been played out successfully many times over, in delivering a system that performs under the bar.

Completing the system design is the beginning of the process, though, and implementation is the rest of the story. Roger Tomlinson's book, the first in this GIS planning series, also ended with one perspective on implementation; mine is another aspect of the same big picture and the second book in the series. Both detail imperative parts of planning for a GIS implementation. The full story of implementation, however, awaits the third book in the series, the one on managing a GIS, to continue the story.

Appendix A

System performance history

There have been many technology changes over the past 20 years that have improved user productivity and performance. Figure A-1 is a graph that shows how display performance has changed over the years as customers took advantage of new technology. In this system performance profile of changes in the 1990s, you can see performance gains resulting from a series of hardware investments. All these investments contributed to improved productivity at the user desktop.

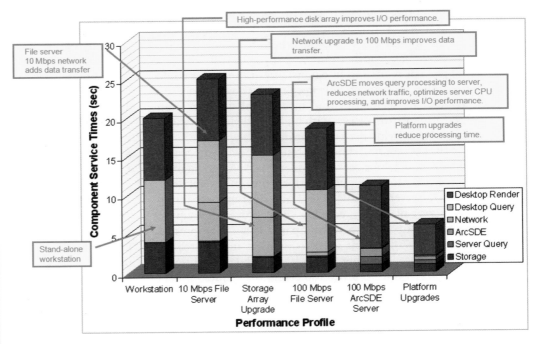

Figure A-1

Advancing technology improved performance.

The first column from the left represents the performance of a stand-alone workstation supporting a typical GIS operation (e.g., requesting display of a new map extent on the user screen). Experience shows that GIS applications tend to be both compute intensive and input/output (I/O) intensive. This means that GIS applications require computer processing and generate network traffic both in amounts large enough to be significant factors in system design. A stand-alone GIS workstation spends about half the processing time on data access (query) and the other half on compute processing (rendering the display).

Figure A-1's second column from the left shows the system performance profile when accessing GIS data from a network file server in place of a local disk. This distributed solution includes additional service time for the network data transfer. Additional system components (network interface cards, network switches, shared network bandwidth, etc.) extend the overall user display response time due to the data transport times. Accessing data from a file server over a 10 Mbps network could increase user display response time by as much as 30 percent. GIS users would keep user workspace on local disks for better performance. This was the case in the past, which changed with improved network bandwidth (see description of the fourth column, below).

The third column shows the result of upgrading the JBOD (just a bunch of disks) configuration on the file server to a high-performance RAID (redundant array of inexpensive disks) storage solution. High-performance RAID storage solutions improved disk access performance by as much as 50 percent.

The fourth column shows the effect of increasing network bandwidth from 10 Mbps to 100 Mbps, a 90 percent reduction in data transfer time. Users could improve productivity by keeping user workspace on remote file servers, instead of on local disks, as high-performance disk access with high-bandwidth networks began to provide better performance.

The fifth column shows the result of moving the spatial data to a database server (geodatabase). A geodatabase server can improve query performance in several areas. DBMS server technology relocates query processing to the server platform, reducing client processing requirements by up to 50 percent. Spatial data is compressed by 30 to 70 percent when supported in an ArcSDE geodatabase, reducing network traffic by an additional 50 percent. The DBMS also filters the requested data layer so only the requested map layer extent is sent over the network to the client, further reducing network traffic. The query processing performed by the ArcSDE server, using DBMS query indexing, data cache, and optimized search functions, will also reduce the processing load on the server to less than half of that used to support client query requests. Moving spatial data to an enterprise ArcSDE geodatabase can significantly improve overall system performance in a distributed computing solution.

The final column on the right shows the effect of upgrading the platform environments to more current models with twice the processing performance, reducing CPU processing time by 50 percent.

Hardware component investments contribute directly to user productivity and the overall productivity of the organization. Computer technology is changing rapidly, and the product of this change is higher performance and improved productivity at the user desktop. Organizations need to budget for this change and make wise investments in their infrastructure portfolio to maintain high productivity in the workplace. A smart investment strategy pays large dividends in supporting successful GIS operations.

System performance testing

Much of the basic understanding of GIS performance and scalability can be gained from focused benchmark testing. The ESRI Enterprise Test Lab was established in 1998 to better understand performance and scalability of ESRI software technology. The test lab has provided independent software performance validation benchmark testing with each ESRI core software release to further refine and improve the understanding of ESRI technology. Over the years, performance models have been improved to support successful deployment of ESRI technology.

A notable series of ArcInfo (ArcINFO through 1997) performance tests was conducted at the Data General Development labs in Westborough, Massachusetts, in July 1998 with the introduction of Microsoft Windows Terminal Server 4.0 edition. This was an update of ARC/INFO performance testing completed by ESRI in 1993. This performance testing established an early foundation for the system architecture design sizing models, and confirmed the models for use in system architecture design consulting services.

The ESRI system performance models are based on an understanding of how computer platforms respond to an increasing number of concurrent ArcInfo batch process loads. An ArcInfo performance benchmark was used to evaluate platform response to increasing user loads. Excessive memory was configured on each server platform (2–4 GB per CPU) to avoid paging and swapping of executables during testing (recommended physical memory requirements are established by separate tests that measure memory allocated to support each application process).

Each benchmark test series was extended to evaluate platform support for four batch processes per CPU. The ArcInfo sizing model is seldom used to identify performance expectations beyond two concurrent batch processes per CPU. The sizing models perform very well at the lower range of these test series.

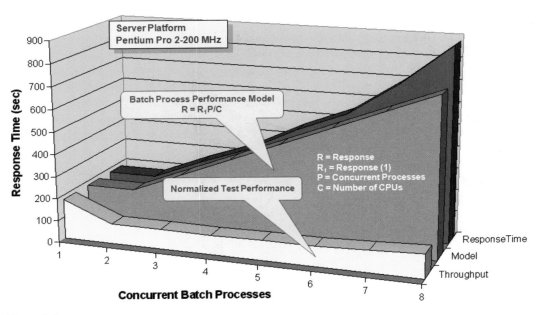

Figure A-2

Early testing of a dual processor: Illustrates the ArcInfo benchmark results from testing a dual-processor Pentium Pro 200 Windows Terminal Server, summarized in a chart. (Note the assumption for all the following charts is that one batch process consumes one CPU.) An individual test was completed for each concurrent process configuration (1 through 8). The third-row graphic plots the average response time measured for each of the concurrent process test runs. The first row shows the rate at which the server platform was processing ArcInfo instructions. The center row shows a plot of the sizing model for concurrent ArcInfo batch processing. The results of this test validate the ArcInfo design model and demonstrate good Windows Terminal Server scaling performance with this platform configuration. (Note that the test in these figures were conducted in 1998.)

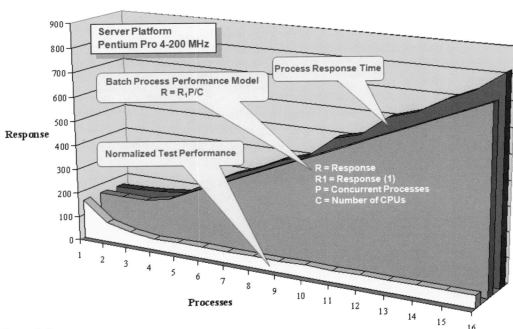

Figure A-3

Early testing of a quad processor: Provides a summary of the ArcInfo benchmark results from testing a quad-processor Pentium Pro 200 Windows Terminal Server. An individual test was completed for each concurrent process configuration (1 through 16). The results of this test validate the ArcInfo design model and demonstrate good Windows Terminal Server scaling performance with this platform configuration.

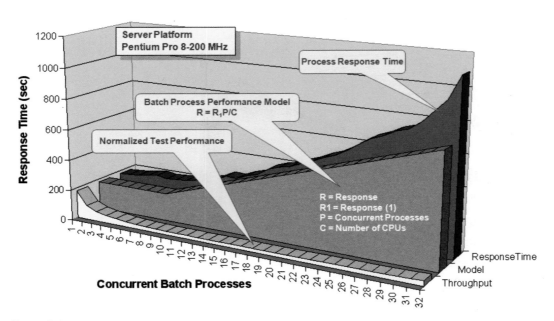

Figure A-4

Early testing of an eight processor: Provides a summary of the ArcInfo benchmark results from an eight-processor Pentium Pro 200 Windows Terminal Server. An individual test was completed for each concurrent process configuration (1 through 32). The results of this test showed reduced performance with larger machine configurations, suggesting optimum application server configurations would be supported by a terminal server farm of several 2- or 4-CPU machines rather than using a more expensive 8-CPU platform to support a large number of concurrent users.

Appendix B

COTS security terms and procedures

Application security encompasses defense measures taken to prevent exceptions in the security policy of applications or the underlying system—vulnerabilities that enter through flaws in the design, development, or deployment of the application. The development of security and control procedures for the custom applications are based on commercial off-the-shelf functionalities provided by Microsoft Windows OS, ArcGIS, RDBMS, and HTTP protocols.

- Windows Access Control List (ACL) provides role-based access control where permissions are assigned to roles and roles are assigned to users.
- Thorough ACLs for file systems Access Control Entries (ACE) can be defined for group rights to specific system objects, such as applications, processes, or files. These privileges or permissions determine specific access rights, such as whether a user can read from, write to, execute, or delete an object.
- ArcGIS controls for client or Web applications are mechanisms implemented either through an ArcGIS out-of-the-box configuration, a custom application enhancement (using ArcObjects), or an ArcGIS Web client.
- Custom control extensions can be used to implement technologies such as Identity Management (IM) and access control. ArcGIS custom control extensions are developed using the ArcObjects development interface. ArcGIS gives the user the ability to restrict ArcGIS client operations (edit, copy, save, print) or it controls users' access to various data assets, based on their role.
- GML is an XML schema used for modeling, transporting, and storing geographic information. ArcObjects, using GML and RDBMS storage functionality, offers a framework and method for auditing controls in ArcGIS multiuser geodatabase environments. A detailed history of GIS workflow activities can be recorded in a GML structure and stored in the RDBMS. In addition to recording who performed the edit, you can supplement activities with comments and notes to provide a traceable, documented, activity log containing before-edit, after-edit, and edit-justification history.
- Integrated operating system authentication and single sign-on (SSO) are two security infrastructures that can be leveraged by ArcObjects applications, to authenticate against and connect to

ArcGIS products using user names and passwords managed in a centralized location. This location can be an encrypted file, an RDBMS table, a lightweight directory access protocol (LDAP) server, or a combination of RDBMS tables and an LDAP server. The primary intent is to insulate users from having to continually authenticate themselves. This technique relies on users' authentication into their desktop workstation (integrated operating system authentication) or the organization's SSO infrastructure.

- Native authentication by ArcSDE and RDBMS: Strong authentication controls can be established between ArcGIS and system components through the use of native authentication, allowing the user to be authorized by downstream systems. Arc-SDE using the direct-connect architecture supports native Windows authentication from the ArcGIS client connecting to the RDBMS. The direct-connect configuration allows ArcGIS clients to leverage RDBMS connectivity functionality. Deployed using two-tier ArcSDE architecture configured with a RDBMS SSL transport layer, native authentication provides an encrypted communication channel between the trusted operating system and the RDBMS.

- SSL is a protocol that communicates over the network through the use of public key encryption. SSL establishes a secure communication channel between the client and server. Encryption functionality of the RDBMS converts clear text into cipher text that is transmitted across the network. Each new session initiated between the RDBMS and the client creates a new public key, affording increased protection. Using ArcSDE in a direct connect configuration eliminates the use of the ArcSDE application tier by moving the ArcSDE functionality from the server to the ArcGIS client. By doing this, you are creating a dynamic link library, and the client application is enabled to communicate directly to the RDBMS through the RDBMS client software. ArcSDE interpretations are performed on the client before communication to the RDBMS. This provides the client application the ability to leverage network encryption controls supplied by the RDBMS client.

- IPSec is a set of protocols that secures the exchange of packets between the ArcGIS client and the RDBMS server at the IP level. IPSec uses two protocols to provide IP communication security controls: authentication header (AH) and encapsulation security payload (ESP). The AH offers integrity and data origin authentication. The ESP protocol offers confidentiality.

- Intrusion detection is available for ArcGIS users: network-based intrusion detection analyzes network packages flowing through the network; or the software can provide a host-based, intrusion-monitoring operation on a specific host.

- Feature-level security implemented in parallel with ArcSDE, for example, allows the data maintenance team to assign privileges at the feature level, restricting data access within the geodatabase object. RDBMS feature-level security is based on the concept of adding a column to a table that assigns a sensitivity level for that particular row. Based on the value in that column, the RDBMS determines, through an established policy, whether the requesting user has access to that information. If the sensitivity level is met, the RDBMS allows access to the data; otherwise, access is denied.

- Data-file encryption can be used by the ArcSDE direct-connect architecture by using a data encryption "add-in" in the RDBMS, which works with ArcGIS products accessing an RDBMS as a data store, custom ArcObjects applications, and custom non-ESRI-technology-based applications using the ArcSDE C and Java APIs to access nonversioned data.

- RDBMS privileges: RDBMS assigns SELECT, UPDATE, INSERT, and DELETE privileges to either a user or role. The ArcSDE command line and ArcCatalog leverage the RDBMS privilege assignment functionality and provide an interface that allows the administrator to assign privileges.

- HTTP authentication is a mechanism by which a method is used to verify that someone is who they claim to be. The standard methods of HTTP authentication integrated with ArcGIS Web applications are the basic, digest, form, and client-certificate methods. Basic authentication involves protecting an HTTP resource and requiring a client to provide a user name and password to view that resource. Digest authentication also involves protecting an HTTP resource by requesting that a client provide user name and password credentials; however, the digest mechanism encrypts the password provided by the client to the server. Form-based authentication is identical to basic, except that the application programmer provides the authentication interface using a standard HTML form. Client-certificate is the most secure authentication method in that it uses the organizational PKI environment to provide and authenticate digital certificates for both client and server.

Acronyms and glossary

Acronyms

ADF	application development framework
AML	ARC Macro Language
API	application program interface
ASC	application server connect
ASP	application service provider
CF	ColdFusion
CIFS	common Internet file services
COTS	commercial off-the-shelf
CPT	Capacity Planning Tool
DC	direct connect
DCOM	distributed component object model
DMZ	demilitarized zone
DS	data source
EDN	ESRI developer network
ETL	extract-translate-load
GB	gigabyte
Gbps	gigabits per second
g.net	(regional) geography network
GIS	geographic information system
HTTP	hypertext transfer protocol
I/O	input/output
ICA	independent computing architecture
IP	Internet protocol
ISP	Internet service provider
JBOD	just a bunch of disks
Kb	kilobit
KB	kilobyte
Kbps	kilobits per second
LAN	local area network
MAC	media access control
Mb	megabit
MB	megabyte
Mbps	megabits per second
NFS	network file server
NIC	network interface card
ODBC	open database connectivity
RAID	redundant array of independent disks
RDP	remote desktop protocol

SAN	storage area network
SDE	Spatial Database Engine
SM	service manager
SMP	symmetrical multiple processor
SOA	service-oriented architecture
SOAP	simple object access protocol
SOC	server object containers
SOM	service object manager
SPEC	Standard Performance Evaluation Corporation
SS	spatial services or Web spatial server
SSL	secure socket layer
SSO	single sign on
TCP	transmission control protocol
WA	Web applications or Web applications server
WAN	wide area network
WSE	Web services extensions
WTS	Windows Terminal Server
XML	extensible markup language

adaptive In the context of system design, a term that describes something that helps you or your organization adapt to technology change; for example, a model that accurately represents the real world is adaptive.

application server A computer platform used to host user application software executables that are managed and deployed from a central computer facility. These servers can include desktop user application sessions (Windows Terminal Servers) or Web user transactions (Web servers, map servers, server object container machines).

Arc07 or Arc08 See *performance baseline.*

bandwidth A data rate measured in bits per sec.; also referred to as network throughput. The digital data rate limit (or network capacity) of a physical communication link is related to its bandwidth in hertz.

batch process A computer program (procedure) or series of procedures executed without user interaction. A batch process would normally execute for a period of time significantly beyond that of an average user display transaction.

batch productivity A measure of the workflow processing throughput of a batch process expressed in terms of peak displays per minute (DPM).

benchmark See *performance benchmark.*

capacity The peak computer platform or network processing throughput. Capacity can refer to a single component or an integrated system of platform and network components. The peak system throughput is limited by the capacity of the weakest system component.

capacity planning A process of defining individual component performance and capacity specifications based on estimated peak user workflow requirements. Capacity planning is about building a distributed hardware and network infrastructure to support enterprise GIS operations.

Capacity Planning Tool (CPT) The Microsoft Office Excel workbook application provided with this book. Readers of this book can use the CPT to collect peak user workflow requirements and establish their own specific target performance and capacity specifications to ensure successful GIS deployment.

client The term used to describe the location of the user computer interface.

component service time The total program execution or network transfer time required for each system component to complete processing of an average work transaction. Component service times are normally discussed in reference to an integrated system of components, where the sum of the component service times plus any additional processing delay or wait times are combined to represent the work transaction response time.

composite workflow Many Web applications today are developed from a group of published Web service components. A custom user workflow created by combining two or more single user workflow software service times is referred to as a composite workflow.

compute-intensive Adjective describing a computer software process that consumes a significant amount of server processing resources; a process whose peak throughput is determined by the processing capacity of the platform CPU.

compute server An IT term used to describe a server located in the data center that is used to host primary application executables; used to distinguish compute servers from data servers in the computer room. Compute is an adjective describing the type of computer or server platform or environment. (In early GIS history, we used "compute server" when referring to UNIX servers that hosted ARC/INFO applications serving terminal clients. We could not use "terminal server" in those days, since that term was used to identify wall network connections for mainframe terminals. Terms have evolved over the years. The mainframe wall network connections have gone away, so this type of server can now be called a "terminal server.")

computer network A group of interconnected computers. Computer network interface cards (NIC) connect computers to network switches that are used as a connection point for groups of computers and supporting network devices (printers, plotters, etc.). Network capacity is established by the capacity of the switch connections (ports).

computer processor chip The small integrated circuit board that supports the computer core processors. Commodity Intel computers normally contain two computer processor chips, with each chip supporting one, two, or four processor cores.

concurrent user model A table used in capacity planning to identify the number of concurrent workflow users that can be supported on a baseline platform performance. When ESRI first delved into capacity planning, a simple capacity relationship was used along with published vendor SPEC benchmarks to translate peak concurrent user loads to other platform environments. Performance baselines and models were updated on an annual basis to address changes in technology. (See *transaction-based sizing model*).

container machine The server object container (SOC). A term used with ArcGIS Server describing the machine hosting the software (ArcObjects) and deployed SOC instances.

core See *processor core*.

data center A central computer facility used to house centrally managed servers and computer data resources. Data centers employ enhanced security and system management software to administer enterprise applications and database server resources, thus providing services to client computers located across the local area network (LAN). The data center is also a central network access point for wide area network (WAN) and Internet user locations.

data server Server platforms that host database software or provide network access to a file system repository.

data source The term used with the Capacity Planning Tool to identify the source of distributed data resources. The following shorthand notation is used to represent standard GIS workflow data sources:

- **SDE DC/DBMS.** ArcGIS applications connecting to a geodatabase using an ArcGIS direct-connect communication architecture.
- **SDE server/DBMS.** ArcGIS applications connecting to a middle-tier ArcSDE remote server, which then connects to a separate geodatabase server.
- **SDE DBMS.** ArcGIS applications connecting to a geodatabase using an ArcGIS SDE communication architecture, with ArcSDE installed on the DBMS server.
- **Small shapefile.** Term used for a shapefile data source with the extent roughly the size of a city neighborhood (9 vector layers, 145,000 features, 140 MB with additional 241 MB raster layer).
- **Large shapefile.** Term used for a shapefile data source with the extent roughly the size of a large city (9 vector layers, 10,000,000 features, 1.2 GB with additional 241 MB raster layer).
- **File geodatabase.** ArcGIS file-based XML data source supporting geodatabase schema and up to a terabyte of geospatial data.

display or workflow display Refers to a unit of work for the purpose of measuring a user workflow throughput (work transaction). A display can refer to an average map display transaction (i.e., Web map transaction, or Web service that returns a map) or a unit of processing time (i.e., average processing loads generated by a power user workflow during a 6-second display cycle time). One way to define an average workflow display would be to collect the CPU processing and network traffic throughput generated by a single power user performing a workflow over a 10-minute period and divide the results by 100.

display cycle time The total time between client workflow displays; it is equal to the display response time plus the user think time.

display queue time Wait time. Any process delay times due to use of shared system resources. Wait times result when component capacity exceeds 50 percent utilization due to random transaction arrival times plus any cumulative travel times identified as system latency.

display response time See also *network response time*. The total amount of time from user display request until the complete display is returned to the user interface. Display response time includes all system component service times plus all queue times.

display service time The total program execution or network transfer time required for a system component (platform or network) to complete processing or transfer of an average work unit.

distributed   Used here to describe operations or system architectures that include shared system components such as networks, application servers (Windows Terminal Server or Web server farms), and a common data source, or geodatabase in GIS parlance. A distributed database allows two or more people to be working on the same data in separate locations. (Capacity planning is about building a distributed hardware and network infrastructure to support enterprise GIS operations.)

file server   Server platforms that provide network access to a file system repository.

floating-point   Floating-point numbers are the set of real numbers that include all negative and positive values, wherein fractional values are represented as decimal numbers. Real numbers are represented by the computer with a fixed field length. The decimal point within the numeric representation is allowed to float to take advantage of the full field length. The accuracy of the real number will depend on the number of digits in the field length and the size of the number (larger numbers will have reduced accuracy due to the fixed number of digits in the numerical representation). The most common floating-point representation in computers is that defined by the IEEE 754 standard.

geodatabase   A database designed to store, query, and manipulate geographic information and spatial data. It is also known as a *spatial database*. Within a spatial database, spatial data is treated as any other data type. Vector data can be stored as point, line, or polygon data types, and may have an associated spatial reference system. A geodatabase record can use a geometry data type to represent the location of an object in the physical world and other standard database data types to store the object's associated attributes. Some geodatabase formats, such as those used by ESRI in its ArcGIS software, also include support for storing raster data.

geodatabase server   The server platform hosting the geodatabase.

geodatabase transition   A term used to distinguish between database replication and geodatabase replication.

high-available configuration   One that continues to perform as a system despite the loss of any individual server. See chapter 4 for high-available Web configurations.

I/O intensive   Refers to a workflow process where peak throughput is limited by the system traffic throughput capacity (input/output) of the system environment. System traffic throughput capacity is a combination of the throughput capacities of both platform traffic and the network. The platform traffic throughput capacity is normally limited by the installed network interface card (NIC). The network throughput capacity is determined by the most limited shared bandwidth segment (network router or switch interface).

image data   Raster or picture data types, where a single or group of picture files are assembled to represent a data layer.

information product   Roger Tomlinson's term for data transformed by GIS into information particularly useful to you or your organization, and delivered via computer, often but not always in the visual form of a map. An example is economic data analyzed in relation to a specific location. See Tomlinson's book, *Thinking About GIS*.

integer   Integer numbers are the set of values (0, 1, 2, 3, etc.) and their negative values (0, -1, -2, -3, etc.) that are represented without a decimal point. Integer numbers represented in a computer with a fixed field length, where the number of significant digits to the right of the decimal point are a fixed value. The accuracy of an integer number will be constant based on the number of significant decimal values. The maximum size of the number will be limited by the fixed field length.

Internet   The world wide public communication network. User access to the public Internet infrastructure is provided by private Internet service providers (ISPs) that charge for the service connection.

local area network (LAN)   Private network infrastructures connecting computers within a relatively small area, such as a building or a campus. Local networks are supported by switched Ethernet protocols that enable high bandwidth communications between computers and shared network devices (printers, plotters, etc.) within the building or group of buildings.

local site   A site that supports computers connected by a LAN. Connections can include campus networks that have several LAN environments connected together over shared switched network connections or network routers, using common LAN protocols.

map server   A computer platform used to host Web mapping server software. This was a common term used in reference to hosting early Web mapping software. See *container machine*.

multicore chips   A computer chip with more than one processor core manufactured on a single chip die. The multicore chips share a level 2 (L2) cache that improves processing performance by reducing time to access program instructions. Multicore chip designs were introduced by computer hardware vendors in the 2004–2006 timeframe, in an effort to continue improving computer performance and capacity while reducing hardware costs. Current AMD and Intel processors provide two- and four-core chip designs, with higher capacity designs (8, 16, etc.) promised in future technology releases. Sun T-series processors already contain eight core processors per chip, and IBM Power PC and HP Itanium processors provide dual-core chips.

multithread process   A popular software execution model that allows multiple threads (independent services) to exist within the context of a single process, sharing the software code resources while providing independent execution. Multithreaded software reduces memory consumption by running the same number of service threads within a reduced number of processes.

network   See *computer network*.

network bandwidth   See *bandwidth*.

network capacity   Network capacity is defined by the capacity (bandwidth) of the router or switch connecting each network component to the client-to-server communication path.

network interface card (NIC)   The network interface card is a network component device used to connect computers to the physical local area network (LAN).

network queue time   Network wait time or display queue time, all three variations are used to identify any transport delay times due to use of shared system resources.

Wait times result when network traffic throughput exceeds 50 percent of the available network bandwidth capacity due to random transaction arrival times and any cumulative travel times identified as system latency.

network response time or display response time   Both are used to identify the total amount of network transfer time from initial data request until the data transfer reaches the user interface. Display response time includes all network transport service times plus all network queue times.

network service time   The term network service time is use to identify the data transfer time of an average work unit (workflow display) over the limiting network connection bandwidth. Estimated data transport time can be calculated by dividing the total traffic per display by the limiting network bandwidth connection.

network throughput   The amount of digital data per time unit that is delivered over a physical or logical link, or that is passing through a certain network node. For example, it may be the amount of data that is delivered to a certain network terminal or host computer, or between two specific computers. The throughput is usually measured in bits per second (bits/s or bps).

network traffic   A common term for network throughput. Usually measured in Mbps.

network workflow traffic   Includes client traffic and database traffic, meaning the data transfer volume of an average work unit (traffic per display). Client traffic per display, measured in Mbpd, is a measure of communication between the client and central data center, and is used to compute client network transport service times. The database (DB) traffic per display, measured in Mbpd, is a measurement of traffic between the servers in the data center, and is used in calculating how much server NIC traffic to anticipate (all server traffic flows through the server NIC interface).

operations   A general term synonymous with "systems"; in the context of system design in particular, it refers to the well-defined user workflows (processes and procedures) in place, underpinning the daily work efforts of an organization. Oil and gas companies have used the term with GIS since the early 1990s; within local governments, the term is used for enterprise applications (permit systems, customer service systems, accounting systems).

parallel processing The simultaneous use of more than one processor core to execute a single program. Most computer programs are executed sequentially, one instruction after the next, just as most human procedures are completed as a sequential series of procedural steps. Parallel processing requires that several instructions of a single procedure be executed in parallel. This in turn requires that the procedure be divided into several parallel threads executed concurrently, then brought back together to compute the final result.

peak workflow throughput As used with the Capacity Planning Tool, this term refers to the number of work units completed during the estimated peak workflow period. Peak workflow throughput is calculated by multiplying the identified peak users by the set peak client displays per minute, and the result is identified in terms of peak displays per minute for each workflow (peak users x DPM/client = DPM). The capacity planning models used in the CPT are based on the peak workflow throughput values.

performance A general term referring to the efficiency of the computer or system to do work. Performance components include processing time (service times and network transport times), queue time (random arrival wait time), travel time (latency), and think time (time for user interaction with the display). Display response time includes the processing time, queue time, and any additional travel time.

performance baseline A specific reference platform performance represented by a SPEC performance benchmark value. Performance baseline values for workflow service times used in the ESRI Capacity Planning Tool are established each year based on the performance of available vendor platform technology. ESRI performance baselines are identified by the term "Arc" followed by the year (i.e., Arc07, Arc08).

performance benchmark A test run of a computer program that takes the measure of the execution performance of that program. Both the test and the measurement of the test are called a performance benchmark. Performance benchmarks are used to identify hardware processing loads generated by the software when executing a defined procedure. Platform vendors publish performance benchmarks to represent the relative processing capabilities of their hardware technology. Performance benchmarks are also useful for identifying changes in software performance

from one release to the next, and can be used as a reference for establishing capacity planning performance targets.

performance budgets The workflow software component service times used to establish a projected performance and scalability baseline. Component software performance should be reviewed during development, prototyping, and deployment to verify that application performance is within the established performance budgets.

performance model A representation of the relationships between the various system performance factors, which can be used for platform sizing and predicting appropriate network bandwidth. In this book, performance models are in the form of a chart, a table, or an Excel interface.

performance targets The workflow software response times used to establish a projected performance and scalability baseline. Component software performance should be reviewed during development, prototyping, and deployment to verify that application performance is within the established performance targets.

platform architecture The hardware platform and software component configuration supporting the production software environment. The platform architecture also defines the protocols used to communicate between the software components and how they are configured.

platform capacity Identifies the peak platform throughput. The Capacity Planning Tool identifies peak platform throughput values based on the selected workflow. In a system environment where multiple workflows are supported on a single platform, the peak platform capacity value is projected based on maintaining the same relative workflow throughput mix.

platform configuration The term used to identify and select a specific platform component configuration. The Capacity Planning Tool identifies the chip manufacture (Intel), the processor model number (5160), the number of processor cores (4 core), the number of chips (2 chip), and the processor MHz and L2 cache [3000(4) MHz]. Platform memory requirements are generated by the CPT using standard guidelines based on system capacity requirements. The platform configuration naming convention is used to identify a specific processor configuration, which may be provided by more than one hardware vendor [i.e., Intel 5160 4 core (2 chip) 3000(4) MHz] platform configuration is provided by Dell, IBM, HP, and Sun.

platform load The total processing time required to support the identified user requirements workflow profile. A composite baseline display service time is computed based on the identified user requirements workflow and allocated to each appropriate platform tier. The computed baseline platform display service time is adjusted for the selected platform performance using the vendor published SPEC benchmark values. The adjusted platform service time is used to calculate resulting platform performance values.

platform node The term used to identify the number of platform servers included in the platform configuration. The CPT will generate the required number of platform nodes for each platform tier based on the identified user workflow requirements and the selected platform configuration. If a fixed number of nodes is identified, the platform performance will be calculated based on the fixed node value.

platform operating system The platform operating system (OS) is the software that manages the sharing of the resources of a computer and provides software programs with an interface used to access those resources. An operating system processes system data and user input, and responds by allocating and managing tasks and internal system resources as a service to users and programs of the system.

platform service times See also *service time*. Refers to the platform processing load required to support the selected workflow configuration. The workflow software component processing loads (service times or the platform performance baseline) are allocated to the assigned platforms to establish the baseline platform service times. The baseline platform service time is translated to the selected platform service time using the SPEC relative benchmark values. The selected platform service time and the user requirements peak workflow are used to calculate the platform performance values.

platform sizing A process informed by the individual platform sizing charts provided in chapter 9 (Platform performance). These charts identify individual platform capacity based on the standard ESRI software workflow models and platform SPEC performance benchmarks provided by vendors. Platform sizing identifies platform capacity requirements and is useful for compar-

ing several platform alternatives on a single display. These charts use the same capacity planning models incorporated in the CPT, and were used for capacity planning by ESRI system design consultants since the mid-1990s.

platform sizing chart The platform sizing chart can be used to identify peak platform throughput based on vendor provided platform SPEC performance benchmarks. Platform sizing charts are provided in chapter 9 for the ArcGIS Desktop Windows Terminal Server, ArcSDE Geodatabase Server, and Web Services (ArcIMS, ArcGIS Server) platform environments. The platform sizing charts do not provide the workflow response times, software component service times, queue times, network transport times, network latency, or network traffic requirements that are included in the Capacity Planning Tool.

platform tier A group of common server platforms configured for a scale-out platform architecture. Web services are deployed in a two-tier (Web server, database server) or three-tier (Web server, container machine, database server) architecture. A single-tier workgroup server tier is provided for a single platform configuration hosting the Web applications and database on the same platform. Windows Terminal Server farms host centralized data center deployment of Windows desktop applications, and are identified as their own platform tier.

platform vendor The hardware vendor that builds and sells the supported host server platforms. Platform vendors also conduct and publish the SPEC performance benchmark results for their own hardware. The vendor-published SPEC benchmark values are used to identify relative platform performance in the Capacity Planning Tool.

platform zone A process for using the CPT for configuring mixed platform tier environments. The example in chapter 10 (Capacity planning) shows how to use the CPT for configuring three separate Web workflows (ArcIMS Map Service, ArcGIS Server Standard Map Service, and ArcGIS Server Advanced Edit Session) sharing a common Windows Terminal Server and database server data center environment.

process queue time The waiting in line for processing time is called queue time. Process queue time is used to identify the total system wait times resulting from the random arrival distribution of individual process transaction requests. When multiple requests arrive for processing by a single service provider (core processor), only one request can be processed at a time and the others must wait in the queue.

processing The work done by the computer in executing the software program instructions. The core processor unit is the computer hardware component that executes the software instructions. The amount of time the core processor is busy executing software instructions is called the processing time, or service time spent by the computer to process the program instructions.

processor chip cache A cache used by the central processing unit of a computer to reduce the average time taken to access memory. The cache is a smaller, faster memory, which stores copies of the data from the most frequently used main memory locations. Hardware vendors design computer chips to provide the optimum performance and capacity by reducing the chip size and optimizing the chip configuration.

processor core The computer processor component used to execute the software program instructions. The core implements optimizations such as superscalar execution, pipelining, and multithreading. Multicore chips provide a single integrated circuit that can include two, four, or more cores on a single integrated silicon chip configuration.

processor MHz The speed (clock rate) at which a microprocessor executes instructions. Computer chips include an internal clock that regulates the rate at which instructions are executed and synchronizes all the various computer components. The core processor design optimizes the number of program instructions that can be executed during a clock cycle. For a specific processor core design, faster clock speeds provide faster program execution. Clock speeds are expressed in megahertz (MHz) or gigahertz (GHz).

processor socket The connector on the computer motherboard that houses a computer processor. Processor sockets use a pin grid array (PGA) where pins on the underside of the processor connect to holes in the processor socket.

queuing theory The mathematical study of waiting lines (or queues). Queuing theory is an academic study of distribution arrival times, and provides a definition of random arrival distributions used in supporting a variety of business decisions. Statistical arrival distributions are difficult to quantify. Mathematical queuing

models are most reliable when estimating large, random-population arrival distributions.

queue time The system wait times resulting from the random arrival distribution of individual process transaction requests. When multiple requests arrive for processing by a single service provider (core processor or shared network port), only one request can be processed at a time and the others must wait in the queue. The waiting in line for processing time is called queue time, and can be an indication of overcapacity conditions.

random arrival times An arrival distribution where all available arrival times are equally likely. Queuing theory can be used to estimate wait times for a random distribution of arrival times. In computer processing, hundreds of communications are transmitted to support a single user display transaction. These display transactions typically occur over a period of less than 5 seconds. For systems supporting many concurrent user workflows, it is quite probable that the resulting transaction arrival times would follow a random distribution profile. Queuing theory can be used to estimate the probability that multiple transaction requests will arrive for processing at the same time, and will have to wait in line for processing. Queuing theory can be used to estimate the wait time (queue time) for capacity planning purposes. An approximation of the queue time (a function that depends on component utilization rates) is included in the CPT to account for these random arrival times.

relative performance The term used in comparing the productivity of two common service providers. The SPEC performance benchmarks were established to provide a fair measure of relative compute performance of one hardware vendor platform configuration compared to another, for capacity planning purposes. A selected platform is identified to represent the performance baseline, and the SPEC published benchmark value represents the performance of the test platform relative to the baseline platform.

remote site User locations that are located over a remote network connection (WAN or Internet). A local site is a user location that is located on the same LAN environment.

requirements analysis The process for identifying peak user workflow requirements. The Capacity Planning Tool includes a requirements analysis module that can be used as a template to identify peak user workflow requirements. User workflow software component service times are identified on the CPT workflow tab, and then used to complete the user analysis within the CPT user requirements module.

response time The total time from initial transaction request to final delivery of the service. For a user map display, this includes all system processing times (service times), data transfer times, queue times, and any transport latency times.

sequential processing Refers to software programs that include a single threaded execution (execute the software procedure one step at a time). Each step (instruction) in the procedure (program) is executed in sequence, and each instruction in the sequence must be completed before moving on to the next instruction.

server A hardware platform used to host a software program that performs a compute service. A server is normally a shared platform environment that hosts multiple concurrent user workflows during a common service period.

service provider Any computer system component (server platforms or communication networks) that process workflow service requests (user displays or program transactions). (The workflow units used in the CPT are called user displays.)

service time See also *platform service times*. The transaction processing time (usually identified in seconds) required to service a single workflow unit of work (called a user display or workflow display). A performance baseline workflow display can be identified as the work of a power user performing a workflow load over a period of six seconds (10 displays per minute).

single-thread processing Refers to the fact that a single processor core is limited to executing a single batch process thread during a single clock cycle. Multiple concurrent processing requests must wait in a process queue and will be executed sequentially by a single processor core. A server platform configuration with 4-core processors can execute up to four concurrent process requests at the same time, each core limited to executing a single process execution.

site bandwidth The bandwidth of the site router or gateway connection to the WAN or Internet communication service provider. The cost of the connection service is usually based on the amount of connection bandwidth provided by the service.

software baseline service times Refers to the performance baseline processing times used to represent the software workflow model. Standard ESRI workflow software service times are provided on the CPT Workflow tab based on performance capabilities of the core GIS software. The ESRI workflow service times can be used as a template to establish specific project workflow performance budgets. The software baseline service times identified for the project workflows establish the performance models used in the CPT system loads analysis. These workflow baseline service times represent the software component performance budget represented by the CPT capacity planning analysis.

software service times Refers to both the software baseline service times used in establishing the workflow performance model and the adjusted software service times (corrected for selected platform performance) represented in the Workflow Performance Summary chart included on the CPT2006 tab.

SPEC baseline performance value The reference SPECrate_int2006 platform performance baseline used for publishing the standard ESRI workflow software service times and all other customized workflows included on the CPT Workflow tab. The SPEC baseline performance value is updated each year to represent the current platform hardware performance published on the SPECint_rate2006 benchmarks.

SRint2000 A shorthand notation for the published vendor SPECrate_int2000 platform benchmarks. The SRint2000 benchmarks were used to represent vendor platform relative performance from year 2000 through December 2006. Hardware platform performance comparisons after 2006 use the SRint2006 benchmarks.

SRint2006 A shorthand notation for the published vendor SPECrate_int2006 platform benchmarks. The SRint2006 baseline benchmark values are used for relative platform performance service time adjustments in the Capacity Planning Tool. SPEC performance values for vendor platforms sold from 2000 through 2006 would be found in the SRint2000 benchmark results. The SRint2006 results can be estimated from the SRint2000 results by dividing by a factor of 2.1, allowing approximate performance comparisons between the older and newer platform environments.

standard ESRI workflows The term used to identify the software technology workflow templates included on the CPT Workflow tab. These workflows can be used as a reference for establishing project-specific workflows during a capacity planning analysis. Workflows with increased software service times represent more conservative performance targets, and workflows with reduced software service times represent more aggressive performance targets. The standard ESRI workflow software service times represent reasonable performance targets for the associated ESRI software technology.

storage A term used to describe a persistent storage media (normally on disk or tape) that retains digitally encoded data. Current platform storage solutions are provided by hard disk devices that provide a persistent storage repository for digitally encoded data. Storage can also include computer tape, compact disk (CD), and memory, which also provide temporary or persistent media storage environments. The term storage used in capacity planning normally refers to either internal server storage, direct attach storage, storage area networks, and network attached storage environments supported by a redundant array of independent disk (RAID) hardware configurations.

stovepipe A term used to describe business systems that support their data on internal disk drives rather than sharing their data on a storage area network (SAN). Hardware funds are invested to support a single project effort, rather than shared to support more optimum corporate-level business needs. Stovepipe systems do not connect to the rest of the community, and it is difficult to share business resources with this type of architecture.

terminal server See *Windows Terminal Server*.

throughput The term used to describe the rate of doing work, expressed as a transaction rate, such as transactions per minute. In communication networks, throughput is the amount of digital data per time unit that is delivered over a physical or logical link, or that is passing through a certain network node. For example, it may be the amount of data that is delivered to a certain network terminal or host computer, or between two specific computers. The throughput is usually measured in bits per second (bits/s or bps), occasionally in data packets per second or data packets per timeslot. The system throughput or aggregate throughput is the

sum of the data rates of delivery to all terminals in a network. The maximum throughput of a node or communication link is synonymous to its capacity.

In computer platforms, throughput is the number of displays per minute (DPM) hosted by the computer platform tier or supported on a certain platform node. For example, throughput could be the amount of processing required by a host computer or by a server platform tier. The throughput is measured in DPM, wherein each display represents a single unit processing load specific for each defined user workflow. The system throughput or aggregate platform throughput is the sum of the workflow processing times, where the platform display (unit of work) is a weighted average of the total platform processing load. The maximum throughput of a node or server tier is synonymous to its capacity.

See also *bandwidth, capacity, network throughput, network traffic, peak workflow throughput, platform capacity, platform sizing chart, workflow capacity.*

transaction The term used to describe a unit of work (for GIS workflows, a transaction may be represented by a map display). Each user workflow may have a different measure of work. The transaction provides a unit for measuring throughput, service time, queue time, and response time—all measurements are relative to the type of workflow transaction.

transaction-based sizing model A more adaptive tool in estimating the platform size required to meet user needs.

Adaptive means that those models have been found to be more accurate in addressing changes in platform performance, allowing you to adapt to future changes more readily. The concurrent user models were not as accurate, and it was important to update the platform sizing charts on a periodic basis for them to be useful. The transaction-based sizing models can be accurate over many years without updating the baseline performance platform. The SPEC benchmarks are an example of a transaction-based sizing metrics.

user productivity The term used to represent the displays per minute (DPM) performed by the average user for each specific workflow included in the CPT requirements analysis. Recommended user productivity for ArcGIS Desktop workflows is 10 DPM, while recommended user productivity for Web application workflows is 6 DPM. User productivity is restricted by workflow display response time and required user think time. Minimum user think time must always be positive, and should provide the user with sufficient user interaction time to review the display and provide appropriate user input. User productivity should be adjusted as needed to accommodate a realistic user think time.

user think time   The human interaction time included in the user workflow to accommodate the required user input for each display cycle. The user think time is calculated by first computing the display cycle time (60 seconds / user productivity) and then subtracting the display response time (total time required for the system to return the user display). For user workflows, the user think time must always be greater than zero. A minimum user think time is included at the top of the DPM/client column in the CPT user requirements module. This minimum time can be adjusted to represent an average minimum user think time, and is used to support the conditional color warnings in the user productivity column (yellow when less than minimum, and red when less than zero). A workflow with zero think time is a batch process (no user interaction time).

user workflow   Used here to identify the configuration of the GIS software technology employed to enable the user work activity. A separate user workflow can be identified for each use-case or information product description involved in quantifying user technology requirements in the user needs assessment. Some workflows may be employed to generate a broad range of information products using the same software technology and data source. A single unit of work is identified as a workflow display, and is quantified by specific baseline component software processing time and the generated network traffic per display (unit of work). For capacity planning purposes, a desktop software workflow display transaction is defined in terms of component software service times (desktop, WTS, SDE, DBMS) and network traffic (Mbpd) generated by a power user during an average period of 6 seconds (10 DPM productivity). A Web software workflow display transaction is defined in terms of component software service times (WA, SOC, SDE, DBMS) and network traffic (Mbpd) generated by a power user during an average period of 10 seconds (6 DPM productivity).

utilization   For hardware platforms, the term used to identify the percentage of time the platform core processors are busy executing the software program instructions. The processing time (service time) is computed by comparing the time the core processors

are busy over the total work unit cycle time (i.e., 10 percent utilization of a single core processor over a 6-second work cycle time would represent 0.6 seconds service time). Platform utilization approaches 100 percent as platform throughput reaches full output capacity. (Another way of putting it is that utilization is a measure of platform capacity consumed by a specific throughput, and is represented as a percentage of platform capacity. Peak platform throughput is achieved at 100 percent utilization.)

For network communications, the term is used to identify the percentage of time the network connection is busy while supporting the required network traffic. Network utilization approaches 100 percent as network traffic throughput reaches bandwidth capacity.

Web application server (WA)   The server platform hosting the Web application software when configured in a three-tier Web architecture. The WA is identified with the platform service times in the Requirements Analysis Module, and the Web application server tier is included as one of the platforms in the Platform Selection Module. The Web application platform service times are included with the Web spatial server, and the Web application software is hosted on the container machine platform tier in a two-tier Web architecture.

Web browser   A client software application that enables a user to display and interact with text, images, maps, videos, and other information typically generated by a Web application over the World Wide Web or a local network.

Web server   Server platform hosting Web applications.

wide area network (WAN)   A computer network that covers a broad area; any network whose communication links cross metropolitan, regional, or national boundaries. Or, less formally, a network that uses routers and public communications links. Contrast with personal area networks (PANs), local area networks (LANs), campus area networks (CANs), or metropolitan area networks (MANs), which are usually limited to a room, building, campus, or specific metropolitan area (e.g., a city), respectively. The largest and most well-known example of a WAN is the Internet.

Windows Terminal Server (WTS)   The term includes a reference to Microsoft Terminal Services and the Citrix XenApp (previous Citrix Presentation Server) technology. Terminal Services is a component of Microsoft Windows (both server and client versions) that allows a user to access applications and data on a remote computer over any type of network. The Microsoft Remote Desktop Protocol, part of Microsoft's Terminal Services, is based on Citrix technology and was licensed from Citrix in 1997. The Citrix XenApp is built on the Independent Computing Architecture (ICA), Citrix Systems' thin client protocol. Unlike traditional frame-buffered protocols like VNC, ICA transmits high-level window display information, much like the X11 protocol, as opposed to purely graphical information. This is possible because the Citrix Display Driver, which is installed in Session Space, is capable of capturing high-level GDI draw commands that can be replayed on GDI-capable clients (for example, Windows-based clients). Clients are available for several operating systems, including Microsoft Windows (CE, 16-bit, 32-bit and 64-bit platforms), Mac OS, Linux, and other Unix-like systems.

work transaction An artificial term used by the Capacity Planning Tool as a measure of an average unit of work. The CPT defines each unique user workflow by identifying the computer processing time (software service times) and data traffic for an average work transaction. For simple GIS map displays, the work transaction may be identified as the work (computer processing and data transport) associated with the average map display. For more complex workflows, the work transaction may be identified as the work (computer processing and data transport) of a power user over a specified period of time (6 seconds). The work transactions for each user workflow are located in the software service times section of the CPT Workflow tab.

workflow capacity The peak workflow throughput of a selected server platform configuration.

workflow performance Any of the performance values (service time, queue time, response time, etc) associated with a specific workflow on a selected server platform configuration.

workflow performance summary An Excel bar chart included in the CPT User Requirements Module that identifies workflow service times, queue times, network transport times, network latency, and total response time based on identified user requirements, network bandwidth, and selected platform architecture.

workflow requirements Information required by the CPT User Requirements Module to identify the peak user workflow loads. This includes the definition of each workflow on the CPT Workflow tab, selection and location of user workflows on the CPT2006 Capacity Planning tab, and identification of peak users and productivity. The CPT can be used as a template for identifying user workflow requirements.

workflow service times The work transaction software service times identified on the Workflow tab that define a user workflow. The CPT applies the workflow service times to the identified user requirements to complete a system loads analysis. The system loads analysis identifies peak network traffic and processing loads on the selected platform configuration.

workstation The client desktop computer.

Index

Related titles from ESRI Press

Thinking About GIS: Geographic Information Planning for Managers, third edition

ISBN 978-1-58948-158-9

ESRI Press publishes books about the science, application, and technology of GIS. Ask for these titles at your local bookstore or order by calling 1-800-447-9778. You can also read book descriptions, read reviews, and shop online at www.esri.com/esripress. Outside the United States, contact your local ESRI distributor.